T0396404

Biological Invasions in Europe and the Mediterranean Basin

Monographiae Biologicae
Volume 65

Series Editors

H. J. Dumont & M. J. A. Werger

Biological Invasions in Europe and the Mediterranean Basin

Edited by

F. di Castri, A. J. Hansen & M. Debussche

Kluwer Academic Publishers
Dordrecht / Boston / London

Library of Congress Cataloging-in-Publication Data

```
Biological invasions in Europe and the Mediterranean Basin / edited by
F. di Castri, A.J. Hansen & M. Debussche.
      p.   cm. -- (Monographiae biologicae ; v. 65)
   ISBN-13:978-94-010-7337-0
   1. Biological invasions--Europe.  2. Biological invasions-
-Mediterranean Region.   I. Di Castri, Francesco.  II. Hansen, A.
J., 1955-   . III. Debussche, M.  IV. Series.
QP1.P37  vol. 65
[QH135]
574 s--dc20
[574.5'24]                                              89-20105
```

ISBN-13:978-94-010-7337-0   e-ISBN-13:978-94-009-1876-4
DOI: 10.1007/978-94-009-1876-4

Published by Kluwer Academic Publishers,
P.O. Box 17, 3300 AA Dordrecht, The Netherlands.

Kluwer Academic Publishers incorporates
the publishing programmes of
D. Reidel, Martinus Nijhoff, Dr W. Junk and MTP Press.

Sold and distributed in the U.S.A. and Canada
by Kluwer Academic Publishers,
101 Philip Drive, Norwell, MA 02061, U.S.A.

In all other countries, sold and distributed
by Kluwer Academic Publishers Group,
P.O. Box 322, 3300 AH Dordrecht, The Netherlands.

Printed on acid-free paper

Drawings for the cover by René Ferris, CNRS, Montpellier.

All Rights Reserved
© 1990 Kluwer Academic Publishers
Softcover reprint of the hardcover 1st edition 1990

No part of the material protected by this copyright notice may be reproduced or utilized in
any form or by any means, electronic or mechanical, including photocopying, recording or by
any information storage and retrieval system, without written permission from the copyright
owner.

Preface

In view of the massive change in the area of distribution of many world biota across classical biogeographical realms, and of the drastic restructuring of the biotic components of numerous ecosystems, the Scientific Committee on Problems of the Environment (SCOPE) decided at its general Assembly in Ottawa, Canada, in 1982 to launch a project on the 'Ecology of Biological Invasions'.

Several regional meetings were subsequently organized within the framework of SCOPE, in order to single out the peculiarities of the invasions that took place in each region, the behaviour of their invasive species and the invasibility of their ecosystems. Most noteworthy among such workshops were one in Australia in August 1984, one concerning North America and Hawaii in October 1984, and one dealing with southern Africa in November 1985.

A leitmotiv of these workshops was that most of the invasive species to those regions were emanating from Europe and the Mediterranean Basin, inadvertently or intentionally introduced by man. It was therefore considered as a timely endeavour to organize the next regional meeting in relation to this region.

The workshop on 'Biological Invasions in Europe and the Mediterranean Basin' was held in Montpellier, France, 21 to 23 May 1986, thanks to the financial support of SCOPE and of the A.W. Mellon Foundation, and the logistic facilities of the Centre National de la Recherche Scientifique (C.N.R.S.). It was not organized as a formal and large conference with presentation of well-structured papers. Rather, a small number of invitees, envisaged as being authors of chapters of the foreseen synthesis book, presented draft introductions on different topics with the purpose of stimulating discussion with a selected group of participants. Therefore, this volume does not represent the proceedings of the Montpellier workshop. All chapters were elaborated long after the workshop, but taking into due account the insights from the discussion. Furthermore, a few other authors were selected later on to fill evident thematic gaps.

The book comprises 27 chapters. After an introduction, 10 chapters deal with plant invasions, two of them from a paleoecological viewpoint; Fox's chapter extends the comparison to all regions of the world with a mediterranean-

v

vi

type climate. The 8 chapters on animal invasions cover different topics: invasions in ancient and recent times, invasions by vertebrates and invertebrates, invasions in terrestrial and in aquatic and marine environments; four chapters refer, to a varied extent, to invasion by parasites. Finally, 8 chapters approach the processes and mechanisms of invasion, from physiological, genetic and ecological viewpoints; while based on examples from Europe and the Mediterranean Basin, their considerations transcend the boundaries of the region.

The core of this volume is definitively on invasions towards and within Europe and the Mediterranean Basin. Nevertheless, many remarks and hypotheses are intended to help explain why species from this region have such a great invasion potential in relation to other continents.

Of course, a comprehensive treatment would have been impossible due to the extent and the heterogeneity of the region, and the multiplicity of taxa. Each chapter has to be considered as a kind of case study, illustrating problems of greater coverage and repercussions.

Finally, it is unquestionable that recent events linked with the increased human impact on ecosystems, the improved transportation systems, the driving forces of an internationally-wide market and trade economy, and also the impending man-made global climatic change will further promote a mix up of new species assemblages in the biosphere. It is hoped that this volume will contribute to the understandig of such complex and far-reaching phenomena.

F. di Castri   A.J. Hansen   M. Debussche

Contents

Preface	v

Part One. Introduction · 1

1. On invading species and invaded ecosystems: the interplay of historical chance and biological necessity
by F. di Castri · 3

Part Two. Plant invasions · 17

2. Plant invasions in Central Europe: historical and ecological aspects
by J. Kornaś · 19

3. History of the impact of man on the distribution of plant species
by K.V. Sykora · 37

4. Recent plant invasions in the Circum-Mediterranean region
by P. Quezel, M. Barbero, G. Bonin and R. Loisel · 51

5. The invading weeds within the Western Mediterranean Basin
by J.L. Guillerm, E. Le Floc'h, J. Maillet and C. Boulet · 61

6. Widespread adventive plants in Catalonia
by T. Casasayas Fornell · 85

7. History and patterns of plant invasion in Northern Africa
by E. Le Floc'h, H.N. Le Houerou and J. Mathez · 105

8. Invasions of adventive plants in Israel
by A. Dafni and D. Heller · 135

9. Man and vegetation in the Mediterranean area during the last 20,000 years
by J.-L. Vernet · 161

viii

10. Plant invasions in Southern Europe from the Paleoecological point of view
by A. Pons, M. Couteaux, J.L. de Beaulieu and M. Reille — 169

11. Mediterranean weeds: exchanges of invasive plants between the five Mediterranean regions of the world
by M.D. Fox — 179

Part Three. Animal invasions — 201

12. The invasion of Northern Europe during the Pleistocene by Mediterranean species of Coleoptera
by G.R. Coope — 203

13. Migratory Phenomena in European animal species
by G. Marcuzzi — 217

14. The bean beetle (*Acanthoscelides obtectus*) and its host, the French bean (*Phaseolus vulgaris*): a two-way colonization story
by V. Labeyrie — 229

15. Some recent bird invasions in Europe and the Mediterranean Basin
by P. Isenmann — 245

16. Of mice and men
by J. Michaux, G. Cheylan and H. Croset — 263

17. Invasions by parasites in continental Europe
by C. Combes and N. Le Brun — 285

18. Human activities and modification of ichtyofauna of the Mediterranean sea: effect on parasitosis
by C. Maillard and A. Raibaut — 297

19. Influence of environmental factors on the invasion of molluscs by parasites: with special reference to Europe
by T.C. Cheng and C. Combes — 307

Part Four. Mechanisms of invasions — 333

20. In search of the characteristics of plant invaders
by J. Roy — 335

21. Biogeographical and physiological aspects of the invasion by *Dittrichia* (ex *Inula*) *viscosa* W. Greuter, a ruderal species in the Mediterranean Basin
by J.P. Wacquant — 353

22. Invaders and disequilibrium
by P.H. Gouyon — 365

23. Species-specific pollination: a help or a limitation to range extension?
by F. Kjellberg and G. Valdeyron — 371

24. Genetic differentiation in beech (*Fagus sylvatica* L.) during periods of invasion and regeneration
by B. Thiebaut, J. Cuguen, B. Comps and D. Merzeau — 379

25. Invasion of natural pastures by a cultivated grass (*Dactylis glomerata* L.) in Galicia, Spain: process and consequence on plant-cattle interactions
by R. Lumaret — 391

26. Introduced and cultivated fleshy-fruited plants: consequences of a mutualistic Mediterranean plant-bird system
by M. Debussche and P. Isenmann — 399

27. Fire as an agent of plant invasion? A case study in the French Mediterranean vegetation
by L. Trabaud — 417

List of contributors — 439

Index of Genera and Species — 443

General index — 458

PART ONE

Introduction

1. On invading species and invaded ecosystems: the interplay of historical chance and biological necessity

FRANCESCO DI CASTRI

Abstract.

The invasion by alien species of new regions and territories is a phenomenon of paramount importance, particularly in the last four centuries after the 'Great Discoveries'. Biological invasion is likely to acquire soon an even greater frequency, because of the current transportation systems and the forthcoming global climatic change.

'Invader' species have diverse sets of ecological, physiological, genetic and morphological characteristics that make them suitable for wide dispersion, colonization and competition. We refer to their intrinsic aptitude and potential for invasion as their 'biological necessity'.

Nevertheless, no one of the various sets of biological characteristics can fully explain success or failure to invade. It is indispensable for invaders to have caught opportunities to leave and to be transported, and to have found at their arrival open spaces, available resources and ecosystems poorly resistant to invasions. This is their 'historical chance'.

Among new patterns of invasion are those associated to the release of genetically designed organisms. Bioengineered organisms cannot be related to any of the existing biogeographical realms. A new world-wide 'anthropogenic realm', with its own peculiar characteristics and trends, is to be considered.

Introduction

It is a truism to say that human activities are changing the face of the world and that very few ecosystems, if any, are completely 'natural' in their species composition and functioning patterns. Acting as a *geological agent*, man has largely modified landscape composition, and is rapidly expanding an atypical ecosystem type, the urban one. As a *biogeographical agent*, by creating new habitats, new barriers and mostly new bridges, man has favoured new mixtures and assemblages of species, often belonging to previously distinct and separated biogeographical realms. Because of his new (and non-desired) role as a *climatic*

F. di Castri, A. J. Hansen and M. Debussche (eds.), Biological Invasions in Europe and the Mediterranean Basin.
3–16. © 1990, *Kluwer Academic Publishers, Dordrecht.*

agent, man is likely to produce the most unprecedented and rapid migration trends (as well as many extinctions) of all biota of the world.

Consequently, in the surroundings of human settlements and agricultural fields (and even in less modified habitats), our neighbouring plants and animals may represent a mosaic-like association of most of the biogeographical realms of the world.

The problem of the biological invasions, illustrated in a so captivating way by Elton in 1958, has already profound implications of high economic importance in such areas as agriculture and weeds control, biological control of pests, aquaculture and epidemiology (e.g. rabies control, see Bacon (1985); even the spread of AIDS can be considered as a peculiar case of biological invasion.

Moreover, exploring the mechanisms and processes related to biological invasions has *per se* high heuristic value, since it implies a 'meeting point' and a close interaction for ecologists, geneticists, population biologists, physiologists and evolutionists, agriculturists, parasitologists, and even historians like Braudel (1979) and Crosby (1986).

In particular, Crosby emphasizes the role played by Europe as a source from which many groups of biological invaders left to aggressively colonize other continents, often displacing local biota. He refers to this phenomenon as 'ecological imperialism'.

What is a biological invader?

A biological invader is a species of plant, animal or micro-organism which, most usually transported inadvertently or intentionally by man, colonizes and spreads into new territories some distance from its home territory. Often, invaders spread from one biogeographical realm to another.

Classical examples of invaders are, for instance, the black rat, *Rattus rattus* (see Michaux *et al.* this volume), or the house sparrow, *Passer domesticus* (see Niethammer 1969), both clearly associated to man. Most likely emanating from Asia, they have progressively colonized Europe and from there all continents, being very common in the vicinity of most human settlements. Invaders can belong to most of the taxa; even large conifers such as *Pinus radiata* from California and *Pinus pinaster* from Europe are aggressive invaders of South African ecosystems (Macdonald *et al.* 1986).

Unavoidably, there are different viewpoints on the notion of a 'biological invader', as exemplified in Figure 1. The central overlapping zone of the four circles symbolizing the approach of biogeographers, ecologists, population biologists and geneticists, and practitioners, is intended to cover the definition and the characteristics of an invader. Nevertheless, no single approach embraces comprehensively the notion of invaders.

A biogeographer is primarily interested by the time of the invasion and the progressive spreading of the areas of distribution, while ecologists and

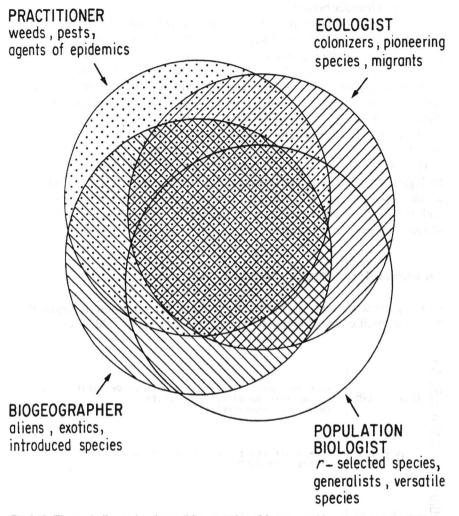

Figure 1. The symbolic overlapping and intercrossing of four approaches to cover the definition and the characteristics of a biological invader (after M. Rejmánek, personal communication, largely modified).

geneticists may concentrate their attention on the mechanisms and processes involved in these phenomena. The practitioner has necessarily in view the economic and health damage provoked by these invasions, as well as the measures of quarantine, preventive control or eradication to be taken as regards a given invader.

As a matter of fact, the 'invasion' term should be primarily used with a biogeographical connotation, but this is not always the case, not even in this volume. Incidentally, 'invader' is by no means synonymous of cosmopolitan species. Most of the invaders are not cosmopolitan; several cosmopolitans have not an invader behaviour.

According to different perceptions and research objectives, biological invasions can be studied at different hierarchical levels of space and time (hopefully by combining different scales), in consideration to the time of occurrence and spreading of the invasion or the distance from the originary territory of the invader. Figures 2 and 3 provide an overview of this scaling problem. Scales of space and time are given only as a very rough approximation. Attention should be drawn to the period of the Great Discoveries which has provoked *de facto* a breakdown of the biogeographical realms as well as a 'revolution' of the food customs all around the world.

The role of human history driving forces in the Old World as related to biological invasions has been treated in greater detail by di Castri (1989a), as has the importance of taking into account hierarchical scales when dealing with this kind of problems (di Castri and Hadley 1988). Practically all scales of space and time are approached in the different chapters of this volume.

The intermingled profile of a potential invader

At the present state of knowledge, it would be inappropriate to argue that there are specific sets of biological characteristics which could be considered

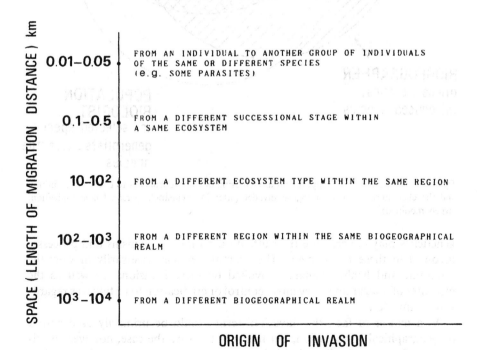

Figure 2. Scale of space as regards the distance from the place of origin of a given invasion.

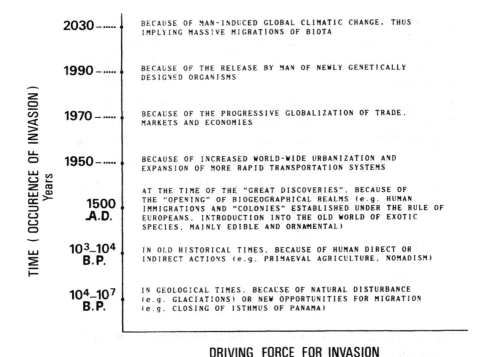

Figure 3. Scale of time as regards the initial driving forces for biological invasions.

strictly peculiar to the condition of being 'an invader'.

Table 1 single out some biological attributes that are likely to facilitate the invasibility by given species of plants or animals. Roy's chapter in this volume discusses in a more thorough and comprehensive way the characteristics of a plant invader (see also Baker 1965, Barrett and Richardson 1986, Newsome and Noble 1986).

Some of the attributes of Table 1 are quite obvious; for instance, morphological characteristics of seeds that facilitate transportation by wind, like the winged seeds of *Fraxinus, Ulmus* or *Acer*, or by animals such as the spiny seeds of *Xanthium* hanging up the skin or wool of large herbivores. In this respect, it is worth mentioning the monumental work of Thellung (1908–1910) who identified hundreds of adventive species near Montpellier, most of them introduced through the import, the hanging out and the drying of wool in Port-Juvénal. Some of these species are not still represented there, so that it would be very interesting to identify those definitively naturalized in the region, in order to quantify the difference between introduction/colonization and naturalization processes.

Admittedly, the attributes listed in Table 1 are very incomplete, some of them could be shifted from one column to another, or could appear in more than a single column. Nevertheless, they give room for some generalization:

8

Table 1. Some biological attributes of a possible invader.

Related to ecology and physiology	Related to morphology and behaviour	Related to genetics and population dynamics
Wide potential niche	Small body size	Subject to *r*-selection
Non-specialized germination and regeneration patterns	High mobility	High fecundity
	High vagility	High population growth
Non-specialized pollination patterns	Apt to phoresis	Short and simple life-cycle
Dormancy	High resistant spores	High genetic variation
Rapid growth	Seed morphology (spiny seeds or burr, plumed seeds, winged seeds or samara) suitable to long dispersal by wind or animals	Uniparental reproduction
High resource allocation to reproduction		Polyploidy
Longevity of seeds able to create seed banks		
Edible fruits and seeds transported by animals		

(a) No one species can possess all these attributes, but an unpredictable proportion of some of them.

(b) As regards possibilities of invasion, there are large differences between plants and highly mobile animals as mammals, birds or fishes, or small vagile organisms like Tardigrada, Rotifera, Nematoda, Fungi and Bacteria.

(c) Nevertheless, there are varied groupings of ecophysiological and genetic characteristics which seem to facilitate successful invasion (but not a single and stable grouping can be recognized).

(d) Conversely, species having many of these attributes have proved (up to now) not to be invaders. Even more, species having opposite attributes (e.g. highly specialized species, *K*-selected strategy species) have shown an invader behaviour.

For instance, in taxonomic groups of the same genus (e.g. *Rattus, Eichornia, Bromus,* etc.), some species have shown to be highly invasive, while other very related ones do not show any invader potential. It is a very challenging research topic to clarify why related species of the same genus reveal such a different behaviour. In other words, it is important not only to investigate why a species is an invader but also, equally relevant, why its closer relative *is not* an invader.

One can also wonder if the potential for invasion to a new territory has some correlation with the dispersal behaviour in the home region. There are examples leading to opposite conclusions. For instance, *Pinus pinaster* and *Pinus radiata* are both strongly invaders in South Africa (MacDonald *et al.* 1986), while only *Pinus pinaster* is slightly invasive in Southern France as compared with close species. *Pinus radiata*, on the contrary, has a quite restricted distribution in its home country, California.

In any event, one can stress that factors extrinsic to species biological attributes (e.g., historical opportunities or characteristics of the recipient ecosystems) are often the most important ones in determining the success of an invasion.

Finally, the mobility of modern society and the globalization of its functioning will probably change the characteristics of the 'ideal invader'.

The chance to be an invader

In line with some of the above paragraphs, this section will deal more with chance and historical opportunities than with intrinsic biological attributes. Disturbance will be a key-word; the meaning of endogenous or natural disturbance and of exogenous or man-made disturbance is that proposed by Fox & Fox (1986).

Table 2 summarizes the three main conditions; a kind of pre-adaptation acquired in the home country; the opportunity and the requisites to afford

Table 2. Conditions facilitating the potential to invade new territories.

Historical conditions in the home territory	Facilities for transportation and/or migration	Local conditions facilitating colonization by new invaders
Geological and evolutionary history of recent natural disturbance (e.g. glaciations, regional tectonic pulses, frosts, droughts)	Intensive exchanges of people and their products because of trade, colonization or war Rapid transportation systems	Existence of open spaces and spare resources Ecosystems subject to frequent natural disturbances
Early man-related history of exogenous disturbance (e.g. grazing pressure by large herbivores, fires)	High vagility of invaders Phoresis (active transportation of small animals by larger insects, birds, etc.)	Man-disturbance of ecosystems similar to that of the home territory
Domestication by man and commensalism	Longevity of seeds, resistance of spores and possibility of long dispersal (by oceanic currents, wind, etc.)	Absence of pathogens, parasites, predators, competitors left behind in the home territory
		Homoclimatic and mostly homocultural (similar land-use patterns) conditions as compared to home territory
		'Insularity' conditions (evolutionary history with isolation patterns) in islands, southernmost tips of continents (South America, South Africa, Australia) and western fringes of continents (e.g. Chile)

a long transportation; and the finding of non hostile conditions at the arrival to the new territories. Expressed in more anthropogenic terms, one can say: 'receiving a basic background open to alternative professions', 'catching the right ship' and 'becoming better competitor than the local species'. In all these steps, there is an interplay of chance and necessity, of human-derived opportunities and of evolutionary heritages.

It is undeniable that some of the conditions of the first and second columns are more applicable to species originating in Europe and the Mediterranean Basin, so that, in general, species from the Old World have a greater potential for invasion than those of the other continents (di Castri 1989a).

Some of the factors included in Table 2 are self-explanatory. I would like to put some emphasis only on a few points.

First of all, when man-made disturbances have similar effects than previous or concomitant natural disturbance in the same region, the pre-adaptive patterns are magnified. This is the case, for instance, of natural fires (and the selection for fire-adapted plants) followed later by man-made fires to open new spaces for agriculture and grazing; the grazing pressure by large wild herbivores, followed by grazing of domesticated animals; the alternative phases of clearing because of tectonic phenomena, killing frosts and extended droughts, followed later on by forest clearing by man to establish primaeval agricultural

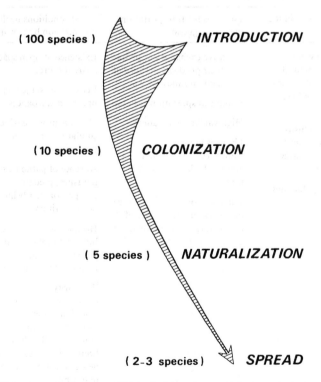

Figure 4. Main steps of a process of biological invasion, and the hypothetical loss of invading species at each step.

fields. The pre-adaptations because of this kind of natural/man-made 'relay' increase considerably the potential for invasion, mainly because the similarity between old and newly-established landscape patterns seems to have a greater relevance for invasion that the simple similarity of climates between the former and the new territories. In conclusion, invaders are favoured in new habitats when those habitats are subjected to disturbances that are novel to native species, but similar to those long experienced by the invading species.

Finally man, by multiplying the 'insularity' conditions throughout the fragmentation of land, further increases the chance for successful biological invasions. Island-like ecosystems are the most prone to be invaded.

Among the factors enumerated in Table 2, commensalism would be worth of a more extended discussion, as one of the most important conditions dealing with species dispersion and invasion related to man. As regards the common house mouse (*Mus musculus domesticus*), ecological and ethological processes are implied in the chromosomic differentiations between wild and commensal populations (Auffray *et al.* 1986, 1988, Bonhomme *et al.* 1983). Such aspects are further discussed in the chapter of Michaux *et al.* (this volume).

In any event, a simple introduction of a species in a given territory does not necessarily implies its naturalization and dispersal. On the contrary, its chance is rather limited as exemplified in Figure 4. It should be stressed, for instance, how few of the arthropods and other small animals falling down from the 'aerial plankton', after long-range transportation by wind, have a real possibility of colonization.

The natural resistance of ecosystems to biological invasions

It has been empirically observed that some regions or ecosystem types are more vulnerable than others as regards biological invasions. In addition, analogous ecosystems situated in different biogeographical regions show a different susceptibility to invasion. For instance, ecosystems from the Northern Hemisphere (mostly from the Old World) are more resistant than those of the Southern Hemisphere, particularly their southernmost parts. Timing of invasion has been also different in the various regions of the world.

Comparing the five regions of the world with a mediterranean-type climate (di Castri 1981, Fox, this volume), that is to say, the Mediterranean Basin, California, Chile, South Africa, South-Western and Southern Australia, ecosystems of the Mediterranean Basin are the most resistant to invasion, and even represent the home-country of most invaders to the other regions.

As a basis for intercontinental and inter-region comparisons on invasibility, see Drake *et al.* 1989, Duffey 1988, Groves & Burdon 1986, Kornberg & Williamson 1987, Macdonald *et al.* 1986, Mooney & Drake 1986, Sauer 1988, Wilson & Graham 1983.

Patterns of disturbance regime and degrees of biogeographical isolation during ecosystems evolution (di Castri 1989b) are the two main factors to explain such differences.

It has been postulated that no invasion can occur without a previous disturbance of the recipient ecosystem (Fox & Fox, 1986, see also Kornas, this volume). I tend to agree, so far it is accepted that there are so subtle alterations that are badly perceivable by man, and that species introduction can represent *per se* a disturbance. According to the disturbance hypothesis, ecosystems subjected to irregular massive disturbances (floods, killing frost, etc.) and to episodic extreme events are very susceptible to invasion; ecosystems with recurrent disturbance are prone to invasion, while those free from disturbance (but this is a biological abstraction) or with low-intensity disturbance are invasion-resistant. Nevertheless, it is important to differentiate between the long-term disturbance regime and the disturbance regime in place at the time of invasion; the most invasion-prone system may be one where disturbance is traditionally mild and infrequent, but where new forces cause intense and frequent disturbances.

Following the species richness hypothesis (Fox & Fox 1986) on the way that rich communities would be less susceptible to invasion, communities with many interacting species and high interconnectedness would be better suited to fully utilize existing resources, and therefore able to prevent new species to become involved. It seems an acceptable speculation, but there are little experimental data to prove or disprove this hypothesis.

As a matter of fact, it happens that species of *Hakea* from Australia as well as Northern Hemisphere *Pinus* (e.g. *pinaster, radiata*) are able to invade natural ecosystems with high diversity in Southern Africa (Macdonald *et al.* 1986). In Europe, close natural (or almost natural) forest ecosystems are very resistant to biological invasions, and very few exceptions can be quoted (see Kornas, this volume). Susceptibility to invasion sharply increases under conditions of stress, for instance, because of acid rains.

Riverine and riparian ecosystems, where intermediate to high disturbance regime is an intrinsic feature, are more easily invasible; fishes from South-Eastern North America such as *Gambusia affinis*, from North America as for instance *Ameiurus nebulosus* and *Salmo gairdneri* (rainbow trout) have been successful invaders, among others, as well as *Ondatra zibethica* (the North American muskrat) and *Myocastor coypus* from South America. Coastal and urbanized environments – the most disturbed ecosystems – show a profusion of invaders.

Most of the biological invasions to Europe, and certainly the most conspicuous ones (*Rattus rattus, Rattus norvegicus, Mus musculus, Passer domesticus*, etc.) come from the East (including the Far East); incidentally, most of the human migrations and invasions in historical times show the same East-West direction. Also neolithization, with all the cortège of associated plants and animals, appeared some 4.000 years earlier in the Eastern Mediterranean, shifting slowly towards the Western Mediterranean (Le Houérou, 1981). Many of the old migrations of Coleoptera (see Marcuzzi, this volume) are in a South-North direction in the Mediterranean.

After the discovery of Americas, many invasions to Europe came from the Northern Eastern part of North America (see Quézel *et al.* this volume), with little or no matching in relation to homoclimatic conditions.

A particular case is that of terrestrial ecosystems of the Mediterranean Basin: they have been and are still subjected to very frequent natural disturbances, mainly droughts and, more rarely, killing frosts, and they have a most ancient history of continuous man-made disturbances who have almost completely shaped the Mediterranean landscape. On the other hand, they lack an history of massive disturbance of the overall territory (like the glaciations of Northern-European regions), and the primaeval human disturbance, through partial clearing, grazing and agriculture, has rather increased the species diversity at the landscape level. In addition, because of their crossroads biogeographical situation, they have been submitted to successive waves of migrations in accordance with the climatic shifts of the glacial and interglacial periods. Mediterranean biota constitute therefore a rich mixture of elements of different biogeographical origin, and also of diverse ecological tolerance in view of the great heterogeneity of the landscape. *In situ* speciation, tolerance to new conditions ('pre-adaptation') including of an anthropic origin, and colonization as an 'escape response' from other ecosystems (cooler or dryer) shape all together the rich mosaic of Mediterranean biota (di Castri 1981, 1989a and b).

Mediterranean ecosystems tend to be resistant to invasion (but new conspicuous succulents have been introduced from Americas, like *Opuntia*, and from South Africa), but their resistance can be formulated in quite a peculiar way: mediterranean ecosystems are rich in species and invasion-resistant because they have been subjected to continuous intermediate disturbances and to very frequent migrations. First of all disturbance was of endogenous kind, but later of both endogenous and exogenous nature in a close feedback. Resistance exists mainly because these ecosystems have been invaded and colonized several times in geological and historical terms by invaders of different biogeographical origin (all available resources having being utilized and all niches filled in), and because a series of early relations with man up to commensalism and domestication have been developed. In other words, since in mediterranean biota there are so many 'old' invaders, they are able to prevent the access of 'new' invaders. In addition, thanks to the old association with human activities, mediterranean taxa are 'preadapted competitors' to colonize new man-modified environments and, transported intentionally or inadvertedly by man, to invade and utilize opportunistically spare resources and open spaces appearing in distant territories due to man impact.

Conclusions

The mechanisms of a biological invasion have not often an unequivocal explanation, because of an almost inextricable mixture of biological (including evolutionary), environmental and anthropic (including historical) factors. Experimental research is not always possible, predictive modelling and risk assessment imply intrinsic difficulties, and preventive measure and adequate legislation are often lacking in several countries (or are inefficiently applied). Biological control, that is to say, researching, finding and spreading a 'natural' enemy of the invader has proved to be successful in a number of cases, sometimes associated with the use of chemicals (herbicides, pesticides).

It remains that the problem of biological invasions has also very great economic repercussions. Possibility of improving our predictive capacity on future trends of biological invasions is even more glooming, because of the current ecological and economic framework.

What will be the behaviour of invaders in a world (or a region) with an impressively declining biological diversity? How invaders will react in face of new disturbances and stresses, such as acid rains or massive eutrophication of freshwater and coastal ecosystems? What will be the consequences on the invasion potential of further fragmentation and homogeneization of landscapes and of vast agglomerations of built-environment? Nobody is now able to reply to these kinds of questions, and important research efforts on these topics should be continued and improved, dealing not only with the successful invaders, but also evaluating failure rates and assessing loosers' behaviour.

Furthermore, the unquestionable importance of genetic bioengineering should be matched by equally important ecological considerations. Attributes of organisms and environments to be considered in risk evaluation before release are numerous and interacting in complex ways (see tables in article of Tiedje *et al.* 1989). These genetically engineered organisms transcend the existing biogeographical realms, and should be viewed as belonging to a kind of 'anthropogenic realm'.

But the main emerging problem is that of the progressive globalization of ecological as well as of economic features. Even the meaning of biological invasions may be interpreted in a different way in a world where trades and markets respond to global driving forces, and where entire ecosystem types are under threat. And to this dynamic and unpredictable 'film' of human current transformations, it should be superimposed that of the impending man-induced climatic change (greenhouse effect, ozone depletion). This change will act simultaneously on all biota of the world, preceeded and in concomitance with higher frequency of extreme climatic events (Wigley, 1985). At that moment, an intensive selective pressure will be exerted world-wide on all biota to change their habitat or their requirements.

They will all become, to a different extent and degree, invaders or loosers. And as in the past, this biological *necessity* to change will have different fate according to the facilities or obstacles of different *chance* offered intentionally or inadvertedly by man's own historical destiny.

References

Auffray, J.-C., Cassaing, J., Britton-Davidian, J. & Croset, H. 1986. Les populations sauvages et commensales de *Mus musculus domesticus*. Implication des structures populationnelles sur la différentiation caryologique. In: *Coll. Nat. CNRS Biologie des Populations*, CNRS, Paris: 279-285.

Auffray, J.-C., Tchernov, E. & Nevo, E. 1988. Origine du commensalisme de la souris domestique (*Mus musculus domesticus*) vis-a-vis de l'homme. *C.R. Acad. Sci. Paris.* 307 (Série III): 517-522.

Bacon, P.J. (ed.) 1985. *Population dynamics of rabies in wildlife.* Academic Press, London, 358 p.

Baker, H.G. 1965. Characteristics and modes of origin of weeds. In: H.G. Baker & G.L. Stebbins (eds.), *The Genetics of Colonizing Species.* Academic Press, New York: 147-169.

Barret, S.C.H. & Richardson, B.J. 1986. Genetic attributes of invading species. In: R.H. Groves & J.J. Burdon (eds.), *Ecology of Biological invasions: an Australian perspective.* Australian Academy of Science, Canberra and Cambridge Univ. Press, Cambridge: 21-33.

Bonhomme, F., Catalan, J., Gerasimov, S., Orsini, P. & Thaler, L. 1983. Le complexe d'espèces du genre *Mus* en Europe centrale et orientale. I-Génétique. *Z. Saügetierkunde,* **48**: 78-85.

Braudel, F. 1979. *Civilisation matérielle, économie et capitalisme, XVe-XVIIIe siècle.* Tome 2. Les jeux de l'échange. Armand Colin, Paris, 600 p.

Crosby, A.W. 1986. *Ecological imperialism. The biological expansion of Europe, 900-1900.* Cambridge Univ. Press, Cambridge, 368 p.

di Castri, F. 1981. Mediterranean-type shrublands of the world. In: di Castri, F., Goodall, D.W. & Specht, R.L. (eds.), *Mediterranean-type Shrublands.* Ecosystems of the World 11, Elsevier, Amsterdam: 1-52.

di Castri, F. 1989a. History of biological invasions with special emphasis on the Old World. In: Drake, J.A., Mooney, H.A., di Castri, F., Groves, R.H., Kruger, F.J., Rejmánek, M. & Williamson, M. (eds.), *Biological invasions. A Global Perspective.* SCOPE 37. John Wiley & Sons, Chichester: 1-30.

di Castri, F. 1989 b. The evolution of terrestrial ecosystems. In: Ravera, O. (ed.), *Ecological Assessment of Environmental Degradation, Pollution and Recovery.* Elsevier, Amsterdam: 1-30.

di Castri, F. & Hadley, M. 1988. Enhancing the credibility of ecology: interacting along and across hierarchical scales. *GeoJournal* 17(1): 5-35.

Drake, J.A., Mooney, H.A., di Castri, F., Groves, R.H., Kruger, F.J., Rejmánek, M. & Williamson, M. (eds.) 1989. *Biological Invasions. A Global Perspective.* SCOPE 37. John Wiley & Sons. Chichester, 525 p.

Duffey, E. (ed.) 1988. Special Issue: Biological Invasions of Nature Reserves. *Biological Conservation* 44(1-2): 1-135.

Elton, C.S. 1958. *The ecology of invasions by animals and plants.* Methuen, London, 181 p.

Fox, B.J. & Fox, M.D. 1986. Resilience of animal and plant communities to human disturbance. In: Dell, B., Hopkins, A.J.M. & Lamont, B.B. (eds.), *Resilience in Mediterranean-type ecosystems.* Dr. W. Junk Publ., Dordrecht: 39-64.

Groves, R.H. & Burdon, J.J. (eds.) 1986. *Ecology of biological invasions: an Australian perspective.* Australian Academy of Science, Canberra and Cambridge Univ. Press, Cambridge, 166 p.

Kornberg, H. & Williamson, M.H. (eds.) 1987. Quantitative aspects of the ecology of biological invasions. *Phil. Trans. R. Soc. Lond.* B 314: 503-742.

Le Houérou, H.N. 1981. Impact of man and his animals on Mediterranean vegetation. In: di Castri, F., Goodall, D.W. & Specht, R.L. (eds.), *Mediterranean-type Shrublands.* Ecosystems of the World 11, Elsevier, Amsterdam: 479-521.

Macdonald, I.A.W., Kruger, F.J. & Ferrar, A.A. (eds.) 1986. *The ecology and management of biological invasions in Southern Africa.* Oxford Univ. Press, Cape Town, 324 p.

Mooney, H.A. & Drake, J.A. (eds.) 1986. *Ecology of biological invasions of North America and*

16

Hawaii. Springer-Verlag, New York, 321 p.

Newsome, A.E. & Noble, I.R. 1986. Ecological and physiological characters of invading species. In: R.H. Groves and J.J. Burdon (eds.), *Ecology of biological invasions: an Australian perspective.* Australian Academy of Science, Canberra and Cambridge Univ. Press, Cambridge: 1-20.

Niethammer, G. 1969. *Some problems connected with the House Sparrow's colonization of the world.* The Ostrich, Suppl. N.8, Proc. IV Pan-African Ornith. Congress: 445-448.

Sauer, J.D. 1988. *Plant migration. The dynamics of geographic patterning in seed plant species.* Univ. of California Press, Berkeley, 282 p.

Thellung, A. 1908-1910. La flore adventice de Montpellier. *Mém. Soc. Nat. Sc. Natur. Math. Cherburg.* 37:57-728.

Tiedje, J.M., Colwell, R.K., Grossman, Y.L., Hodson, R.E., Lenski, R.E., Mack, R.N. & Regal, P.J. 1989. The planned introduction of genetically engineered organisms: ecological considerations and recommendations. *Ecology* 70(2): 298-315.

Wigley, T.M.L. 1985. Impacts of extreme events. *Nature* 316: 106-107.

Wilson, C.L. & Graham, C.L. (eds.) 1983. *Exotic plant pests and North American agriculture.* Academic Press, New York, 522 p.

PART TWO

Plant invasions

2. Plant invasions in Central Europe: historical and ecological aspects

JAN KORNAŚ

Abstract

Since the introduction of agriculture to Central Europe ca. 7,000 years ago, man-accompanying (synanthropic) plants have increasingly expanded in this area. Both native and introduced species were involved, most of them having a 'general purpose' genotype which seems to result from the 'r' type of selection. At present, permanently established aliens constitute 10–20% of the local floras in Central Europe. In this paper, groups of aliens of various immigration ages and degrees of naturalization are defined and illustrated with case histories. Four stages are distinguished in the process of naturalization of alien plants: (1) introduction of propagules and emergence of first individuals, (2) establishment in heavily disturbed sites, (3) colonization of less disturbed sites, (4) invasion into undisturbed sites, each subsequent stage being more difficult to achieve than the previous one. Consequently, most of the aliens occur only in man-made ruderal and/or segetal plant communities, and very few of them were able to penetrate into undisturbed natural vegetation.

Introduction

Central Europe is among those parts of the world, that have been profoundly changed by human activities. Since the Neolithic Period (i.e. for ca. 7,000 years) sedentary peoples have practiced agriculture and cattle-raising here, and exerted a growing pressure upong the living and non-living environments. Consequently, an ever increasing alteration of the plant cover resulted – a phenomenon called synanthropization (Faliński 1966, 1975, Kornaś & Medwecka-Kornaś 1967, 1974, 1986, Kornaś 1982, 1983). It consists of several processes as specified in Table 1.

In the present Chapter I shall concentrate mainly upon one of the floristic aspects of synanthropization, namely the expansion of man-accompanying (synanthropic, hemerophilous) taxa. Changes in the flora, however, are closely correlated with those in the vegetation, and therefore, when discussing the

F. di Castri, A. J. Hansen and M. Debussche (eds.), Biological Invasions in Europe and the Mediterranean Basin.
19–36. © 1990, *Kluwer Academic Publishers, Dordrecht.*

20

Table 1. Processes of synanthropization of plant cover (Kornaś 1982).

Changes in the Flora
- expansion of taxa positively affected (man-accompanying = synthropic, hemerophilous plants)[a]
- decline of taxa negatively affected (retreating = hemerophobous plants)[b]
- evolutionary changes of the taxa themselves

Changes in the Vegetation
- expansion of hemerophilous plant communities
- decline of hemerophobous plant communities
- formation of new (anthropogenous) plant communities

[a] Linkola 1919, Jalas 1955, Kornaś 1966.
[b] Linkola 1919, Kornaś 1966.

floristic consequences of human activities, I shall often be referring to the processes at the community level. I shall be using Central European examples to demonstrate the general principles concerning the invasions of synanthropic plants. I shall start with a series of case histories, and close with an attempt towards conclusions of a more universal significance. I shall be dealing mainly with the vascular plants, but most of my conclusions may be applied also to non-vascular plants and animals (especially invertebrates).

Sources of information

Thanks to the activity of many generations of botanists there is a wealth of factual material concerning man-made changes in the flora of Central Europe. This information has not yet been sufficiently synthesized. It is contained in two main sources: the paleo (-ethno) -botanical data, based on subfossil plant remains and dating back from both historic and prehistoric times (Willerding 1978), and the floristic data (either published or included in herbarium collections) which usually go back no more than 150–200 (– 400) years. Only the latter are reasonably complete and enable us to produce dot maps of changes in the distribution of individual taxa (Figures 1, 3, 4, 5). Therefore, they will form the main subject of my considerations.

Diversity of man-accompanying plants

The synanthropic plants (Table 2) consist of both native (indigenous) taxa, occurring spontaneously on river banks, land slides, sand dunes, etc., which have secondarily spread into man-made habitats (apophytes), and of aliens which have emerged in a given area thanks only to man (anthropophytes). The history of expansions of native synanthropic plants is still rather poorly known, although there exists some paleobotanical evidence of these events (Ralska-Jasiewiczowa 1981, Trzcińska-Tacik & Wasylikowa 1982). At present,

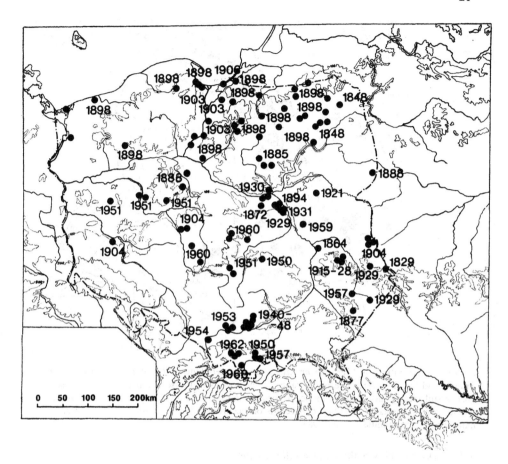

Figure 1. Expansion of *Elsholtzia ciliata* in Poland based on literature and herbarium records (dates of earliest collections indicated). *E. ciliata* is a species of temperate Asiatic origin which invaded Poland in the 19th century: it remains strictly limited to ruderal habitats in villages thus representing the group of epoecophytes. Świeboda 1963.

more than a half of the total number of plant species occurring in manmade habitats of arable fields, ruderal places, urban sites and industrial environments in Central Europe are natives, and some of them, e.g. *Elymus repens* (*Agropyron repens*), *Calamagrostis epigejos*, *Tussilago farfara*, etc., proved to be exeptionally resistant against adverse conditions of air and soil pollution, water stress, mechanical disturbance, and so on. These species are of particular interest for recultivation of areas devoid of vegetation, e.g., industrial waste heaps.

We are much better informed about the invasions of aliens (anthropophytes). The general patterns of these invasions are discussed by Thellung 1915, 1918–1919, Ridley 1930, Meusel 1943, Lousley 1953, Andersson 1956, Elton 1958, Kornaś 1966, 1982, 1983, Kornaś & Medwecka-Kornaś 1967, 1986, Sukopp

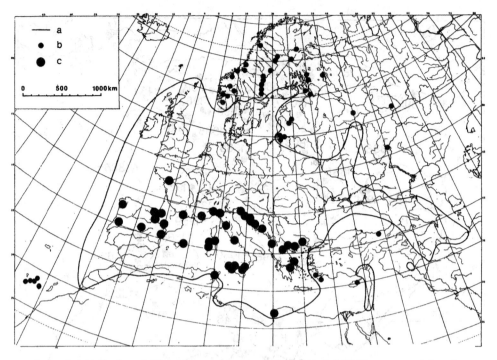

Figure 2. Distribution of *Sherardia arvensis* – an archaeophyte of Mediterranean origin – in Eurasia and North Africa. The initial range of the species in the southern part of the present area is still clearly indicated by its occurrence in natural vegetation (a – limit of continuous distribution, b – isolated localities, c – records on occurrence of *S. arvensis* in natural or semi-natural plant communities). Zając 1983.

Table 2. Diversity of synanthropic plants (Kornaś 1982).

Natives (apophyta)
- permanently established in man-made habitats (eu-apophyta).
- introduced temporarily (apophyta ephemera)
- escaped from cultivation (oekiophyta)

Aliens (anthropophyta)
* Permanently Established (metaphyta)
  older immigrants, before 1500 A.D. (archaeophyta)
  - introduced (archaeophyta adventiva)
  - man-made (archaeophyta anthropogena)
  - survived in man-made habitats only (anthropophyta resistentia)
  newcomers, after 1500 A.D. (kenophyta = neophyta sensu Meusel 1943)
  - established only in ruderal and/or segetal communities (epoecophyta)
  - established in semi-natural communities (hemiagriophyta)
  - established in natural communities (holoagriophyta = neophyta sensu Thellung 1915)

* Not Permanently Established (diaphyta)
  - introduced temporarily (ephemerophyta)
  - escaping from cultivation (ergasiophygophyta)

For discussion and further references see: Kornaś 1966, 1968a, 1978, 1982, 1983, Trzcińska-Tacik 1979, Zając 1979, Kornaś & Medwecka-Kornaś 1986.

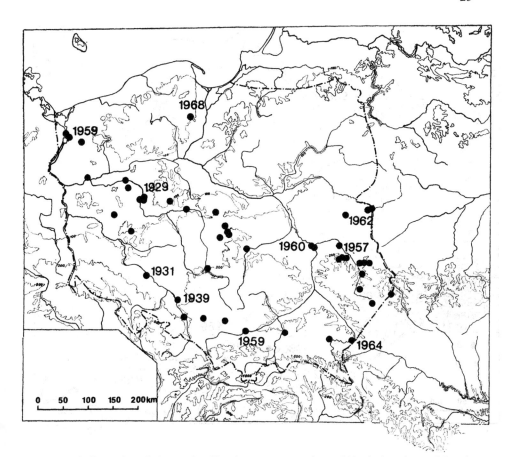

Figure 3. Expansion of *Amaranthus blitoides* – an epoecophyte of North American origin – in Poland (dates of earliest collections indicated). The species remains strictly limited to ruderal habitats and seems not to be able to penetrate to the northeastern part of the country, most probably because of climatic barriers. Frey 1974.

1969, 1972, Perring 1970, Baker 1972, Hawksworth 1974, Hill 1977, Fukarek 1979. Further references to case histories from Central Europe are to be found in Krawiecowa 1951, Sukopp 1962, 1966, Kornaś 1966, 1968b, 1971a, Kornaś & Medwecka-Kornaś 1968, Faliński 1968, 1971, 1972, Olaczek 1976, Trzcińska-Tacik 1979, Zając 1979).

The very beginning of the expansion of aliens to Central Europe was obviously connected with the introduction of agriculture ca. 7,000 years ago. Since then their immigration has continued at a variable pace until the present. We therefore need to distinguish groups of various immigration age among the anthropophytes (Table 2). The major division is that between the older immigrants (archaeophytes) and the newcomers (kenophytes = neophytes sensu Meusel 1943), with the demarcation point set usually at the beginning of the

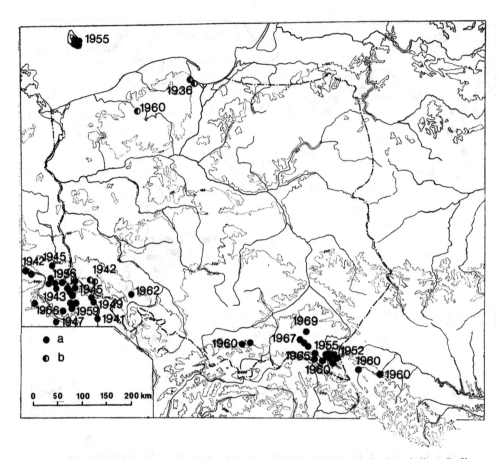

Figure 4. Expansion of *Veronica filiformis* in Poland (dates of earliest collections indicated). *V. filiformis* is native to the mountains of Anatolia and the Caucasus; introduced to Europe as an ornamental, it became a hemiagriophyte established in semi-natural plant communities of pastures and meadows, but only in montane and maritime areas with cool and humid climate (a – records from semi-natural plant communities, b – records from man-made habitats of gardens and parks). Pietras 1970.

Modern Era (A.D. 1,500), when – after the discovery of America – the first New World weeds appeared in Europe.

The older immigrants

At least 140 species of archaeophytes occur in Poland at present (Zając & Zając 1975, Zając 1979), and numbers of a similar magnitude are also indicated from other Central-European countries. Approximately 70% of these species originated in the Mediterranean Region and the Near East, i.e., in the same areas from which agriculture had been introduced to Central Europe (Figure

2). Many of them certainly arrived at our territory together with the cultivated crops (Zając 1979, 1983, 1986). All unquestionable archaeophytes in Central Europe remain strictly limited to arable fields and/or ruderal sites; they never occur in undisturbed natural vegetation. We unfortunately cannot determine whether there also were older immigrants which were able to penetrate into natural vegetation and to become fully established there. If there were such cases, we would not be able at present to distinguish these fully naturlized immigrants from the native non-synanthropic species.

Paleobotanical records on the earliest known occurrences of individual species of archaeophytes in Central Europe are spread over the whole range of periods from the Neolithic (ca. 6,500 years ago) till the Middle Ages. The following exemplaes from Poland (Wasylikowa & Gluza 1977) clearly illustrate this situation:

(a)*Neolithic* (ca. 6,500–3,700 years B.P.): *Agrostemma githago, Avena fatua, Bromus secalinus, Chenopodium urbicum, Buglossoides arvensis (Lithospermum arvense), Bilderdykia convolvulus (Polygonum convolvulus), Sinapis arvensis*;

(b)*Halstatt* (ca. 2,700–2,400 years B.P.): *Anagallis arvensis, Hyoscyamus niger, Lolium temulentum, Malva neglecta, M. sylvestris, Setaria pumila, S. viridis, Thlaspi arvense*;

(c)*Roman* (ca. 2,000–1,570 years B.P.): *Echinochloa crus-galli*;

(d)*Early Mediaeval* (ca. 1,350-700 years B.P.): *Anthemis arvensis, Consolida regalis, Euphorbia helioscopia, Galium tricornutum, Geranium dissectum, Leonurus cardiaca, Neslia paniculata, Onopordum acanthium, Urtica urens, Verbena officinalis*;

(e)*Late Mediaeval* (ca. 700–500 years B.P.): *Anthemis cotula, Silene gallica*.

Recently, a marked decline of many archaeophytes has been noticed. It is caused by improvements of agricultural methods, which eliminate more and more ancient segetal weeds, and by the urbanization of rural areas, which leads to the extinction of the old-fashioned ruderal flora in the villages (Kornaś 1971b, Sukopp & Trautmann 1976, Sukopp *et al.* 1978, Hilbig 1982, Weeda 1985, Svensson & Wigren 1986).

The newcomers: case histories

There is a list of several hundred successsfully established recent synanthropic immigrants to Central Europe, and many of them have been studied in much detail. They may be divided into three groups according to the degree of their naturalization:

(1)those occurring only in man-made (ruderal and/or segetal) plant communities (epoecophytes),

(2)those established also in semi-natural plant communities (of meadows, pastures, disturbed river banks, lake shores, etc.) (hemiagriophytes), and

(3)those which succeeded in establishing themselves in entirely undisturbed natural vegetation (holoagriophytes).

Let me now discuss a few typical case histories for each of these groups, based mainly on the data from Poland.

The epoecophytes

The majority of permanently established newcomers in Central Europe are strictly limited to ruderal sites and/or arable fields. This is true of a large group of species of temperate North American origin, such as *Amaranthus retrofluxus* (Frey 1974), *Conyza canadensis (Erigeron canadensis)* (Wein 1932, Wagenitz 1964–1979) and *Oxalis europaea* (Hegi 1924, Hantz 1979), all established in Poland for more than a hundred years, or *Amaranthus albus* (Frey 1974), *A. blitoides* (Figure 3) and *Lepidium densiflorum* (Kobendza 1950, Latowski 1982), introduced more recently and spreading at present as ruderal plants in towns. An especially interesting case is that of *Galinsoga parviflora*, an annual of South American origin which escaped from cultivation from various European botanical gardens and appeared in Poland in the early 19th century (Majdecka-Zdziarska 1929, Wagenitz 1964–1979). It is presently one of the most common weeds of ruderal places, gardens and root-crop cultures. The rate of growth of its European populations was exponential, as demonstrated for the Netherlands by Van Soest (1941), perhaps because of the lack of competition in the open habitats it had colonized. *G. parviflora* is a subtropical plant, very sensitive to low temperatures. In spite of this, it was able to become naturalized in temperate regions thanks to its very short life-cycle of a few weeks, and a very high production of seeds which withstand winter frost well.

The second important group of epoecophytes in Central Europe, comes from continental regions of Eastern Europe and Asia. It is represented, for example, by *Senecio vernalis*, which has its original home in the steppes of southern and central Russia (Meusel 1943, Kornaś 1966). It penetrated into Poland on a wide front in the 19th century, moving very rapidly from the east westwards. This migration went on chiefly through the agency of wind, which carried the light-plumed achenes. Man contributed to the migration only indirectly, by creating – through the destruction of natural vegetation – suitable habitats for the new arrival on rubble, fallow land, embankments, etc. It is very symptomatic that *S. vernalis*, a markedly continental species, followed in its expansion essentially the same routes along which the native xerotherms (plants capable of withstanding drought and heat) previously migrated. Almost all weeds of steppe origin used the same migrational routes.

The hemiagriophytes

As a representative of the newcomers which were able to settle down in semi-

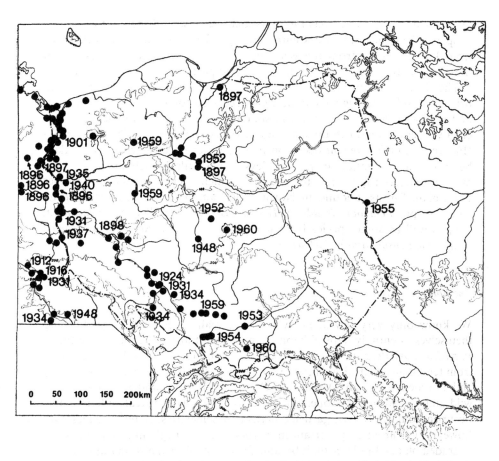

Figure 5. Expansion of *Bidens frondosa* – an hemiagriophyte of riverside habitats – in Poland (dates of earliest records indicated). This species, of North American origin, penetrated from the north-west to the south and east using both river valleys and railroad tracks as migratory routes. Trzcińska-Tacik 1961.

natural habitats, I should like to mention *Veronica filiformis* (Fig. 4), a mountain species growing naturally in Anatolia (Turkey) and Caucasus (USSR). It spreads very successfully in Western and Central Europe, in regions with humid maritime or mountain climates, but markedly avoids drier continental areas. It is a perennial of garden lawns, pastures and meadows, and reproduces mainly vegetatively by broken fragments of shoots. In Poland it has recently appeared in the Sudetic Mts. (since 1942) and the Carpathian Mts. (since 1952, Pietras 1970).

Another example of this group of newcomers is *Juncus tenuis*, of North American origin. It was introduced into Europe accidentally in the first half of the 19th century, undoubtedly in connection with the transport of goods by sea, in as much as its first stations were always in the neighbourhood

28

of ports (Meusel 1943, Kornaś 1966). Its sticky seeds were dispersed by vehicles, people and animals all over Central Europe. The first mention of *J. tenuis* from Poland comes from the year 1901, and at present the species occurs most frequently on wet paths in meadows and pastures and on forest roads.

The degree of naturalization of several other aliens of North American origin is similar. *Epilobium adenocaulon* occurs on forest edges and clearings, and is now invading large areas in southern Poland. *Erigeron annuus* grows chiefly at edges of shrub-thickets and in riverside meadows. *Bidens frondosa* (Figure 5) and *B. connata* occur on muddy water edges of rivers and lakes. Semi-natural vegetation in river valleys is particularly rich in newcomers. Willow thickets (*Salix* sp.), for example, which occupy the sites of former flood-plain forests, abound in North American species, e.g., *Acer negundo, Aster* sp. pl., *Echinocystis lobata, Helianthus tuberosus, Rudbeckia laciniata*, etc., as well as some species from temperate Asia, e.g., *Reynoutria japonica (Polygonum cuspidatum), Sorbaria sorbifolia*, etc.

The holoagriophytes

We know only very few aliens which have really succeeded in establishing themselves in entirely natural habitats in Central Europe. From among them, *Impatiens parviflora* deserves a special attention (Trepl 1984). It was brought from temperate Central Asia to the botanical gardens of Geneva (Switzerland) and Dresden (German Democratic Republic) in 1837, and very soon went wild and spread over Europe. For many years it was a typical ruderal plant, occurring only in towns, gardens, parks and cemeteries. A few decades ago, however, it started to penetrate into woods, at the beginning only to badly degraded places, but later on it became firmly established also in quite natural stands of deciduous forests (Figure 6). It is now the most serious invader in this type of habitat, even in places apparently not disturbed by man.

Another similar example is that of *Acorus calamus* which came to Central Europe from Western Asia through the agency of the Turks in the 16th century, and at the beginning of the 17th century succeeded in becoming completely established in entirely natural communities of reedswamps, so that now it gives the impression of being a fully native member of the flora (Wein 1939, 1940, 1942).

A quite exceptional case in which the newcomer proved to be so aggressive that it succeeded in displacing native species in apparently natural habitats, and formed a new plant community type is that of *Solidago gigantea* subsp. *serotina*. This North American plant, introduced again as an ornamental and escaped from cultivation, invaded river valleys in most parts of Poland and penetrated into various types of habitats, including those of shrub thickets and riverside forest (Wagenitz 1964–1979, Guzikowa & Maycock 1986). It often forms extensive single species stands from which the native plants are nearly completely excluded. However, we still do not know for sure whether

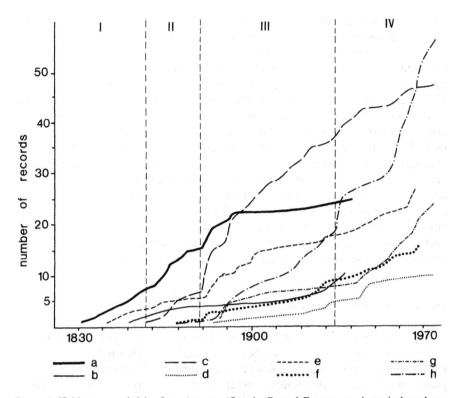

Figure 6. Habitats occupied by *Impatiens parviflora* in Central Europe: a – botanical gardens, b – culture plots, c – parks and cemeteries, d – industrial plants and timber yards, e – gardens and fencerows, f – ruderal places, g – railway tracks, h – natural forests. During its expansion in the last 150 years this species passed through all four successive stages of naturalization (I–IV) and became fully established in apparently undisturbed communities of deciduous forests. Trepl 1984, Kowarik 1985, Kowarik & Sukopp 1986.

this would have been the case, if the river valleys had not previously been modified, especially with respect to their water regime.

The introduced aquatic aliens seem to include species which do penetrate more easily than the terrestrial ones into undisturbed stands of indigenous vegetation. A well known case of this kind is that of the invasion of *Elodea canadensis* to Europe (Meusel 1943, Kornaś 1966).

Temporarily introduced aliens

A large body of evidence indicates that many recently introduced aliens (ephemerophytes) are not at all capable of becoming established under the new conditions, not to mention their spreading and colonizing further stations (Sukopp 1976, Kowarik 1985, Kowarik & Sukopp 1986). In some of them

30

seedlings do not develop into mature plants but perish soon after germination; others remain sterile or do actually flower but bear no fruits, or the seeds produced are not viable. Some of these alien species persist in their stations for a long time, or even reproduce there, and then suddenly disappear. Enormous numbers of such ephemeral arrivals were encountered in places with a constant influx of plant diaspores from remote areas of the Earth: in the sea-ports and river-ports, in the vicinity of large railway junctions, and in industrial centres, especially near textile factories, flour mills and oil mills. From among a total of 799 species of newcomers found in Montpellier (France) and its vicinity, no less than 692 turned out to be ephemerophytes (Thellung 1912). A list containing 1,633 species of 'wool adventives' was compiled by Probst (1949) from data for various centres of textile industry in Western and Central Europe. All these species, of course, do not play any role in the future of the local flora, but they certainly are worthy of studying because they very clearly demonstrate the difficulties which the aliens have to overcome when becoming established in a new area.

The newcomers: statistical data

Let me now try to evaluate the share of various groups of introduced species in the entire flora of a country, on the example of Poland (Kornaś & Medwecka-Kornaś 1967, 1968). The Polish flora consists of nearly 2,300 species of vascular plants, more than 2,000 species being native and about 250–300 species introduced by man and already well established in the country. Thus, the proportion of introduced and established aliens considerably exceeds 10%, similarly as in other European countries (Federal Republic of Germany 16%, British Islands 16%, Finland 18% – Sukopp 1972). No less than 531 species of temporarily introduced ephemerophytres have been recorded in Poland until now, and this figure may easily be enlarged through further observations. However, the ephemerophytes, being unable to maintain themselves permanently, cannot be regarded as true members of the flora of a country, and therefore they will be omitted in my further considerations.

More than 100 introduced and permanently established aliens are believed to be recent immigrants in Poland, having arrived since the beginning of the 16th century. The occurrence of 100 newcomers in plant communities has been carefully analysed. Fifty-seven species proved to grow only in ruderal places and/or arable fields (epoecophytes); 33 species are found also in semi-natural vegetation (hemiagriophytes), while only 10 species are reported to grow permanently in undisturbed vegetation (holoagriophytes). The distribution of 33 species established in semi-natural vegetation may be summarized as follows:
- various types of disturbed riverside habitats 17
- forest edges and felled forest areas 7
- dry secondary grasslands 4

31

– meadows, pastures and grassy foot-paths 3
– heaths and clearing in conifer forests 2

Of the 10 species of holoagriophytes which penetrated into undisturbed stands of natural plant communities, only 4 species (discussed already in the previous paragraphs) are really widely distributed throughout the country (*Impatiens parviflora, Acorus calamus, Solidago gigantea* subsp. *serotina, Elodea canadensis*). The remaining 6 species occur rarely or very rarely on river banks (2), along small rivulets (2), on maritime sandy beaches (1) and on limestone rocks (1). Only very few other doubtful or extremely rare examples could possibly be added to this list.

The above data clearly show that in Poland (and the same is true of all other Central European countries) introduced plant species most often occur only in man-made habitats and very seldom establish themselves as true members of the natural vegetation. Undisturbed natural plant communities seem to be highly resistant against the penetration of newcomers. This is especially evident in communities with a closed plant cover (forests, grasslands, bogs, etc.), but even some initial plant communities with many open spaces, e.g., on sand dunes, rocks, etc., offer only little opportunity for the newcomers to become established. The semi-natural types of closed vegetation (e.g., meadows and pastures) are similar in this respect. On the contrary, the open synanthropic vegetation of arable fields and ruderal places is a very suitable environment for invaders: to a lesser extent this is also true of semi-natural secondary communities with much bare ground, especially in river valleys, where more than 2/3 of all naturalized species of newcomers were found.

The newcomers: stages of naturalization, biological features of the plants

When analysing the case histories of recently introduced aliens in Central Europe, we inevitably come to the conclusion that the process of their immigration and naturalization usually follows a similar scenario (Kornaś 1966, 1982, 1983, Baker 1972, Trzcińska-Tacik 1979, Kowarik 1985, Kowarik & Sukopp 1986). In stage one, the propagules of the alien are accidentally introduced or deliberately imported to a new area, and the first individuals shoot up. In stage two, the newcomer becomes permanently established in one or more strongly disturbed sites. In stage three, the immigrant begins to colonize also such sites which are only slightly disturbed. Finally, in stage four, it can succeed in penetrating into wholly undisturbed sites (Figure 6). Many newcomers go through the first and second stages being cultivated (as crop plants, ornamentals, in botanical gardens, etc.). Each subsequent stage appears much more difficult than the preceding one. That is why many immigrants do not pass beyond stage one, i.e. beyond the status of temporarily introduced ephemerophytes. Much less numerous are those which become permanently settled in man-made (ruderal and/or segetal) vegetation (epoecophytes), and only a small fraction succeed to penetrate semi-natural

32

vegetation (of meadows, pastures, river banks, etc. – hemiagriophytes). This is even more true of those newcomers which are able to settle in completely natural communities (holoagriophytes). Only quite exceptionally a recently introduced alien immediately starts to colonize such undisturbed sites, 'jumping over' the stages two and three (as e.g., *Elodea canadensis* has done). Consequently, the numbers of aliens in more advanced stages of naturalization are always much lower than of those in the initial stages.

The situation can easily be explained by an action of twofold set of selective barriers which a newcomer has to overcome: that of climatic factors, which are acting more or less similarly in every place of a given area and eliminate most of the ephemerophytes, and that of biotic factors which follow a very intricate local pattern and prevent most of the species established in man-made habitats from penetrating into closed natural communities (Sukopp 1962, 1966, Harper 1965, Kornaś & Medwecka-Kornaś 1967, 1968, Faliński 1968). The importance of climatic selection is particularly emphasized by the fact that nearly all successful recent immigrants originated in regions with climates closely resembling that of the newly colonized area (those introduced to Central Europe mostly in temperate latitudes of North America and Asia). On this principle, a method has been even proposed for discovering equivalent climates in various parts of the world (Kerguélen 1963).

Many attempts have been undertaken to find out what is the biological background of expansiveness in man-accompanying plants. Various morphological, ecological, physiological and genetical features have been pointed out as helpful in this mode of life (Baker 1965, 1974, Baker & Stebbins 1965, Grant 1967, Mulligan & Findlay 1970, Hill 1977, Grime 1979, Kornaś 1982).

The most important and most common of them seems to be the wide range of ecological tolerance which enables the plants to survive in unstable and unpredictable environments. Its genetical expression is a 'general purpose' genotype as defined by Baker, connected with the characters which are singled out by the 'r' type of selection: early production of seeds in the life-history, short life duration (often an annual habit combined with self-compatibility), allocation of large proportion of resources to reproductive activities, production of large amounts of small seeds with highly modifiable dormancy period, effective seed dispersal, and very low competitive ability. Most of these characteristics of colonizing species ('ruderals' sensu Grime 1979) were noticed already as early as 1894 by the botanist Mac Leod (1894), who very aptly called this group of plants 'proletarians' (Hermy & Stieperaere 1985).

The reproductive strategies sorted out by the 'r' type of selection greatly assist the newcomers to succeed in the initial stages of naturalization. They do not suffice, however, to overcome the biotic barriers in stage four, when features of 'K'-plants ('competitors' sensu Grime 1979, 'capitalists' sensu Mac Leod 1894) are much more serviceable: high competitive ability based on effective vegetative reproduction, robust growth form, perennial habit, etc. Very few synanthropic plants (as e.g., *Solidago gigantea* subsp. *serotina*) are able to combine the two contrasting groups of characteristics necesary to pass over both the early and the late stages of naturalization. This is probably the main reason why only so few really successful invaders (holoagriophytes) exists among hundreds of newcomers.

Perspectives

What are the prospects of future plant invasions in Central Europe, as inferred from the present knowledge of these phenomena? It seems that there has already been a marked decrease in new arrivals in the recent years, possibly because most of the major regions of the world did already exchange the majority of their perspective synanthropic taxa with Europe. This trend will most probably continue also in the future and therefore we have to expect a rather decreased influx of aliens to our area. We are, of course, unable to forsee which species are to be expected as new arrivals. The most probable candidates for this are plants with wide distributional areas in their home countries (Forcella & Wood 1984, Forcella 1985). Annuals with 'r'-strategy are most likely to continue to play the role of epoecophytes, while perennial 'K'-strategists may render a few further most aggressive invaders (holoagriophytes). The future share of aliens in the flora of Europe will depend upon the degree of disturbance of the existing plant cover, especially of the still remaining patches of semi-natural and natural vegetation. This is unpredictable in details, but it certainly will be more and more increasing in the course of time. At any rate, we have to regard both the present and future migrations of synanthropic plants as the effect (and not the cause) of disturbance in the existing plant cover.

References

Anderson, E. 1956. Man as a maker of new plants and new plant communities. In: W.L. Thomas (ed.), *Man's role in changing the face of the earth*, The Univ. of Chicago Press, Chicago, pp. 763–777.

Baker, H.G. 1965. Characteristics and modes of origin of weeds. In: H.G. Baker & G.L. Stebbins (eds.), *The genetics of colonizing species*, Academic Press, London, pp. 147–172.

Baker, H.G. 1972. Migrations of weeds. In: D.H. Valentine (ed.), *Taxonomy, phytogeography and evolution*, Academic Press, London, pp. 327–347.

Baker, H.G. 1974. The evolution of weeds. *Ann. Rev. Ecol. Syst.* 5: 1–24.

Baker, H.G. & Stebbins, G.L. (eds.) 1965. *The genetics of colonizing species*. Academic Press, London. XV, 588 pp.

Elton, Ch.S. 1958. *The ecology of invasions by animals and plants*. Methuen, London. 181 pp.

Faliński, J.B. 1966. Antropogeniczna roślinność Puszczy Białowieskiej jako wynik synantropizacji naturalnego kompleksu leśnego. (La végétation anthropogène de la Grande Forêt de Białowieża comme résultat de la synanthropisation d'un complexe forestier naturel). *Rozpr. Uniw. Warszwask.* 13: 1–256.

Faliński, J.B. (ed.) 1968. Synantropizacja szaty roślinnej. I. Neofityzm i apofityzm w szacie roślinnej Polski. (Synanthropization of plant cover. I. Neophytism and apophytism in the flora of Poland). *Mater. Zakładu Fitosocjol. Stosowanej Uniw. Warszawsk.* 25: 1–229.

34

Faliński, J.B. (ed.) 1971. Synantropizacja szaty roślinnej. II. Flora i roslinność sysnantropijna miast w związku z ich warunkami przyrodniczymi, dziejami i funkcją. (Synanthropization of plant cover. II. Synanthropic flora and vegetation of towns connected with their natural conditions, history and function). *Mater. Zakładu Fitosocjol. Stosowanej Uniw. Warszawsk.* 27: 1–317.

Faliński, J.B. (ed.) 1972. Synantropizacja szaty roślinnej. IV. Synantropizacja szaty roślinnej w parkach narodowych i rezerwatach przyrody. (Synanthropization of plant cover. IV. Synanthropization of plant cover in national parks and reservations). *Phytocoenosis* 1(4): 223–305.

Faliński, J.B. 1975. Anthropogenic changes of the vegetation of Poland. *Phytocoenosis* 4(2): 97–115.

Forcella, F. 1985. Final distribution is related to rate of spread in alien weeds. *Weeds Res.* 25: 181–191.

Forcella, F. & Wood, J.T. 1984. Colonization potentials of alien weeds are related to their 'native' distributions: implications for plant quarantine. *J. Austral. Inst. Agric. Sci.* 1984: 35–40.

Frey, A. 1974. Genus *Amaranthus* in Poland. *Fragm. Florist. Geobot.* 20(2): 143–201.

Fukarek, F., (Hrsg.) 1979. *Pflanzenwelt der Erde*. Urania, Leipzig. 290 pp.

Grant, W.F. 1967. Cytogenetic factors associated with the evolution of weeds. *Taxon* 16(4): 283–293.

Grime, J.P. 1979. *Plant strategies and vegetation processes*. J. Wiley & Sons, Chichester. XI, 222 pp.

Guzikowa, M. & Maycock, P.F. 1986. The invasion and expansion of three North American species of goldenrod (*Solidago candensis* L. sensu lato, *S. gigantea* Aiton and *S. graminifolia* (L.) Salisb.) in Poland. *Acta. Soc. Bot. Polon.* 55(3): 367–384.

Hantz, J. 1979. The genus *Oxalis* L. in Poland. *Fragm. Florist. Geobot.* 25(1): 65–112.

Harper, J.L. 1965. Establishment, aggression and cohabitation in weedy species. In: H.G. Baker & G.L. Stebbins (eds.), *The genetics of colonizing species*, Academic Press, London, pp. 243–265.

Hawksworth, D.L. (ed.) 1974. *The changing flora and fauna of Britain* (Systematics Association Special Volume No. 6). Academic Press, London, XIII, 461 pp.

Hegi, G. 1924. Oxalidaceae. In: G. Hegi (Hrsg.) *Illustrierte Flora von Mitteleuropa*, 1 Aufl. Bd. IV, Teil 3, Hanser, München, pp. 1644–1656.

Hermy, M. & Stieperaere, H. 1985. Capitalists and proletarians (Mac Leod): an early theory of plant strategies. *Oikos* 44(2): 364–366.

Hilbig, W. 1982. Preservation of agrestal weeds. In: W. Holzner & M. Numata (eds.), *Biology and ecology of weeds* (Geobotany 2), W. Junk, The Hague, pp. 57–59.

Hill, Th.A. 1977. *The biology of weeds*. E. Arnold, London. IV, 64 pp.

Jalas, J. 1955. Hemerobe und hemerochore Pflanzenarten. Ein terminologischer Reformversuch. *Acta Soc. Fauna Fl. Fenn.* 72: 1–15.

Kerguélen, M. 1963. Possibilité de détection des climats homologues par des méthodes floristiques. *Ann. Amélior. Pl.* 13(3): 277–289.

Kobendza, R. 1950. Aperçu critique de certaines espèces du genre *Lepidium* R. Br. et les nouvelles espèces pour la flore polonaise. *Acta Soc. Bot. Polon.* 2(2): 439–453.

Kornaś, J. 1966. Influence of man and his economic activities on the vegetation of Poland. The synanthropic flora. In: W. Szafer (ed.), *The vegetation of Poland*, Pergamon Press-PWN, Oxford-Warszawa, pp. 97–137.

Kornaś, J. 1968a. A geographical-historical classification of synanthropic plants. *Mater. Zakładu Fitosocjol. Stosowanej Uniw. Warszawsk.* 25: 33–41.

Kornaś, J. 1968b. A tentative list of recently introduced synanthropic plants (kenophytes) established in Poland. *Mater. Zakładu Fitosocjol. Stosowanej Uniw. Warszawsk.* 25: 43–53.

Kornaś, J. 1971a. Changements récents de la flore polonaise. *Biol. Conservation* 4(1): 43–47.

Kornaś, J. 1971b. Recent decline of some synanthropic plant species in Poland. *Mater. Zakładu Fitosocjol. Stosowanej Uniw. Warszawsk.* 27: 51–64.

Kornaś, J. 1978. Remarks on the analysis of a synanthropic flora. *Acta Bot. Slovaca*, Ser. A

3: 385-393.

Kornaś, J. 1982. Man's impact upon the flora: processes and effects. *Memorabilia Zool.* 37: 11-30.

Kornaś, J. 1983. Man's impact upon the flora and vegetation in Central Europe. In: W. Holzner, M.J.A. Werger & I. Ikushima (eds.), *Man's impact on vegetation*, W. Junk, The Hague, pp. 277-286.

Kornaś, J. & Medwecka-Kornaś, A. 1967. The status of introduced plants in the natural vegetation of Poland. *IUCN Publ.*, New Series 9: 38-45.

Kornaś, J. & Medwecka-Kornaś, A. 1968. The occurrence of introduced plants in natural and semi-natural plant communities in Poland. *Mater. Zakładu Fitosocjol. Stosowanej Uniw. Warszawsk.* 25: 55-63.

Kornaś, J. & Medwecka-Kornaś, A. 1974. The vegetation of Cracow. *Folia Geogr., Ser. Geogr.-Phys.* 8: 153-169.

Kornaś, J. & Medwecka-Kornaś, A. 1986. *Geografia roslin. (Plant geography).* PWN, Warszawa (in Polish). 528 pp.

Kowarik, I. 1985. Zum Begriff 'Wildpflanzen' und zu den Bedingungen und Auswirkungen der Einbürgerung hemerochorer Arten. *Publ. Natuur-Hist. Genootsch. Limburg* 35(3-4): 8-25.

Kowarik, I. & Sukopp, H. 1986. Unerwartete Auswirkungen neu eingeführter Pflanzenarten. *Universitas* 41(483): 828-845.

Krawiecowa, A. 1951. Analyse géographique de la flore synanthropique de la ville de Poznań. *Poznańskie Towarz. Przyjac. Nauk, Wydz. Mat.-Przyr., Prace Komis. Biol.* 13(1): 1-132.

Latowski, K. 1982. Taxonomic carpological study of the Eurasian species of *Lepidium* L. genus. *Uniw. A. Mickiewicza w Poznaniu, Ser. Biol.* 23: 1-105.

Linkola, K. 1919. Studien über den Einfluss der Kultur auf die Flora in den Gegenden nördlich vom Ladogasee. I. Allgemeiner Teil. *Acta Soc. Fauna Fl. Fenn.* 45(1): I-VII, 1-429.

Lousley, J.E. (ed.) 1953. *The changing flora of Britain.* Bot. Soc. Brit. Isles, Arbroath. 203 pp.

Mac Leod, J. 1894. Over de bevruchting der bloemen in het Kempisch gedeelte van Vlaanderen. Deel II. *Bot. Jaarb. Dodonaea* 6: 119-511.

Majdecka-Zdziarska, E. 1929. *Galinsoga parviflora* Cav. et *Galinsoga hispida.* Benth. *Bull. Int. Acad, Polon. Sci., Cl. Sci. Math., Sér. Bl. Bot.* 1929: 105-139.

Meusel, H. 1943. *Vergleichende Arealkunde.* I, II. Gebr. Borntraeger, Berlin. 466, 92, 90 pp.

Mulligan, G.A. & Findlay, J.N. 1970. Reproductive systems and colonization in Canadian weeds. *Canad. J. Bot.* 48(5): 859-860.

Olaczek, R. 1976. Changes in the vegetation cover of Poland since the middle of 19th century. *Zesz. Problemowe Postępów Nauk Roln.* 177: 369-408.

Perring, F.H. (ed.) 1970. *The flora of changing Britain.* Classey, Middlesex. 157 pp.

Pietras, B. 1970. Current distribution of *Veronica filiformis* Sm. in Poland. *Fragm. Florist. Geobot.* 16(2): 311-316.

Probst, R. 1949. *Wolladventivflora Mitteleuropas.* Verlag Vogt-Schild, Solothurn. VII, 193 pp.

Ralska-Jasiewiczowa, M. 1982. Prehistoric man and natural vegetation: the usefullness of pollen evidence in interpretation of man-made changes. *Memorabilia Zool.* 37: 31-45.

Ridley, H.N. 1930. *The dispersal of plants throughout the world.* L. Reeve, Ashford. XX, 744 pp.

Soest, J.L. van 1941. De verspreiding van *Galinsoga* in Nederland. *Ned. Kruidk. Arch.* 51: 288-301.

Sukopp, H. 1962. Neophyten in natürlichen Pflanzengesellschaften Mitteleuropas. *Ber. Deutsch. Bot. Ges.* 75(6): 193-205.

Sukopp, H. 1966. Neophyten in natürlichen Pflanzengesellschaften Mitteleuropas. In: R. Tüxen (Hrsg.) *Anthropogene Vegetation.* Ber. Int. Symp. Int. Vereinigung Vegetationsk. 1961, W. Junk, Den Haag, pp. 275-291.

Sukopp, H. 1969. Der Einfluss des Menschen auf die Vegetation. *Vegetatio* 17(1-6): 360-371.

Sukopp, H. 1972. Wandel von Flora und Vegetation in Mitteleuropa unter dem Einfluss des Menschen. *Berichte über Landwirtschaft* 50(1): 112-139.

Sukopp, H. 1976. Dynamik und Konstanz in der Flora der Bundesrepublik Deutschland.

36

Schriftenreihe Vegetationsk. 10: 9–26.

Sukopp, H. & Trautmann, W. (Hrsg.) 1976. Veränderungen der Flora und Fauna in der Bundesrepublik Deutschland. *Schriftenreihe Vegetationsk.* 10: 1–409.

Sukopp, H., Trautmann, W. & Korneck, D. 1978. Auswertung der Roten Liste gefährdeter Farn- und Blütenpflanzen in der Bundesrepublik Deutschland für den Arten und Biotopschutz. *Schriftenreihe Vegetationsk.* 12: 1–138.

Svensson, R. & Wigren, M. 1986. A survey of the history, biology and preservation of some retreating synanthropic plants. *Acta Univ. Upsal.* – Symb. Bot. Upsal. 25(4): 1–74.

Świeboda, M. 1963. Distribution of *Elsholtzia Patrini* (Lep.) Garcke in Poland. *Fragm. Florist. Geobot.* 9(2): 239–243.

Thellung, A. 1912. La flore adventice de Montpellier. *Mitt. Bot. Mus. Univ. Zürich* 58: 1–728.

Thellung, A. 1915. Pflanzenwanderungen unter dem Einfluss des Menschen. *Beibl. Englers Bot. Jahrb.* 53, Beibl. Nr. 116: 37–66.

Thellung, A. 1918–1919. Zur Terminologie der Adventiv- und Ruderalflora. *Allg. Bot. Z. Syst. Karlsruhe* 24: 36–42.

Trepl, L. 1984. Über *Impatiens parviflora* D.C. als Agriophyt in Mitteleuropa. *Diss. Bot.* 73: 1–400.

Trzcińska-Tacik, H. 1961. Studies on the distribution of synanthropic plants. 1. *Bidens melanocarpus* Wieg. in Poland. *Fragm. Florist. Geobot.* 7(1): 161–168.

Trzcińska-Tacik, H. 1979. Flora synantropijna Krakowa. (Synanthropic flora of Cracow). *Rozpr. Habilitacyjne Uniw. Jagiellońskiego* 32: 1–278 (in Polish).

Trzcińska-Tacik, H. & Wasylikowa, K. 1982. History of the synanthropic changes of the flora and vegetation of Poland. *Memorabilia Zool.* 37: 47–69.

Wagenitz, G. 1964–1979. Compositae I. In: G. Hegi (Hrsg.) *Illustrierte Flora von Mitteleuropa*, 2 Aufl. Bd. VI, Teil 3, Parey, Berlin, pp. 1–366.

Wasylikowa, K. & Gluza, I. 1977. Flora of the Pleistocene and the Holocene in archaeological excavations. In: E. Rühle (ed.), *Geology of Poland*, Vol. II, Catalogue of fossils, Part 36; Cainozoic, Quaternary, Publ. House Wydawnictwa Geologiczne, Warszawa, pp. 105–122.

Weeda, E.J. 1985. Changes in the occurrence of vascular plants in the Netherlands. In: J. Mennema, A.J. Quené-Boterenbrood & C.L. Plate (eds.), *Atlas van de Nederlandse Flora 2*, Scheltema & Holkema, Utrecht-Bohn, pp. 9–47.

Wein, K. 1932. Die älteste Einführungs- und Einbürgerungsgeschichte des *Erigeron canadensis*. *Bot. Arch.* 34: 394–418.

Wein, K. 1939, 1940, 1942. Die älteste Einführungs- und Ausbreitungsgeschichte von *Acorus calamus. Hercynia* 3, 5, 6.

Willerding, U. 1978. Die Paläo-Ethnobotanik und ihre Stellung im System der Wissenschaften. *Ber. Deutsch. Bot. Ges.* 91: 3–30.

Zając, A. 1979. The origin of the archaeophytes occurring in Poland. *Rozpr. Habilitacyjne Uniw. Jagiellońskiego* 29: 1–213.

Zając, A. 1983. Studies on the origin of archaeophytes in Poland. Part I. Methodical considerations. *Zesz. Nauk. Uniw. Jagiellońskiego, Prace Bot.* 11: 87–107.

Zając, A. 1986. Studies on the origin of archaeophytes in Poland. Part II. *Zesz. Nauk. Uniw. Jagiellońskiego, Prace Bot.* 14: 7–50.

Zając, E.U. & Zając, A. 1975. The list of archaeophytes occurring in Poland. *Zesz. Nauk. Uniw. Jagiellońskiego, Prace Bot.* 3: 7–16.

3. History of the impact of man on the distribution of plant species

K.V. SYKORA

Abstract

Since the earliest times man has introduced plants to regions outside their original area of distribution. Some species have been imported before (archaeophytes) and some after the year 1500 (neophytes), whereas the origin of several species is unknown. Exotics have been transported unintentionally with ship ballast, military transport, foreign grains, foreign fertilizers, wool and cotton, straw and hay. Plants have also been intentionally introduced as medicinal herbs, crop plants and for horticultural reasons. In view of the origins of the species introduced, a European, Oriental, Canadian-Virginian, Cape, North American Forest and Nieuw-Hollandse (Australian) period can be distinguished.

The habitat and vegetation in which neophytes occur are discussed in this paper. Exotic species especially appear to occur in plant communities strongly influenced by man. Although European plant communities generally appear to be resistant to an invasion of exotic species, the same is not true for other continents. Examples are given.

Introduction

Since the earliest times man has, intentionally or unintentionally, carried plants over great distances and has created new environments in which certain species were able to establish themselves. In this way, he has exercised an influence on the natural distribution patterns of plant species and in some cases this has had unfortunate consequences for the original floras. The evidence seems to indicate that man is by far the most effective dispersal agent that ever has existed. His influence on the floras of the world during only the last 500 years has been greater than any other single agent occurring manyfold that length of time (Fosberg 1959).

As a result of man's extensive travels, the natural barriers preventing many species from spreading have been circumvented. Consequently, several species

F. di Castri, A. J. Hansen and M. Debussche (eds.), Biological Invasions in Europe and the Mediterranean Basin.
37–50. © 1990, *Kluwer Academic Publishers, Dordrecht.*

38

have become established in Europe following the influence of man.

In this respect we may distinguish two categories. Archaeophytes (Kreh 1957) were imported before the end of the Middle Ages, while neophytes (Thellung 1912) became part of the flora after the year 1500. The oldest known representatives of imported plants are the weeds on arable land such as *Centaurea cyanus, Agrostemma githago* and *Lolium temulentum*. These species already appeared in Neolithic settlements (Thellung 1915). Other species are *Avena fatua, Bromus secalinus, Ranunculus arvensis, Adonis aestivalis, Delphinium consolida, Legousia speculum-veneris*. Many species, including the original wild forms of our cereals, originated in the steppes and semi-deserts of the Near-East, while others come from the Mediterranean area.

It is often difficult to determine exactly when a species was imported. Indeed, it is not always clear whether a species is native or introduced (Sykora & Westhoff 1977). Various neophytes were initially regarded as native plants, such as *Bidens connatus, Epilobium adenocaulon,* and *Oenothera biennis* all of which came from North America (Sukopp 1972). Archaeophytes are particularly difficult to distinguish from the original species, since the original and secondary areas of distribution continually overlap, and also because the introduction of these species dates far back in time. On the basis of ecological factors, vegetation, reproduction, biology, and the bioclimate, Sukopp and Scholtz (1968) stated that it was highly likely, for instance, that *Poa bulbosa* was not a native plant of western Europe, as was generally thought, but an archaeophyte which has been spread by the raising of sheep.

It has now been established that the European flora is affected by man more than was previously thought. Even in the case of several plants which are a normal part of our grasslands, such as *Festuca pratensis, Trisetum flavescens, Phleum pratense, Daucus carota, Pastinaca sativa and Trifolium pratense*, it is not certain that they are native to Central Europe. Plants may continue to exist in places where they were taken by man after they have disappeared from their original homes (Fosberg 1959). For some plants (e.g. *Alcea rosea*, a cosmopolitan species) the place of origin remains unresolved.

Introduction of species

Many introduced species have been found in the vicinity of ports, some of which were brought in with shipballast (sand from the vicinity of the port of export) (Jehlik 1981). Military transport is also responsible for the spread of various plants. Following the Napoleontic wars, plants from southern Russia such as *Bunias orientalis* grew on a number of sites where Cossacks had camped. As a result of the Franco-Prussian war a 'flora obsidionalis' (siege flora) developed around Paris, consisting of plants from the south of France and Algeria (mainly grasses and Leguminosae), which had come with the forage for domestic animals. In the vicinity of granaries, grinding mills and breweries, weeds from cereal fields are often found having been imported with foreign grains. As early as 1762 the presence of exotics was reported in the vicinity of mills near Montpellier (southern France) (Thellung 1915). In earlier times some hundreds of species were imported with highly contaminated barley from Asia Minor and Morocco. Nowadays, the North American species *Iva xanthiifolia* is imported with cereals. *Artemisia sieversiana* is found with Russian wheat (Jehlik & Hejny 1974). A large number of foreign species are also imported with grass seed. Examples are *Rudbeckia serotina* from North America and *Agrostis castellana* from southern Europe. In the past the importation of foreign fertilizers (seaweeds) has been responsible for the introduction of exotics. Also many species used for birdseed now occur in the wild state, in the same way as the weeds with which they were mixed. This group of adventives includes species of *Panicum* and *Setaria*, *Phalaris canariensis* from the Canary Islands, *Cannabis sativa* from southeast Russia, *Linum usitatissimum*, *Helianthus annuus* from North America and *Guizotia abyssinica* from east Africa.

A large number of adventives were imported through the wool and cotton industries (wool scouring, cotton spinning and other textile processes). The imported wool, in particular, contained several fruits with prickles or thorns. *Xanthium spinosum* was even known as the 'shepherd's plague'. A classic example is the wool scouring mill of Port-Juvénal, near Montpellier (southern France), where more than 500 exotic species of wool adventives have been collected since 1813 (Thellung 1915). A large number of these were, in fact, described by the botanists in Montpellier before the world of botany discovered them in their place of origin.

In the past, fruit from the Mediterranean basin was packed and protected from frost in rye straw and hay from the meadows along the coast. A large number of filed weeds and species found in meadows were unintentionally imported in this way including *Alopecurus myosuroides*, *Vulpia myuros*, *Kochia scoparia*, *Herniaria hirsuta*, *Fumaria micrantha*, *Coronopus didymus*, *Conringia orientalis*, *Reseda lutea*, *Medicago polymorpha*, *Vicia bithynica*, *Oxalis corniculata* and *Plantago lagopus* (Scheuermann 1948). It is assumed that gypsies brought *Datura stramonium* and *Euphorbia uralensis* from southern Europe and western Asia to Europe.

It is not only the higher plants that are imported. A large number of cryptogams have been spread by man and his cattle outside their original areas. *Campylopus introflexus* is now invading the coastal and inland sand dunes of the Netherlands. In the British Islands it commonly occurs on peatbogs. Cattle were responsible for the increased spread of the coprophilic mosses (*Splachnum*) and fungi (*Pilobolus* and *Coprinus*). *Puccinia malvacearum*, *Plasmopara viticola*, and *Sphaerotheca mors-uvea* are examples of genuine adventive parasites. The first species occurs in most parts of Europe as a parasite on *Malva, Althaea,* and *Lavatera*. It probably comes from South America and was first recorded in Spain in 1869. *Plasmopara viticola*, a parasite on vines, is native to North America, and was first recorded in April 1878 as a disease on the European grape (Thellung 1915). *Sphaerotheca mors-uvea* was imported

40

into Europe via Ireland and Russia around 1900.

In the forests of eastern North America, American chestnut (*Castanea dentata*) was infected by the fungus *Endothia parasitica* introduced from Asia on nursery plants in the year 1900. Within thirty or forty years the proportion of the chestnut in the woods dropped from 40 to 1% (Gleason & Cronquist 1964; Sukopp 1979). In 1911, *Endothia parasitica* had spread to at least ten states, in 1926 it was still spreading southwards and by 1950 most chestnuts were dead except in the extreme south and now it has reached the Pacific coast. In 1938, the blight appeared in Italy and later also in Spain (Elton 1977).

The Blister Rust (*Cronartium ribicola*) was introduced from Europe to North America by the importation of *Pinus strobus* seedlings. Here it has widely destroyed or seriously damaged *Pinus strobus* forests (Hämet-Ahti 1983).

Plants imported intentionally

Plants imported intentionally include crop plants, medicinal herbs, species which have escaped from botanical gardens and plants which have been deliberately cultivated by (amateur) botanists (Büttner 1883).

Examples of naturalized medicinal plants are *Acorus calamus, Lupinus polyphyllus, Althaea officinalis, Parietaria officinalis, Aristolochia clematitis, Chenopodium bonus-henricus, Helleborus viridis, Chelidonium majus, Verbena officinalis, Marrubium vulgare and Leonurus cardiaca* (Preuss 1928). *Castanea sativa* and *Juglans regia*, from southern Europe, spread over central and western Europe in Roman times (Welten 1963). Ornamental plants which have escaped from gardens and are now very widespread include *Saponaria officinalis, Robinia pseudacacia*, all species of *Solidago, Sagittaria latifolia, Galinsoga parviflora, Matricaria matricarioides, Elodea canadensis, Stenactis annua* and *Erigeron canadensis* (Büttner 1883).

Plants which have been intentionally planted so as to grow wild include *Cymbalaria muralis* from southern Europe, *Vaccinium macrocarpon* which occurs in North America and on peat land in Europe and *Kalmia angustifolia* from North America (Sukopp 1962).

Origins

In view of the origins of the species planted in botanical gardens a number of periods of time may be distinguished: namely, the European, Oriental, Canadian-Virginian, Cape, North American forest and Nieuw-Hollandse (Australian) periods (Kraus 1894).

The *European period* ended about 1560. The first botanical garden, in Padua (Italy), contained almost exclusively native plants or plants which had become naturalized before the Middle Ages. During the *Oriental period* (1560–1620), a large number of plants came from southern and southeastern Europe and the Near East into European gardens. For the first time *Hyacinthus, Muscari, Tulipa, Narcissus, Fritillaria, Lilium, Anemone* and exotic *Ranunculus* were cultivated. A number of flowering bushes were also included (e.g., *Philadelphus coronarius* and *Syringa*). *Laburnum anagyroides* and *Syringa vulgaris* were introduced to Leiden (The Netherlands) by Clusius in 1594 and 1601 respectively (Kraus 1894). *Diospyros kaki*, from Japan and Korea, was cultivated in the 16th century in Padua (Italy) as a plant against veneral disease and it soon naturalized in the Mediterranean region. The most famous centre for these botanical activities was the Habsburg monarchy (Austria), which gave a great impetus to botany. Mattioli and Dodoens, two eminent botanists, were also personal physicians to the Emperor. Carolus Clusius was at the Viennese court from 1573 to 1587. In the beginning of 1588 Clusius received potatos sent to him as 'taratoufli' by missionaries (de Wit 1982).

In the 16th century seeds of Tulips were sent to Europe by Busbecquius, the envoy of king Ferdinand of Austria to the Sultan of Turkey. Shortly afterward, species from the Mediterranean region and from southern Russia were imported and soon they were cultivated in the Netherlands where the tulip was introduced by de l'Ecluse (= Clusius). New cultivars were used for speculations. During the so called tulipomania these speculations were so extraordinary that the Dutch government had to interfere. As prices went down, many people became bankrupt. At present there are some thousands of cultivars.

During the *Oriental period* the first American plants came to Europe via Spain. However, these were almost exclusively South American species from high-altitude areas, as demonstrated by the incorporation of the terms *indica, hispanica* and *peruana*. Examples of these are the Sunflower, Tobacco, *Agave americana, Opuntia*, Tomato, Potato and *Tagetes*. There was a great influx of American species following the development of the areas in America with a temperate climate; that is, after the English occupied Virginia and the French occupied Canada. During this *Canadian-Virginian period* (1620–1687), a number of species appeared including *Robinia pseudacacia, Oenothera biennis, Helianthus tuberosus* and some species of *Aster* and *Solidago*. According to Kraus (1894) *Robinia pseudacacia* was planted for the first time in Europe in the Jardin des plantes (Paris, France) in 1636. In 1697 it was planted in Montpellier (France).

At that time no truely tropical species had yet been imported; these could not survive either the long sea journey or the cold European winters. As more plants were imported from warm countries, the difficulty of preserving them during the winter increased. It was mainly the arrival of plants from the Cape (South Africa) which hastened the development of greenhouses; Leiden (The Netherlands) is generally considered to be the first botanical garden to have had such an enclosure. In the Index Stirpium from 1594 a number of thermophilic species are mentioned like *Aloe, Opuntia, Bambusa* and *Saccharum* (Karstens & Kleibrink 1982). A solarium (hibernaculum) was built

42

there in 1599.

Attempts have long been made to intentionally naturalize exotics. At the end of the 17th century and during the 18th century Nisolle, Gouan, Amoreux, and others, sowed a great number of exotic species in the neighbourhood of Montpellier in order to enrich the flora and to delight their colleagues with a rich botanical discovery. Because this was done with insufficient ecological knowledge, the species sown did not establish. Since that time, the intentional sowing of exotic and especially also of native wild plant species, is a regularly reoccurring issue.

During the *Cape period* (1687–1772) a large number of species from the Cape province in South Africa became available. The main plants imported were succulents in the widest sense and also *Euphorbiaceae, Mesembryanthemum, Aloe* and *Stapelia*. Bulbous plants and *Pelargonium, Protea* and *Erica* were also imported. Most of these spread from Dutch gardens to the other botanical gardens in Europe.

The period of the *North American forest plants* reached its highest point in the middle of the 18th century, when parks were being created in England on a large scale. When these picturesque parks were being laid out such trees as American maples (*Acer*), birches (*Betula*), oaks (*Quercus*) and hawthorns (*Crataegus*) were used. *Prunus serotina* and *Amelanchier lamarckii*, both of which are now naturalized species in the Netherlands, were also imported during this period.

On the 11th of June 1771 Captain James Cook, after a three-year voyage in Endeavour, returned to England with the scientist Joseph Banks and his assistant Dr. Daniel Solander on board. They presented the then newly founded botanical gardens at Kew (Great Britain) with a large number of Australian and New Zealand plants. Well known plants from this *Australian period* are the *Acacia, Eucalyptus, Casuarina, Banksia, Grevillea* and many others.

The influx of tropical plants reached a high point during this period. In Leiden and Amsterdam where heated greenhouses were available, a large number of tropical species were propagated for the first time. In the 19th century, journeys to the tropics became more frequent and less dangerous, and the number of scientific collectors rose year by year. A number of beautiful plant families such as tree fern (*Cyathea*), *Pinus, Zingiberaceae*, tropical *Arum, Orchidaceae* and *Begonia* were imported at this time. In 1891, the number of plant species and varieties in cultivation at Kew was already 19,888, 2,443 of which were of tropical origin (Kraus 1894).

Many Chinese plants have been introduced to Europe via Japan. The influx of Chinese plants into Japan, already demonstrated to exist in prehistoric times, showed a new revival during the Edo-period (1615–1867). The East-Asiatic flora was introduced to Europe comparatively late, which can be attributed to the closed borders of the Chinese imperium (till the Opium war 1839–1842) of Japan till 1854 and of Korea till 1876. Around the year 1700 exotics from America and from the Cape were more numerous in European gardens than species from East Asia. The English already started the cultivation of Chinese plants at the end of the 18th century. *Ginkgo biloba* was planted in European gardens in the 18th century. Only after the foundation of a commercial company (Siebold & Co.) and a Jardin d'acclimation by von Siebold after his return from Japan in 1830, the importation of Japanese plants reached a certain level. The first ship with several plant species arrived in Holland in 1829. The second ship was captured by rebellious Belgians and the cargo was used to found a nursery garden in Gent (Belgium) (Muntschick 1983). *Helianthus annuus* originating from America was first pictured in Japan in 1666.

The 19th century saw the establishment of a variety of exotic plants in European gardens. Today paradoxically, the emphasis is on native plants and increasing numbers of nurserymen are meeting this demand.

Naturalization and habitat

In order to determine the effect of exotic species on the native vegetation, we should know in which communities they have established themselves.

Neophytes especially appear to occur in plant communities strongly influenced by man; they preferably establish themselves in pioneer situations and in open perennial vegetations. Most frequently they are found in the surroundings of cities, on wastelands, road verges, rubble (ruderal places), stationyards, river banks and in coastal communities. Consequently, most neophytes are r-strategists highly adapted to quickly changing environments. On these sites strong competition with native species is avoided.

On the other hand, hardly any neophytes are found in relatively well developed undisturbed vegetation (Sukopp 1979, Kornas 1982). From the about 100 neophytes occurring in Poland, 57 grow on ruderal places and or on cultivated grounds, 34 in semi-natural vegetation (e.g. grasslands) and only 9 in undisturbed more or less natural vegetation (Kornas 1982). Only two of the last mentioned species have an extensive distribution (Kornas & Medwecka-Kornas 1967). The proportion of introduced aliens exceeds 10% in Poland; data for West Germany, the British Isles and Finland are respectively 16%, 16%, and 18% of the total flora. The number of neophytes in a certain region is correlated to the intensity of human influence. The proportion of native species is increasingly reduced under urbanization while the proportion of neophytes increases (Sukopp & Werner 1983). In Berlin, for instance, the number of neophytic ruderal species increased from 20 in 1787 to 51 in 1884 and 79 in 1959. The decrease in the number of native species is likewise related to human activity (Sukopp 1968).

A new European species, *Spartina anglica*, came into existence by a doubling of the number of chromosomes of *Spartina x townsendii*, a crossing between the European species *S. maritima* and the introduced American species *S. alterniflora*. On salt marshes along the coast of the North Sea, *Spartina anglica* is locally ousting the originally occurring *Salicornia stricta* and *Puccinellia*

44

maritima and is building an own plant community, the *Spartinetum*. In less than 55 years after its introduction in 1942, *S. anglica* succeeded in almost crowding out the original *Spartina maritima* (Adema & Mennema 1979).

Lactuca tatarica originates from the steppes in Eastern Europe (Jehlik 1980) where it grows on river banks, steppes and on saline soils. In the southeastern part of the Soviet Union it acts as a noxious weed. It is drought and salt resistant. The oldest discovery in Germany was in 1902. In Great Britain it was discovered as early as 1884. From here it spread along the Baltic Sea, to the Netherlands and Ireland. According to some authors it spread with the help of the Pallas Sandgrouse *Syrrhaptes paradoxus* (Jehlik 1980). According to others it accompanied wool or cereals. In Rotterdam it was found at a locality near to a granary (Kuhbier 1977). It is naturalized in the *Atriplicetum littoralis* and on sparcely vegetated dunes where it is propagating almost exclusively by vegetative means.

Oenothera biennis occurs on disturbed sites in dunes rich in calcium. *O. parviflora* var. *ammophila* is characteristic of the *Elymo-Ammophiletum*.

Cotula coronopifolia, presumably originating from South Africa, is locally frequent on slightly saline moist soils along the Dutch coast. *Cotula coronopifolia* is common in the southern hemisphere (Cape Province, St. Helens (Great Britain), Australia, New Zealand, the southern part of South America), and is spreading along the coasts of both the North Atlantic and North Pacific oceans. It was found for the first time near Emden, northwest Germany, as early as 1739; in 1835 and 1839 it was found near Amsterdam. This species is mainly found in coastal areas in the *Armerion maritimae*, the *Lolio-Potentillion* and the *Puccinellietum distantis* (van der Toorn 1980).

Nicandra physalodes, a garden escape introduced from Peru in 1759, is now a characteristic species of the *Onopordion acanthii*, a community from ruderal disturbed places (Westhoff & Den Held 1969).

Cyperus esculentus was introduced unintentionally to the Netherlands together with commercial bulbs imported from abroad. Nowadays because of its strong vegetative propagation by small tubers, it is one of the top twenty weeds in the world (Groenendael & Habekotte 1988).

Some Asiatic species, escaped from gardens and parks in which they have been introduced in the 19th century, strongly expanded in road verges (e.g. *Reynoutria japonica*, *R. sachalinensis*, *Polygonum polystachyum* and *Heracleum mantegazzianum*) (Bakker & Boeve 1985). Also *Senecio inaequidens* originating from Natal and Transvaal (South Africa) is expanding along roads and railways. It occurs in *Sisymbrietea, Melilotetea, Onopordion acanthii, Artemisietea, Sedo-Scleranthetalia* and disturbed *Nardo-Galion* and *Nardo-Calunetea* (Kuhbier 1977, 1978; Hülbusch & Kuhbier 1979).

It is along roadsides that many adventive species first colonize from crop, pasture, ornamental garden or other origin. In Hawaii, introduced species occur in greatest numbers and with greatest frequency within one metre from the paved surface. In marked contrast, native species were rare or absent within one metre from the road, increased in the second metre and predominated in the third (Wester & Jurvik 1983).

Nowadays road verges are often seeded with a seed mixture in which sometimes exotic species occur like *Sanguisorba muricata* and *Agrostis castellana* or cultivars like tall growing *Lotus corniculatus* (Trautmann & Lohmeyer 1975). In the Netherlands road verges are important refugia for species threatened by modern agricultural practices.

Almost half of all neophytes naturalized in Europe occur along inland waterways, canals and rivers (Sukopp 1972). Two thirds of the Polish neophytes can be found in river valleys. The great number of exotic species in river bordering communities is caused by the facts that: rivers are effective ways of transportation for many species, river banks are kept open by the action of the water, resulting in the permanent presence of sites with low competition, and the considerable anthropogenic influence in these habitats.

Introduced species from open, nutrient-rich, river-bordering communities include: *Bidens frondosus, B. connatus, Xanthium strumarium* subsp. *riparium, Amaranthus retroflexus, Erigeron canadensis, Matricaria matricarioides, Veronica peregrina, Lycopersicon esculentum, Helianthus annuus, Nicotiana sp.*.

Many neophytes escaped from gardens occur on organic drift material in the communities belonging to the *Senecionion fluviatilis* (e.g. *Solidago gigantea, S. canadensis, Aster salignus, A. novi-belgii, A. lanceolatus, A. tradescantii, Impatiens glandulifera, Helianthus tuberosus, Rudbeckia laciniata, Stenactis strigosa, Reynoutria japonica, Armoracia rusticana*) (Sukopp 1962). In a number of cases these species form new communities like the *Impatienti-Solidaginetum* and the *Rudbeckio-Solidaginetum. Impatiens glandulifera*, like *Reynoutria japonica* and *Helianthus tuberosus* is locally suppressing and endangering native species (Trautmann & Lohmeyer 1975). The *Impatienti-Solidaginetum* (*Convolvulion sepii*) is composed of tall growing neophytes with a high biomass production. The expansion of these neophytes is highly correlated with the anthropogenic destruction of the natural riverbank communities (Kopecky 1966). *Impatiens glandulifera* originates from the western Himalayas, from an altitude between 1,800 and 3,000 metres, where it grows along brooks. In 1839, it was introduced to England from where it spread over Europe. Populations of this species now occur on *Bidention, Phalaridion, Convolvulion, Filipendulo-Petasition* and *Salicion albae* sites. On these sites only perennial native species with a strong vegetative propagation can withstand *I. glandulifera*, e.g., *Urtica dioica, Phalaris arundinacea, Calystegia sepium, Petasites officinalis*, and *Poa trivialis*. In river forelands *I. glandulifera* obstructs the natural rejuvenation of woody species (Lhotska & Kopecky 1966).

Open water is also easily colonized by certain neophytes. Examples are *Elodea canadensis* in Europe and New Zealand, *Egeria densa* in the USA, Europe, New Zealand and Japan and *Hydrilla verticillata* in the USA. *Eichhornia crassipes* and *Pistia stratiotes* are widespread in tropical waters. In Japan, adventive macrophytes with wide native ranges in North and South America, such as *Cadomba caroliniana, Eichhornia crassipes, Egeria densa, Elodea nuttallii, Myriophyllum brasiliense* etc., have become a nuisance in lakes, ponds, irrigation canals and some paddy fields (Ikusima 1983). *Elodea canadensis*, presently is a characteristic species of the *Parvopotametalia*. The same applies for *Elodea nuttalii* observed in the Netherlands for the first time in 1941. In the beginning shipping was seriously impeded by the massive increase of the first species. As natural enemies turned up, *Elodea canadensis* rapidly decreased again till an equilibrium was reached.

Azolla filiculoides became extinct in Europe in the glacial period before last. It was reintroduced to botanical gardens from America in the 19th century. It is locally common and forms an own community the *Lemno-Azolletum filiculoidis*, occurring in very nutrient rich, more or less polluted, shallow waters. *Azolla caroliniana* (*Azolletum carolinianae*) introduced from North America, considerably decreased after an initial increase and was replaced by the former association probably because of water pollution (Westhoff & Den Held 1969).

Several species colonized the sea shores. *Asparagopsis armata* (*Falkenbergia rufolanosa*), a red algae, originating from Australia, Tasmania and New Zealand and introduced by shipping, was first noticed in the extreme south of the French Atlantic Coast in 1923 from where it spread along the European Coast (Feldman & Feldman 1942, Horridge 1951, Walker *et al.* 1954). In 1947 *Sargassum muticum* was discovered in North America where it was presumably introduced with Japanese oysters about 1940. It is now naturalized along the whole Atlantic Coast. In 1972 Japanese oysters were imported to the French Atlantic Coast and a few weeks later *Sargassum muticum* was noticed. In 1973 it occurred along the south coast of England, in 1977 it was found in the Netherlands and in 1982 it reached the Frisian islands (Prud'homme van Reine 1977, Kremer *et al.* 1983).

Acorus calamus is the only neophyte that established itself in reed-zones and is now a characteristic species of the *Phragmition*. It was first mentioned in a naturalized state from Rosenbach in 1621. In the 18th century, the number of locations considerably increased. In 1737, Linnaeus wrote 'it grows luxuriant along the Dutch canals'. *Acorus calamus* was imported to Prague and Vienna via Constantinople. *A. calamus* var. *vulgaris* introduced into Europe is a native of the Himalayas (Wein 1932, 1939, 1941 & 1942).

Only a few exotic species established themselves in undisturbed woods. *Quercus rubra, Robinia pseudacacia* and *Prunus serotina* mainly occur in commercial woods (Sukopp 1962). In continental and submediterranean dry woods *Robinia pseudacacia* is changing the whole species composition and is thus building a new community. Locally it is replacing *Betula*. Nitrogen accumulation by *Robinia* promotes nitrophilous species. *Robinia* changes the chemical, physical and biological characteristics of the soil and is accompanied by a typical flora consisting of *Galium aparine, Veronica hederifolia, Chelidonium majus, Urtica dioica, Geranium robertianum, Impatiens parviflora, Galeopsis tetrahit* and *G. bifida, Bromus sterilis, Agrostis tenuis, Sambucus nigra, Crataegus* and *Rubus* species and *Ribes uva-crispa* (Kohler & Sukopp 1964; Kohler 1968). In the Netherlands and Belgium *Prunus serotina* occurs in the *Fago-Quercetum* as a differential species against the *Querco-Roboris betuletum*. In Dutch woods

Hyacinthoides non-scripta, Eranthis hyemalis and *Vinca minor* are relics from cultivation. *Hyacinthoides non-scripta* is especially found in oakwoods occurring in the Dutch dunes (*Convallario-Quercetum dunense*). In Ireland *Rhododendron ponticum* is replacing *Ilex aquifolium* extensively in the oakwoods of Killarney. It was introduced to Britain by 1763 and is now well established on sandy and acid soils (Warburg 1953).

Man introduced numerous plant species into the boreal zone, even in remote areas. Most of these neophytes were able to invade disturbed habitats (like shores) or seral communities only, the undisturbed boreal forest seems to be very resistant against invaders (Hämet-Ahti 1983).

Besides exotics (e.g. *Vaccinium macrocarpon* from North America) have been found on nutrient-poor, drained bogs.

Introduced species also occur in hedges, e.g. *Lycium barbarum, Spiraea salicifolia, Amelanchier lamarckii.*

Besides exotics occur in dry grassland communities, on clearings (*Epilobietea angustifolii*), in small sedge-dominated vegetation of mires (*Parvocaricetea*), in moist heaths (*Ericion tetralicis*), on walls and scree (e.g. *Cheiranthus cheiri* and *Cymbalaria muralis*).

The majority of the neophytes fit into the structure of certain plant communities. Only in a number of cases the establishment results in a suppression of native species; e.g., this is the case with the American *Aster, Solidago* and *Bidens* species. In Europe the native *Bidens* species have partly been ousted by the North American species *B. frondosus* and *B. connatus* which are now characteristic of the nitrophilous *Bidention* communities.

In general one can conclude that in Europe, introduced species only occupy artificial habitats and they seldom establish themselves as a fixed element of natural vegetation. Here, the natural plant communities apparently are very resistent against an invasion of exotic species. This does not apply to all regions on earth. Notably, the open communities of dry climatic zones and the vegetation of biologically isolated regions like islands are even very susceptible to invasions. The percentage of alien species in the flora of many islands is more than 50% e.g., South Georgia, Kerguelen Islands, Tristan da Cunha (Moore 1983). In some other parts of the world the introduction of exotics resulted in a strong impoverishment of the native flora.

Apparently there are both competitive and susceptible floras. In some countries the flora is susceptible to such an extent that the native species have been totally ousted by invading species over extensive areas, within a few decades. The flora of extensive parts of Chili, for instance, largely consists of introduced species. In New Zealand, two thirds of the land surface contain a flora very dissimilar from the original flora; the majority of introduced species being European pasture plants. The flora contains 1,700 (58.6%) alien species (Moore 1983). Half of the Hawaiian plant species, 90% of which are endemic, is endangered by exotics. In this way many species disappeared from St. Helens (Great Britain). In Tahiti, *Rubus moluccanus* dominates almost the whole surface of the island and suffocates the native vegetation. Recent studies,

48

at least in temperate areas, have concluded that the persistence of aliens in island floras is dependent upon the activities of man and his grazing animals (Moore 1983).

Because of the increasing number of nurseries of wild plant species, seeds of these species are nowadays available in large quantities for low prices. Unfortunately, the wild plant species offered by these nurseries are often of foreign origin and seeds are imported on a large scale, for example from nurseries in the north of Africa.

Conclusions

In Europe, neophytes especially appear to occur in vegetation strongly influenced by man (pioneer situations and open perennial vegetations). Here, competition with native species is avoided. The majority of the neophytes fit into the structure of certain plant communities. Only in a few cases native species are suppressed. The number of neophytes in local floras is strongly related to human activity. In some other parts of the world, the vegetation has more severely been altered by invading species.

References

Adema, F. & Mennema, J. 1979. De Nederlandse slijkgrassen. *Gorteria* 9(10): 330–334.

Bakker, P. & Boeve, E. 1985. *Stinzenplanten*. Terra, Zutphen. 168 pp.

Büttner, R. 1883. Flora advena marchica. *Verhandlungen des Bot. Vereins der Provinz Brandenburg.* 25(I): 1–59.

Elton, Ch.S. 1977. *The ecology of invasions by animals and plants.* Science Paperbacks, John Wiley & Sons, New York. 181 pp.

Feldman, J. & Feldman, G. 1942. Recherches sur les Bonnemaisoniacées et leur alternance de générations. *Ann. Sci. Nat., Bot. Ser.* 11(3): 75–175.

Fosberg, F.R. 1959. Man as a dispersal agent. *The Southwestern Naturalist* 3: 1–6.

Gleason, H.A. &, Cronquist, A. 1964. *The Natural Geography of plants*, New York. 420 pp.

Groenendael, J.M. van, & Habekotte, B. 1988. Cyperus esculentus L. – biology, population dynamics, and possibilities to control this neophyte. *Z. Pflanzenkrankh. PflSchutz*, Sonderh. XI: 61–69.

Hämet-Ahti, L. 1983. Human impact on closed boreal forest (taiga). In: W. Holzner, M.J.A. Werger & I. Ikusima (eds.), *Man's impact on vegetation*, Junk, The Hague, pp. 201–211.

Horridge, G.A. 1951. Occurrence of Asparagopsis armata Harv. on the Scilly Isles. *Nature* 167: 732–733.

Hülbusch, K.H. & Khubier, H. 1979. Soziologie von Senecio inaequidens DC. *Abh. Naturw. Verein Bremen* 39: 47–54.

Ikusima, I. 1983. Human impact on aquatic macrophytes. In: WE. Holzner, M.J.A. Werger & I. Ikusima (eds.), *Man's impact on vegetation*, Junk, The Hague, pp. 69–82.

Jehlik, V. 1980. Die Verbreitung von Lactuca tatarica in der Tschechoslowakei und Bemerkungen zu ihrem Vorkommen. *Preslia* 52: 209–216.

Jehlik, J. 1981. Beitrag zur synanthropen (besonders Adventiv-Flora des Hamburger Hafens. *Tuexenia* 1: 81–97.

Jehlik, J. & Hejny, S. 1974. Main migration routes of adventitious plants in Czechoslovakia.

Folia Geobot. Phytotax. 9: 241-248.

Karstens, W.K.H. & Kleibrink H. 1982. *De Leidse hortus.* Waanders, Zwolle. 191 pp.

Kohler, A. 1968. Zum ökologischen und soziologischen Verhalten der Robinie (Robinia pseudoacacia L.) in Deutschland. In: R. Tüxen (ed.), *Pflanzensoziologie und Landschaftsökologie,* W. Junk, Den Haag, pp. 402-412.

Kohler, A. & Sukopp, H. 1964. Über die soziologische Struktur einiger Robinienbestände im Stadtgebiet von Berlin. *Sitzungsberichte der Gesellschaft Naturforschender Freunde zu Berlin* NF 4, H.2: 74-88.

Kopecky, K. 1967. Die Flussbegleitende Neophytengesellschaft Impatienti-Solidaginetum in Mittelmähren. *Preslia* 39: 151-166.

Kornas, J. 1971. Changements récents de la Flore polonaise. *Biological Conservation* 4(1): 43-47.

Kornas, J. 1982. Man's impact upon the flora: processes and effects. *Memorabilia Zoologica* 37: 11-30.

Kornas, J. & Medwecka-Kornas, A. 1967. The status of introduced plants in the natural vegetation of Poland. *IUCN Publications New Series* 9: 38-45.

Kraus, G. 1894. *Geschichte der Pflanzeneinführungen in die europäischen botanischen Gärten,* Leipzig. 73 pp.

Kreh, W. 1957. Zur Begriffsbildung und Namengebung in der Adventivfloristik. *Mitt. flor. soz. Arbeitsgem.* NF 6/7: 90-95.

Kremer, B.P., Kuhbier, H. & Michaelis, H. 1983. Die Ausbreitung des Brauntanges Sargassum muticum in der Nordsee. Eine Reise um die Welt. *Natur und Museum* 113(5): 125-130.

Kuhbier, H. 1977. Der Tatarenlattich Lactuca tatarica (L.) C.A. Meyer auf der Tegeler Plate bei Dedesdorf an der Niederweser. *Drosera* 77(1): 14-20.

Kuhbier, H. 1977. Senecio inæquidens DC- ein Neubürger der nordwestdeutschen Flora. *Abh. Naturw. Verein Bremen* 38(21): 383-396.

Lhotska, M., & Kopecky, K. 1966. Zur Verbreitungsbiologie und Phytozönologie von Impatiens glandulifera Royle an den Flussystemen des Svitava, Svratka und oberen Odra. *Preslia*38: 376-385.

Moore, D.M. 1983. Human impact on island vegetation. In: W. Holzner, M.J.A. Werger & I. Ikusima (eds.), *Man's impact on vegetation,* Junk, The Hague, pp. 237-246.

Muntschick, W. 1983. Die floristische Erforschung Japans um 1700 und Kaempfers Bedeutung für die Kenntnis Japanischer Pflanzen in Europa. In: E. Kaempfer, 1712 (ed.), *Flora Japonica.* Reprint, Steiner Verlag GMBH, Wiesbaden. 315 pp.

Preuss, H. 1928. Das anthropophile Element in der Flora des Regierungsbezirkes Osnabrück. *Veröffentlichungen des Naturwissenschaftlichen Vereins zu Osnabrück* 21: 17-165.

Prud'homme van Reine, W.F. 1977. De reis van een bruinwier rond de wereld. *Gorteria* 8(10/11): 212-216.

Scheuermann, R. 1948. Zur Einteilung der Adventiv- und Ruderalflora. *Ber. der Schweizerischen Botanischen Gesellschaft,* B.58, Bern: 268-276.

Sukopp, H. 1962. Neophyten in natürlichen Pflanzengesellschaften Mitteleuropas. *Ber. d. Deutschen Bot. Gesellschaft* LXXV: 193-205.

Sukopp, H. 1968. Der Einfluss des Menschen auf die Vegetation und zur Terminologie anthropogener Vegetationstypen. In: R. Tuxen (ed.), *Pflanzensoziologie und Landschaftsökologie,* W. Junk, The Hague, pp. 65-74.

Sukopp, H. 1972. Wandel von Flora und Vegetation in Mitteleuropa unter dem Einfluss des Menschen. *Berichte über Landwirtschaft Hrsg. Bundesministerium f. Ernährung, Landwirtschaft und Forsten.* Bd.50, H.1: 112-139.

Sukopp, H. 1979. Florenwandel und Vegetationsveränderungen in Mitteleuropa während der letzten Jahrhunderte. In: R. Tüxen ed. *Gesellschaftsentwicklung (Syndynamik),* Cramer, Vaduz, 469-489.

Sukopp, H. & Scholz, H. 1968. Poa bulbosa L., ein Archäeophyt der Flora Mitteleuropas. *Flora Abt.* B. 157: 494-526.

Sukopp, H. & Werner, P. 1983. Urban environments and vegetation. In W. Holzner, M.J.A.

50

Werger & I. Ikusima (eds.), *Man's impact on vegetation*, Junk, The Hague, pp. 247-260.

Sykora, K.V. & Westhoff, V. 1977. Een nieuwe vindplaats van Campanula latifolia L.; een inheemse soort? *Gorteria* 8 (10/11): 187-193.

Thellung, A. 1912. La flore adventice de Montpellier. *Mem. Soc. nation. Sci. nat. math. Cherbourg* 389: 57-728.

Thellung, A. 1915. Pflanzenwanderungen unter dem Einfluss des Menschen. *Beiblat Nr. 116 zu den Botanischen Jahrbüchern Bd.* 53: 37-66.

Toorn, J. van der, 1980. On the ecology of Cotula coronopifolia L. and Ranunculus sceleratus L. *Acta Bot. Neerl.* 29(5/6): 385-396.

Trautmann, W. & Lohmeyer, W. 1975. Zur Entwicklung von Rasenansaaten an Autobahnen. *Natur und Landschaft* 50(2): 45-48.

Walker, M.I., Burrows, E.M. & Lodge, S.M. 1954. Occurrence of Falkenbergia rufolanosa in the Isle of Man. *Nature* 174: 315.

Warburg, E.F. 1953. A changing flora as shown in the status of our trees and shrubs. In: J.E. Lousley (ed.), *The changing flora of Britain,* Abroath, pp. 171-180.

Wein, K. 1939, 1941, 1942. Die älteste Einfürungs und Ausbreitungsgeschichte von Acorus calamus 1-3 Teil. *Hercinia* Bd. 1-3: 367-455, 72-128, 241-291.

Welten, M. 1963. Zur Chronologie der insubrischen Vegetationsgeschichte. *Ber. des Geob. Inst. der Eidg. Techn. Hochsch. Stiftung Rübel* 34: 77-81.

Wester, L. & Juvik, J.O. 1983. Roadside plant communities on Mauna Loa, Hawaii. *J. of Biogeography* 10: 307-316.

Westhoff, V. & den Held, A.J. 1969. *Plantengemeenschappen in Nederland*, Thieme, Zutphen. 324 pp.

Wit, H.C.D. de, 1982. *Ontwikkelingsgeschiedenis van de biologie*. Deel 1, PUDOC, Wageningen. 415 pp.

4. Recent plant invasions in the Circum-Mediterranean region

P. QUEZEL, M. BARBERO, G. BONIN and R. LOISEL

Abstract

We propose a terminology to distinguish from others those plant species whose natural distributions were changed by human activities (voluntary or not). Such plants may be native or non-native. We also develop a typology of the different ways by which they become integrated or not into the indigenous flora.

Plants associated with man usually increase their distribution. Impressive and rapid invasions of non-native plants often occur in the circum-Mediterranean region, mainly where artificial or artificialized ecosystems prevail. Contrarily, their impact is limited where well-balanced natural ecosystems are dominant. Only when the natural balance is disturbed (e.g. by fires, overgrazing, abandonment of traditional cultivation methods, reafforestation) is the distribution of the native flora severely modified. Invading species with sexual reproduction alone must be separated from those showing asexual reproduction abilities, whether natural or linked to human activities. The importance of seed adaptation for dispersal is emphasized. These plants associated with man in the Circum-Mediterranean region originate mainly from temperate North America and Tropical Africa, but in the Western Mediterranean region the impact of Eastern Mediterranean and Irano-Touranian element is also important. The contribution of plants from the other Mediterranean climatic zones is generally very weak, with some remarkable exceptions.

Introduction

The historical and biogeographical interpretation of the distribution of plant species whose introduction and spread are linked to human activities is an important problem. Clarifying the reasons why native species modify their distribution or ecological preferences in a specified zone is also important. Man has been active in the Mediterranean region for more than seven or

F. di Castri, A. J. Hansen and M. Debussche (eds.), Biological Invasions in Europe and the Mediterranean Basin. 51–60. © 1990, *Kluwer Academic Publishers, Dordrecht.*

52

eight millennia and has strongly influenced local plant distribution. Unfortunately, historical documents on native plants are very scarce and recent studies on cultivated plants (e.g. Renfrew 1979, Zohary & Hope 1973, Van Zeist 1980) raise more questions than answers. Palynology has been of great help in determining the ecological role of many tree species since the end of the last glaciation (e.g., Pons & Suc 1980, Reille *et al.* 1980), however, many gaps are yet to be filled, especially when herbaceous species are concerned. We provide in this chapter a general picture of the ecological, biogeographical and biological importance of recent invading species around the Mediterranean Sea.

Definitions

We have selected a simple terminology (Table 1) to define the plant categories relative to biological invasions. Many attempts have already been made at definitions (Rikli 1903, Thellung 1911–12, Debray & Sanay 1939, Kornas 1966, 1978, Holub 1971, Rousseau 1971, Jovet 1984), resulting sometimes in complicated nomenclatures which are difficult to use.

The species able to adapt to a variety of human activities to ensure their growth and often to extend their natural distribution can be defined as '*synanthrope*'. There are two main groups in the *synanthropes*: the '*apophytes*' and the '*anthropophytes*'.

Table 1. Terminology of the definitions used in the text.

SYNANTHROPE SPECIES: species linked to the voluntary or involuntary action of man which generally modify their natural distribution by extension.

SYNANTHROPE SPECIES OF NATIVE ORIGIN: *Apophytes*
- spread by cultivation methods: *Cultigen Apophytes*
- spread by development of marginal zones: *Ruderal Apophytes*
- spread due to important modifications in ecological or biotic factors, or to accidental causes:
  * fires: *Pyrophyte Apophytes*
  * grazing: *Zoogen Apophytes*
  * logging or voluntary extension: *Substitution Apophytes*

SYNANTHROPE SPECIES OF FOREIGN ORIGIN, VOLUNTARILY OR INVOLUNTARILY INTRODUCED: *Anthropophytes*
- introduced before the end of the 15th century: *Archaeophytes*
- introduced later   : *Kenophytes*
Ephemerophytes: plants appearing episodically
Subspontaneous: species which have escaped from other cultures and survive at least for a certain period without human help
Adventive species introduced involuntarily which survive for a certain period without human help
Naturalized or *Neophytes*: species introduced involuntarily whose survival in the native flora appears to be definitive

Apophytes

The apophytes are native species which have the ability to profit from human activities. Apophytes may be divided into several groups, the most important of which we examine as follows.

The cultigen apophytes are adapted to cultivation methods and have had a wide distribution in the Mediterranean region as long as traditional cultivation techniques were in use. Examples include *Agrostemma githago, Centaurea cyanus, Vaccaria pyramidata, Androsace maxima* with cereal crops and *Silene linicola* and *Camelina alyssum* with flax crops. Mechanization and use of chemicals have deeply disturbed these species, some of them being now extremely rare and others being extinct (Barbero *et al.* 1984). Modern agriculture in orchards and vineyards have driven *Tulipa agenensis* and *T. praecox* to the rim of extinction, and at the same time has allowed the invasion of plants like *Diplotaxis erucoides* and *D. viminea*.

The ruderal apophytes grow mainly in artificial habitats such as vacant lots, roadsides, and rubble. The extension of these environments due to an increase in human activity, particularly in and around cities, has contributed to the spread and diversification of these plants.

A very significant group of apophytes is closely linked to important changes in biotic factors, often of accidental origin, such as the pyrophyte, zoogen and substitution apophytes.

Pyrophyte apophytes have benefited from the increase in forest fires in the Mediterranean region, as shown by the extension of certain matorral species in the past few decades: *Cistus, Ulex, Calicotome, Rhus coriaria,* etc. (Loisel 1976). However, unlike in California, for example, there are practically no annual or perennial species adapted to this particular ecological environment around the Mediterranean sea. The annual species which grow after a fire are generally nitrophilous anthropophytes (Babour & Major 1972).

The zoogen apophytes result from well-known dispersal processes, particularly epizoochory and endozoochory, which normally take place during colonization. However, two examples related to overgrazing, mainly on the southern shores of the Mediterranean, are worth examining. The first example is the spreading of unpalatable toxic and spiny species. In North Africa the natural distribution of *Peganum harmala*, many spiny *Astragalus, Launaea* and *Zilla macroptera* have increased many fold in pre-saharan zones (authors' obs.). The second example, not as widely recognized, but nonetheless important, is the replacement of plants characteristically associated with forest and preforest communities, essentially hemicryptophytes and chamaephytes by therophytes of little ecological value. This phenomenon has led not only to a standardization of the flora, but also to soil deterioration. The therophytes invasion in many communities in North Africa and the Middle East is alarming on biological as well as economical grounds.

The substitution apophytes are also worth mentioning. These are usually woody plants whose natural distribution has been voluntarily modified by

54

human activities, especially by those of foresters. These activities include development of priviliged tree species (mostly conifers) and reafforestation. One of the most remarkable examples is the extension of *Pinus brutia* in Southern Anatolia (Akmann *et al.* 1978). Worth noting also is the reafforestation of the Mediterranean region with Mediterranean species such as *Pinus nigra*, *Cedrus atlantica* and *Abies*, which are now naturally regenerating in degraded and preforested ecosystems. This type of conifer reafforestation, generally covering large areas, is often followed, in old woodlots, by the installation of many species growing on acid humus: *Pyrola* and several Orchidaceae, indigenous species whose natural distribution has increased over the past few decades.

Anthropophytes

The Anthropophytes are synanthrope species which were voluntarily or involuntarily introduced by man. Like the apophytes, they can be separated into different groups. We will discuss successively their origin, their habitat and lastly their particular biological adaptations.

Most of the authors mentioned earlier agree on a classification separating the archaeophytes, introduced before the end of the 15th century, from the kenophytes which were introduced later. This division corresponds to the discovery of the New World. They agree as well on a classification separating the anthropophytes into several groups related to their degree of integration within the native flora.

The ephemerophytes appear for short amounts of time without any human help. Many garden species are included in this category. Their development may last for a few decades before they disappear, unable to compete with the indigenous flora or because of unfitted ecological factors. Numerous Liliaceae and Amaryllidaceae are subspontaneous species in the Mediterranean region.

The adventive species have the same characteristics, but they have been involuntarily introduced and do not originate from cultivated fields (e.g. rice fields) or gardens.

The naturalized species, often called 'neophytes', are plants almost always involuntarily introduced and completely integrated into the native flora. Examples are numerous: *Nothoscordum inodorum, Chenopodium ambrosioides, Phytolacca americana, Oenothera, Solanum, Datura, Aster, Conyza, Erigeron, Artemisia*.

Ecology and abundance of anthropophytes

Abundance in the Mediterranean region

An exhaustive evaluation of all the anthropophytes existing around the Mediterranean is impossible. First of all, the ephemerophytes cannot be reasonably taken into account. The importance of the adventive and sub-spontaneous species is sometimes fairly great, but it is nevertheless impossible to tell at this time if they could eventually be naturalized. An approximate evaluation can be made in which only the neophyte species and a few members of the adventive and sub-spontaneous species will be considered. Therefore, the number of non-native plants in the Mediterranean region can be estimated at close to 250. Considering the 23,000 species existing in the region, this figure is very low, around 1% of the total. In California, as an example of another Mediterranean region, Jepson (1925) considered that 50 to 75% of the flora came from invasions. This figure was actually too high, and Raven & Axelrod (1978) have estimated it at closer to 20%.

Geographic origin of invasive plants

The origin of invasive plants is an important, but sometimes difficult, topic of study, particularly for archaeophytes, and the progressive introduction of cultivated plants will not be discussed here (for a review see Van Zeist 1980). Among the other types of species, we will examine those from Mediterranean climatic zones around the world, temperate climatic zones and zones with tropical flora.

Species from Mediterranean regions

Surprisingly, their importance is limited. However, around the Mediterranean itself, some migrations can be seen. They occur mainly from east to west. Even if this extension is generally of human origin, species like *Cercis siliquastrum, Styrax officinalis, Fraxinus ornus, Phlomis fruticosa, Platanus orientalis, Punica granatum* are constantly expanding westwards. Species of western Mediterranean origin which have naturally developed on the east side of the Mediterranean are very rare.

Another group comes from the Irano-Touranian zone. It is mostly composed of anthropophilous plants related either to ancient cultivated fields (e.g. *Calepina irregularis, Camelina sativa, Neslia paniculata*) or to local fields and currently widely expanding (e.g. *Isatis tinctoria* and *Salvia aethiopis*). *Artemisia annua* is of dubious origin. This plant had a very important place in the famous 'Flora Juvenalis' of Montpellier, southern France (Thellung 1911–12).

Few Mediterranean species come from the Cape Peninsula. Some were voluntarily spread and are now naturally adapted, as is the case for some

Carpobrotus species or *Polygala myrtifolia*. More remarkable still are the invasions of *Oxalis pes-caprae* and *Cotula coronopifolia*, the latter of which is not limited to the Mediterranean region and both of which are particularly exuberant.

From the Australian Mediterranean region, the genus *Acacia* was voluntarily introduced and is now well adapted to the Circum-Mediterranean flora: *A. retinoides, A. dealbata, A. cyanophylla, A. melanoxylon*, etc. Worth noting also are *Atriplex semibaccata* and perhaps some *Eucalyptus* species.

From the American Mediterranean region, very few species are present around the Mediterranean, and they are usually temporary or present only in very limited areas. From California, *Phacelia tanacetifolia* can be cited. This is rather peculiar since the non-Mediterranean American flora provided a significant number of invading species in the Circum-Mediterranean region.

Species from temperate climatic zones

The number of species coming from Old World temperate zones is relatively large. Among others are trees and shrubs such as *Alianthus altissima, Broussonetia papyrifera, Buddleja davidii* and herbaceous plants such as *Impatiens glandulifera, I. parviflora, Artemisia verlotiorum*.

However, the number of species originating from the North American temperate climatic zones is certainly the largest of all invading species around the Mediterranean, among others are: *Panicum capillare, Paspalum dilatatum, Yucca, Amaranthus, Phytolacca americana, Opuntia, Argemone mexicana, Amorpha fruticosa, Robinia pseudacacia, Oenothera, Oxalis, Datura, Solidago, Erigeron, Aster, Galinsoga, Ambrosia, Xanthium*. From the South American temperate climatic zone came *Amaranthus deflexus, Solanum, Salpichroa, Tagetes minuta*.

Species from intertropical climatic zones

The Old World and Tropical Africa provided an important number of species to the Mediterranean region, mostly cultigen summer annual Gramineae such as *Chloris, Echinochloa, Panicum, Paspalum, Pennisetum, Sorghum*.

Only a few species came from Tropical America: some species of *Lippia, Amaranthus, Solanum, Datura, Nothoscordum inodorum*, and *Nicotiana glauca*.

The anthropophytes in relation to their environment

Even a quick look at the anthropophytes shows that they do not thrive in any kind of environment. Most of them have developed in unstable and artificialized habitats; very few can be found in natural environments.

Anthropophytes in unstable and artificialized habitats

These habitats can be separated into two groups: 1) cultivated fields, orchards, gardens,... and 2) unstable zones where nitrates are abundant, such as urban and periurban wastelands, rubble heaps, roadsides, animal resting grounds and fire zones.

Most of the anthropophytes can be found in these environments. The species of the Secalinion alliance occurring in cereal crops for instance, are mostly Gramineae and Cyperaceae of tropical origin. In food crops and orchards many species are characteristic of the Diplotaxidion alliance: about 12 species of *Amaranthus*, 3 to 6 of *Chenopodium*, and *Oxalis, Oenothera, Ambrosia, Aster, Erigeron, Conyza, Artemisia*,... Many of these species also grow in nitrogen-rich environments (Chenopodion alliance) along with some *Beta, Phytolacca americana, Argemone mexicana, Heliotropium, Solanum, Datura*,... Adventive species and ephemerophytes are numerous on roadsides (Onopordion alliance): *Lepidium virginicum, Salpichroa, Xanthium, Galinsoga, Cotula coronopifolia*, etc. *Broussonetia papyrifera, Buddleja davidii, Ailanthus altissima, Robinia pseudacacia* can be found in vacant lots.

Anthropophytes in natural habitats
Plants spread by man are very seldom found in natural habitats. However, there is one exception to this rule: in the Mediterranean region, forest communities along rivers show a large number of anthropophytes, such as *Hemerocallis, Commelina, Amorpha fruticosa, Oenothera, Impatiens, Lippia, Solidago, Heliopsis*, and also *Acer negundo, Lonicera japonica, Reynoutria, Begonia evandsiana*, plus a large number of adventive of naturalized species as well. This peculiarity is linked to the special microclimatic conditions in this type of ecosystem where summer drought is absent.

Anywhere else their importance is limited. In forested and preforested communities, not taking into account some natural regeneration on non-native species voluntarily introduced (mostly Californian conifers), only *Acacia dealbata* has been able to develop a local climax on acid soils in Provence (Loisel 1976). However, its distribution was drastically reduced because of the dramatically cold winters of 1985 and 1986. Neither in Mediterranean matorrals nor in natural grasslands did we find any well-adapted anthropophytes.

A few succulents of North American origin, introduced for ornamental purposes, are now fairly abundant in the Mediterranean warmest zones: *Agave americana* and some *Opuntia* species, for example, are now a part of the Mediterranean flora, as are *Carpobrotus* species from the Cape Peninsula (South Africa).

Biological problems
The anthropophytes successful development could generally be explained through a normal sexual reproduction process. This occurs often, particularly for the therophytes (including some biennials) located in a very disturbed habitat: all the Graminae (*Paspalum, Echinochloa, Panicum*,...), some Amaranthaceae (*Amaranthus*), Chenopodiaceae (*Chenopodium, Atriplex*), Cruciferae (*Lepidium, Neslia, Bunias, Camelina*,...), Oenotheraceae (*Oenothera*),

58

Balsaminaceae (*Impatiens*), and the majority of the Solanaceae (*Datura, Physalis, Solanum*) and Compositae (*Ambrosia, Xanthium, Aster, Erigeron, Conyza, Cotula* and *Artemisia*). Even though precise studies are scarce, this is also probably the case for shrubs such as *Amorpha fruticosa, Lippia* and *Buddleja*, which are strong invaders especially along the rivers.

The spread of the anthropophytes can be related in many cases to asexual reproduction (runners, natural cuttings, bulblets, stump division, etc.). This process is mostly responsible for the extension of *Hemerocallis, Yucca, Broussonetia, Carpobrotus, Oxalis, Solidago, Heliopsis*, etc.

Some species use both strategies: trees such as *Acacia* and more particularly *A. dealbata, Robinia pseudacacia* and *Ailanthus altissima*, but also succulents such as *Agave* (*A. americana* mostly) and several *Opuntia* species whose distribution has been considerably helped by man-made cuttings.

The evaluation of the seed adaptation of anthropophytes to dispersal is not very conclusive and it appears that most of the invading species do not have any particular adaptation at all: *Beta, Chenopodium, Amaranthus, Oenothera, Oxalis, Impatiens, Heliotropium, Datura, Buddleja, Cotula, Artemisia,* Cruciferae and Leguminosae.

Wind dispersal must be important for species with bristle bearing seeds such as some Compositae (*Erigeron, Conyza, Aster, Galinsoga*) and perhaps *Pennisetum*. Animal dispersal is also an important consideration, with ectozoochory found in *Ambrosia* and *Xanthium* and in some species of Gramineae. Endozoochory exists for *Phytolacca americana* and *Solanum*.

Life forms
Even approximate, the percentage of life form types of the anthropophytes brings out certain evident characteristics. The therophytes prevail and account for up to 60%, which is far above normal figures (40–50%) outside the Mediterranean region (Loisel 1976). The phanerophytes come second (14–15%), also far above the usual percentages in the native flora. This can be explained by numerous voluntary introductions of this life form. The geophytes reach 7–8% which is the same as that found in the indigenous flora. Chamaephytes and hemicryptophytes are poorly represented with only 10–11 and 7–8%.

Conclusion

The analysis of the synanthrope species in the Mediterranean region brings out many important facts. These species do not represent a significant component of the Mediterranean flora. Naturalized, adventive and subspontaneous taxa account for only 250 species, about 1% of the total native flora vs. 654 species with 15% of the native flora in California (Raven *et al.* 1977).

Some species have a more important role than others. In the apophyte group many Eastern species (from the Eastern Mediterranean basin and the

Irano-Touranian region) conquered the Western Mediterranean basin, though the reverse is not true. On the other hand, the other Mediterranean floras of the world have provided only a small amount of anthropophytes, except those voluntarily introduced from Australia (*Acacia, Eucalyptus, Atriplex*). The largest amount of anthropophytes comes from the extra-tropical New World.

The therophytes are the most important group, far ahead of others, and the phanerophytes are proportionally over-represented because of deliberate selections. The expansion of many phanerophytes and geophytes, as well as hemicryptophytes is favored by vegetative multiplication.

The apophytes expand their distribution when a natural ecosystem is destabiiized, such as therophytes and toxic or spiny species which spread on the southern side of the Mediterranean when preforest or matorral species invade its northern side. The anthropophytes thrive on artificialized habitats where nitrates are abundant and where there is little compotetion with native plants. The hygrophilous communities on river banks are a special case: because of very favorable ecological conditions, many anthropophytes, as well as apophytes, have settled and thrived. In fact, very few new ecosystems have managed to develop around the Mediterranean. *Acacia* communities and perhaps also *Phoenix dactylifera* oases alone have succeeded, with the exception of some voluntarily spread forest communities which are neither invading nor yet ecologically structured.

References

Akman, Y., Barbero, M. & Quezel, P. 1978. Contribution à l'étude de la végétation forestière d'Anatolie méditerranéenne. *Phytocoenologia* 5(I): 1–79.

Babour, M.G. &. Major, J. 1977. *Terrestrial vegetation of California*. John Wiley & Sons, New York, London, Sidney, Toronto, 1002 pp.

Barbero, M., Loisel, R. & Quézel, P. 1984. Incidences des pratiques culturales sur la flore et la végétation des agrosystèmes en région méditerranéenne. *C.R. Soc. Biogéographie* 59(4): 463–473.

Debray, M. & Sanay, L. 1939. Classification de Thellung. *Bull. mens. Soc. Linn. Seine Maritime* 18: 33–36.

Jepson, W.L. 1925. *A manual of the flowering plants of California*. Univ. California Press, 1238 pp.

Jovet, P. 1984. Trois classifications des plantes synanthropes. *C.R. Soc. Biogéographie* 60(3): 107–119.

Kornas, J. 1966. Influence of man and his activities on the vegetation of Poland. In: Sfafer Wl. (ed.), *The vegetation of Poland*. pp. 97–137.

Kornas, J. 1978. Remarks of the analysis of a synanthropic flora. *Acta Botanica Slovaca* ser. A, 3: 386–393.

Loisel, R. 1976. *La végétation de l'étage méditerranéen dans le sud-est français*. Thèse Doctorat ès Sciences, Aix-Marseille III, 384 pp.

Pons, A. & Suc, J.P. 1980. Les témoignages de structures de végétation méditerranéenne dans le passé antérieur à l'action de l'homme. *Naturalia Monspeliensis*, Colloque de Montpellier, numéro spécial: 69–78.

Raven, P.H. & Axelrod, D. 1978. *Origin and relationships of the Californian flora*. University of California Press, Botany, 72, 134 pp.

Reille, M., Triat-Laval, H. & Vernet, J.L. 1980. Les témoignages de structures de végétation méditerranéenne durant le passé contemporain de l'action de l'homme. *Naturalia Monspeliensis*, Colloque de Montpellier, numéro spécial: 79–87.

Renfrew, J.M. 1966. A rapport on recent finds on carbonized cereals grain and seeds from prehistoric Thessaly. *Thessalica* 5: 21–36.

Rousseau, C. 1971. Une classification de la flore synanthropique du Québec et de l'Ontario. *Le Naturaliste canadien* 98: 530.

Thellung, G. 1911–1912. Flore adventice de Montpellier. *Soc. Sc. Nat. Math. Cherbourg* 38: 57–728.

Van Zeist, W. 1980. Aperçu sur la diffusion des végétaux cultivés dans la région méditerranéenne. *Naturalia Monspeliensis*, Colloque de Montpellier, numéro spéciale: 129–145.

Zohary, D. & Hoff, M. 1973. Domestication of pulses in the Old World. *Science* 182: 887–894.

5. The invading weeds within the Western Mediterranean Basin

J.L. GUILLERM, E. LE FLOC'H, J. MAILLET, and C. BOULET

Abstract

Weed invasions within the Western Mediterranean Basin are related to historical changes in agricultural practices. Some species are used as examples to point out the diverse strategies of weed invaders.

Introduction

There is seldom invasion of natural communities without disturbance, and there is a trend to greater invasion with more prolonged, frequent or intense disturbance (Fox & Fox 1986). A ubiquitous class of human disturbance across the biosphere is associated with agriculture. From the earliest cultivation, new ecosystems have been created and managed to provide food resources. This chronic disturbance for plant cultivation has led to spread of co-evolved weeds species prone still today to episodic invasion processes.

Invasive plants often show the ability to migrate more or less quickly within one area, or into new areas or continents, and to fill new open spaces. The success of many invasions is the inevitable consequences of the plant abilities to react to changes in the agricultural practices, occurring over millennia, particularly all around the Mediterranean Basin.

The manifold source of invasives, the antiquity of the weed flora, the diversifying weed shifts and invasions in the Circummediteranean area are more or less known. Many papers have been produced on these topics (e.g. Barralis 1966, Dafni & Heller 1980, 1982, Greuter 1971, Guillerm 1990, Guillerm & Maillet 1982, 1984, Holzner 1978, 1982, Jovet 1971, Kosinova 1974, Kornas 1983, Montégut 1974, 1984, Pignatti 1983, Pinto Da Silva 1971, Protopapadakis 1985, Yannitsaros & Economidou 1974, Zohary 1949–50, 1983).

Any new interference within the existing vegetation may modify the species relationships, introduce an imbalance, and allow invasion by a local species or by newcomers. An invading plant enlarges its habitat by the colonization

F. di Castri, A. J. Hansen and M. Debussche (eds.), Biological Invasions in Europe and the Mediterranean Basin, 61–84. © 1990, *Kluwer Academic Publishers, Dordrecht.*

62

of an open space corresponding to its ecological niche, or by a life cycle adjustment to a new situation.

The life cycle adaptations and the reproductive strategies allowed the species to take up residence centuries ago (naturalized species, subspontaneous species, or archeophytes), in the latter time (neophytes or present newcomers), or to be only fugitive species.

The dominant, recessive or neutral position of the species within the community determines their spread (cosmopolitan, regional or local, endemic, relic or pioneer species).

The main weeds of the Western Mediterranean Basin

The weed community may include sparse, infesting and invading weeds. The sparse or low-density species are named minor weeds. Infesting species are relatively abundant and harmful for the cultivated plant. Invading weeds are the most harmful, having high densities and often becoming dominant. The infesting and invading weeds are the major weeds (Baker 1974). The invader's status may fluctuate in space and time.

The main weeds infesting or invading the cultivated fields in the Western Mediterranean Basin are listed in Table 1. Some of these species are harmful on a global scale, others, only on a regional or local scale. As any given list is open to criticism, Table 1 is a starting point for further informations, precise details, discussion and comparison with other lists as in Moreira *et al.* (1986).

The weed invasions related to the changes in agricultural practices

The weed communities are the result of ecological changes introduced by man with the recurrent and alternating agricultural practices. It was more or less 9000 B.C. that the Neolothic agricultural interference occurred over the Mediterranean Basin. The most important sources of information are the plant remains from the ancient Egyptian tombs (Kosinova 1974), where about 50 weeds have been recorded from the Neolithic to the Coptic periods, passing by the Predynastic and the Dynastic eras of the Pharaonic Egypt (about 3000 B.C. to 640 A.D.). Most of them are, today, commonly distributed all over the Mediterranean region.

From the Neolithic to the 17–18th century, the weed communities changed little in species composition with the progressive adaptation of colonizing species to these agricultural open spaces, and the emerging of the weed flora specialized to the different ecological environments, and to the diverse types of cultivation.

Nevertheless, from the Middle Ages, the European, East, and Far-East trades favoured the introduction of some Eurasian species, nowaways integrated

Table 1. Invading and/or infesting weed species in the Western Mediterranean Basin (Authors's data).

Species	i/e	Biol. type	Origin
Amaranthaceae			
Amaranthus blitoides S. Watson	e	Th	N. Am.
Amaranthus deflexus L.	e	Th	S. Am.
Amaranthus retroflexus L.	i	Th	N. Am.
Amaryllidaceae			
Gladiolus italicus Miller	i	G	Med.
Araceae			
Arisarum vulgare Targ.-Tozz.	i	G	Med.
Arum italicum Miller	i	G	Med.
Borraginaceae			
Heliotropium curassavicum L.	e	H/Ch	N. Am.
Heliotropium europaeum L.	i	Th	Med.
Caryophyllaceae			
Gypsophila pilosa Hudson	i	Th	Eur.
Silene vulgaris (Moench) Garcke	i	G	Eur.
Stellaria media (L.) Vill.	i	Th	Cosm.
Vaccaria pyramidata Medicus	e/i	Th	Eur.
Chenopodiaceae			
Atriplex patula L.	i	Th	Cir. Bor.
Chenopodium album L.	i	Th	Cosm.
Compositae			
Ambrosia artemisifolia L.	e	Th	N. Am.
Artemisia verlotiorum Lamotte	e	G	As.
Aster squamatus (Sprengel) Hieron.	e	Th	C.S. Am.
Carduncellus eriocephalus Boiss.	i	Th/H	Med.
Carduus getulus Pomel	i	Th	Med.
Carthamus lanatus L.	i	Th	Med.
Chamaemelum fuscatum (Brot.) Vasc.	i	Th	Med.
Chamomilla suaveolens (Pursh) Rydb.	e	Th	As./N. Am.
Chondrilla juncea L.	e	H	Eur/Med.
Chrysanthemum coronarium L.	i	Th	Med.
Chrysanthemum segetum L.	i	Th	Med.
Cirsium arvense (L.) Scop.	i	G	Eur./As.
Conyza albida Willd	e/i		
Conyza blakei (Cab.) Cabrera	e	Th	S. Am.
Conyza bonariensis (L.) Cronq.	e/i	Th	C. Am.
Conyza canadensis (L.) Cronq.	e/i	Th/H	N. Am.
Dittrichia viscosa (L.) W. Greuter	e	Ch	Med.
Galinsoga ciliata (Rafin) S.F. Blake	e	Th	S. Am.
Galinsoga parviflora Cav.	e	Th	S. Am.
Lactuca serriola L.	i	Th	Eur.
Onopordon espinae Cosson ex Bonnier	i	H	Med.
Onopordon nervosum Boiss	i	H	Med.
Otospermum glabrum (Lag.) Willk.	i	Th	W. Med.
Picris echioides L.	i	Th/H	W. Med.

64

Table 1. (Continued).

Species	i/e	Biol. type	Origin
Senecio inaequidens DC.	e	Ch	S. Af.
Senecio vulgaris L.	i	Th/H	Eur./As.
Sonchus asper (L.) Hill	i/e	Th/H	Cosm.
Sonchus oleraceus L.	i/e	Th	Cosm.
Convolvulaceae			
Calystegia sepium (L.) R. Br.	e	G	Eur.
Convolvulus althaeoides L.	i	G	Med.
Convolvulus arvensis L.	i	G	Cosm.
Convolvulus tricolor L.	i	Th	Eur.
Crassulaceae			
Sedum sediforme (Jacq.) Pau	e	G	Med.
Cruciferae			
Cardaria draba (L.) Desv.	i	G	Eur./Med.
Diplotaxis catholica (L.) DC.	i	Th	Eur.
Diplotaxis erucoides (L.) DC.	i	Th	Med.
Diplotaxis simplex Sprengel	i	Th	Med.
Diplotaxis tenuifolia (L.) DC.	e	H/Ch	Eur.
Raphanus raphanistrum L.	i	Th/H	Cosm.
Rapistrum rugosum (L.) All.	i	Th/H	Med.
Roripa sylvestris (L.) Besser	i	G	Med.
Sinapis arvensis L.	e	Th	Eur.
Cyperaceae			
Cyperus esculentus L.	i/e	G	Paleot.
Cyperus rotundus L.	i/e	G	Paleot.
Equisetaceae			
Equisetum arvense L.	i	G	Cir. Bor.
Equisetum ramosissimum Desf.	i	G	Eur.
Equisetum telmateia Ehrh.	i	G	Eur.
Euphorbiaceae			
Acalypha virginica L.	e	Th	N. Am.
Chrozophora obliqua (Vahl) A. Juss. ex Spren.	i	Th	Med.
Chrozophora tinctoria (L.) A. Juss.	i	Th	Med.
Euphorbia nutans Lag.	e	Th	N. Am.
Euphorbia serrata L.	e	Th/G	Med.
Graminae			
Avena sterilis L.	i	Th	Med.
Brachiaria eruciformis (Sm) Griseb.	e	Th	Paleosubt.
Bromus wildenowii Kunth	e	H	S. Am.
Bromus diandrus Roth	e	Th	Cosm.
Bromus madritensis L.	e	Th	Med.
Bromus sterilis L.	e	Th	Eur.
Cenchrus echinatus L.	e		
Cynodon dactylon (L.) Pers.	i	G	Cosm.
Dactyloctenium aegyptium (L.) Beauv.	e	Th	Trop.
Dichanthium ischaemum (L.) Roberty	e	H	Cosm.

Table 1. (Continued).

Species	i/e	Biol. type	Origin
Digitaria sanguinalis (L.) Scop.	i/e	Th	Cosm.
Echinochloa crus-galli (L.) Beauv.	i/e	Th	Eur.
Lolium multiflorum Lam.	i	Th	Med.
Lolium rigidum Gaudin	i	Th	Paleot.
Panicum capillare L.	e	Th	Cosm.
Panicum dichotomiflorum Michx	e	Th	Cosm.
Panicum miliaceum	e	Th	As.
Panicum repens L.	e	G	Paleot.
Paspalum dilatatum Poiret	e	G	Néotrop.
Paspalum paspalodes (Michx) Scribner	e	G	Cosm.
Phalaris paradoxa L.	e	Th	Med.
Poa annua L.	i	Th	Cosm.
Setaria gracilis H.B.K.	e	H	Cosm.
Setaria verticillata (L.) Beauv.	i	Th	Cosm.
Setaria viridis (L.) Beauv.	i	Th	Cosm.
Labiatae			
Lamium amplexicaule L.	i	Th	Eur./As.
Leguminosae			
Melilotus indica (L.) All.	e	Th	Med./As.
Melilotus segetalis (Brot.) Ser.	i	Th	Med.
Scorpiurus muricatus L.	i	Th/H	Med.
Liliaceae			
Allium polyanthum Schultes & Schultes	i	G	Med.
Allium vineale L.	i	G	Eur.
Ornithogalum narbonense L.	e	G	Med.
Malvaceae			
Abutilon theophrasti Medicus	e	Th	As.
Onagraceae			
Epilobium tetragonum L.	e	H	Med.
Orobanchaceae			
Orobanche crenata Forskal	i	par.	Med.
Oxalidaceae			
Oxalis corniculata L.	e	G	Med.
Oxalis latifolia Kunth.	e	G	S. Am.
Oxalis martiana Zucc.	e	G	S. Am.
Oxalis pes-caprae L.	e	G	S. Af.
Papaveraceae			
Papaver rhoeas L.	i	Th	Eur.
Phytolacaceae			
Phytolaca americana L.	e	Ch	N. Am.
Polygonaceae			
Bilderdykia convolvulus (L.) Dumort	i	Th	Eur.
Emex spinosa (L.) Campd.	i	Th	Med.
Polygonum amphibium L.	e	G	Cosm.

66

Table 1. (Continued).

Species	i/e	Biol. type	Origin
Polygonum aviculare L.	i	Th	Cosm.
Polygonum lapathifolium L.	e	Th	Cosm.
Reynoutria japonica Houtt.	e	G	As.
Rumex bucephalophorus L.	i	Th/H	Med.
Rumex crispus L.	i	H	Cosm.
Portulacaceae			
Portulaca oleracea L.	i/e	Th	Med.
Ranunculaceae			
Clematis vitalba L.	e	Np	Eur.
Rosaceae			
Rubus caesius L.	e	Ch	Eur.
Rubus ulmifolius Schott	i	G	Eur.
Rubiaceae			
Galium aparine L.	i	Th	Eur.
Galium tricornutum Dandy	i	Th	Med.
Rubia peregrina L.	e	G	Med.
Santalaceae			
Thesium humile Vahl	i	par.	Med.
Scrophulariaceae			
Veronica hederifolia L.	i	Th	Eur.
Veronica persica Poiret	i	Th/H	As.
Solanaceae			
Solanum eleaegnifolium Cav.	e	G/H/Ch	S. Am.
Solanum luteum sp. alatum (Moench) Dostat	i/e	Th	Cosm.
Solanum nigrum L.	i/e	Th	Cosm.
Umbelliferae			
Ammi majus L.	i	Th	Med.
Daucus carota L.	i/e	Th/H	Paleotep.
Foeniculum vulgare Miller	e	H	Med.
Ridolfia segetum Moris	i	Th	Med.
Torilis arvensis (Hudson) Link	i	Th	Med.
Torilis nodosa (L.) Gaertner	i	Th	Med.

Status		Origin				
i	= infesting	Afr.	= Africa	Neotrop.	= Neotropical
e	= invading	S. Af.	= South Africa	Paleosubt.	= Paleosubtropical
		N. Am.	= North America	Paleotep.	= Paleotemperature
Biological type		C. Am.	= Central America	Paleot.	= Paleotropical
Th	= therophyte	S. Am.	= South America		
G	= geophyte	As.	= Asia		
H	= hemicryptophyte	Cosm.	= Cosmopolist		
Ch	= chamephyte	Eur.	= Europe		
Np	= nanophanerophyte	Med.	= Mediterranean		
par	= parasit		region		

within the Mediterranean flora, according to their winter and spring cycles, as *Veronica persica, Anagallis arvensis, Sonchus oleraceus, Picris echioides, Cardaria draba.*

The more drastic shifts occurred a little later with species coming from the New World, as *Conyza bonariensis* and *Conyza canadensis* which are now present all over the Old World. For these two species, climatic differentiation favoured *Conyza bonariensis* in the eastern part of the Mediterranean Basin, and *Conyza canadensis* within the western part. From the New World colonization time results also the invasion by some tropical weeds occurring with the introduction of summer cycle plants, as *Zea mais, Sorghum, Solanum lycopersicum, Solanum tuberosum.*

These new tropical weeds became good invaders in the new biotopes, represented by the summer cultivations, where they strongly compete with the local weed species. In the cultivated fields of France, the difference between the floristic richness of the winter and spring flora and the summer flora is considerable (Le Maignan 1981). Within 2000 records including 909 species, about 45% are winter and spring therophytic species, and only 10% are summer therophytic species. The other 45%, in majority out of summer species, are hemicrytophytes (25%), geophytes (12%), chamephytes and phanerophytes (8%). Within the 10% of summer therophytic species, 1.5% are tropical therophytic species. Similar percentages are obtained in considering the species present upper 10% of the records (550 species).

From the 19th century and the beginning of the 20th century, cultivation spread and intensified with mechanization and fertilization practices favoring nitratophilous weeds as *Polygonum* spp., *Lamium amplexicaule, Stellaria media* or perennial weeds, as for example *Cirsium arvense*, and *Convolvulus arvensis.*

These more uniform weed communities reflected the more homogeneous environmental conditions. The intensified agriculture caused many sensitive species to diminish their ranges by retreating towards its centre. In the same time, improved seed cleaning eradicated a number of specialized species, with short-lived seeds, which depend on being sown with the crop, and are not able to grow outside arable lands.

Another drastic alteration in the composition of weed communities occurred with the use of herbicides. In winter and spring cereals, the sensitive species decreased or disappeared, as *Agrostemma githago, Asperula arvensis, Rhagadiolus stellatus* (Maillet 1981).

Ploughing cut the regenerative organs of plants into many parts. Both vegetative propagules and seeds were transported by the agricultural machines from field to field and over large distance. Some of them remained in the soil seed bank. The huge size of this seed bank enabled species, as *Papaver rhoeas, Centaurea cyanus, Sinapis arvensis*, to reappear casually or suddenly from an inadequate herbicide use or application, or be driven back to refugial sites.

The outcome was that species resistant to herbicides suffered less competition from other weeds. The resistant species have now became able to grow in

68

greatly increased densities, and with single individuals being larger than before. These resistant species are able also to enlarge their range of distribution and to fill the niches of the eliminated species, conquering new areas where they were not able to compete before. This development is evidence of the dependance between the ecological behaviour of the species and the competition of other species, and vice-versa for the competitive power of the species depending on the environmental conditions (Ellenberg 1968).

In the cereals of the Western Mediterranean Basin, some species became resistant to a specific herbicide, and only a new chemical composition is able to eliminate them.

The tolerant weeds are early autumnal and short-lived species, eradicated with difficulty in the beginning development of the cereal, as *Lolium* spp., *Avena* spp., *Galium* spp., *Veronica* spp. They are also spring weeds, indifferent to the automnal application of herbicides, as *Chenopodium album, Avena fatua, Convolvulus arvensis,* diverse Umbellifereae.

The main invasions occur in the summer annual cultivations where the local species are less adapted to these new conditions, in the perennial cultivations, including large open spaces (vineyards, orchards, groves), and in market gardening.

In tomato fields, inadequate herbicide applications favoured *Solanum nigrum, Solanum luteum, Solanum sarrachoides, Solanum sublobatum.*

In maize, triazine applications promote invasions by *Digitaria* spp., *Setaria* spp., *Echinochloa crus-galli,* and more recently by *Panicum capillare* and *Panicum dichotomiflorum* which metabolize triazineas maize. Resistant species as *Amaranthus retroflexus* and *Solanum nigrum* (susceptible before) occur now. The enlarged interow favoured newcomers as *Abutilon theophrasti,* in France and *Acalypha virginica,* and *Galinsoga parviflora* in Italy.

Irrigation also favoured the extent of *Paspalum dilatatum, Paspalum paspalodes, Polygonum persicaria, Arisarum vulgare, Oxalis pes-caprae, Oxalis corniculata, Cyperus rotundus, Panicum repens, Phragmites australis, Equisetum telmateia, Artemisia verlotiorum* in Mediterranean orchards. The developments of resistant lianas also occurs as *Bryonia cretica, Solanum dulcamara, Clematis vitalba, Humulus lupulus* (Guillerm & Maillet 1982, Maillet & Zaragoza-Larios 1976, Moreira 1976).

Many of the mentioned species are C4 photosynthesis species which are good competitors and spread quickly.

In the opening of the perennial cultivation (as vineyards, orchards), the herbicides associated with the no-tillage practice enlarged the spread of the surrounding species into the cultivated lands (Guillerm & Maillet 1984). Species with a great seed dispersal (anemochory, ornithochory), a strong regenerative system, and with thicked leaves (a waxy cuticule decreasing the herbicide penetration) are favoured. This is emphasized by the alteration of the herbicide product in autumn, enabling their spread in spring time.

These main species, as *Rubia peregrina, Smilax aspera, Hedera helix, Sedum sediforme,* occur from the surrounding garrigue, or are newcomers, as *Oxalis pes-caprae, Oxalis latifolia, Senecio inaequidens, Solanum eleaegnifolium.*

In rice fields, the main shifts in the weed communities occur also from the changes in the agricultural practices (hand weeding to chemical control, direct seedling or bedding rice plant). The present main infesting or invading weeds are (Vasconcellos 1959, Foury *et al.* 1954, Bolos & Masclans 1955, Pignatti 1957, Tallon 1958, Tomaselli 1958, Maghami 1966, Tinarelli 1973, Sutisna 1980, Espirito Santo & Rosa 1981, Guillerm & Sutisna 1983, Guillerm *et al.* 1989, Sanon 1986) *Echinochloa crus-galli* var. *hostii* and var. *longisetum, Cyperus difformis, Lindernia dubia* (the late one from the N.E. America), the 'wild rices' including diverse *Oryza* (a complexe group originated from Asia and Africa) and common species of the environmental communities enlarging their niches in the rice fields as *Scirpus mucronatus, Scirpus maritimus, Phragmites australis, Typha* spp., *Alisma plantago-aquatica*, and *Chara* spp.

The invading weed strategies

The following species, related to different invading periods, illustrate the diverse strategies for weeds to invade the cultivated fields. These strategies, results of the genetic and somatic variabilities of the species, are reflected by the life form system adaptations to survive the unfavourable disturbances.

Sinapis arvensis, Avena fatua, and *Avena sterilis* are former colonizing species, invading Mediterranean Europe and Africa since the Neolithic period. *Sinapis arvensis* presently decreases within intensive cultivation by anti-dicotyledoneous herbicide uses, but it remains likely adaptations to new environmental conditions. The two mentioned species of *Avena* are, today, increasing as they are cropmimics (Baker 1974).

The neo-tropical *Solanum nigrum* introduced 3 or 4 century ago, is a usual summer facies in perennial cultivations, but it is very increasing in summer cultivations.

Oxalis pes-caprae, originated from the Cape province of South-Africa, introduced at the beginning of the last century, progressively become an invader, specially in irrigated fields.

Solanum eleaegnifolium, a neo-tropical species, recently introduced, causes now serious damages, particularly in Morocco. It is an excellent example of life-cycle adaptation to the crop rotation.

Senecio inaequidens is a new invader in Southern France and Italy, originated from the Cape province of South Africa also. This species stays a long time into ruderal sites, and recently increases in fields, favoured by chemical weed control.

Sinapis arvensis

A large colouring's heterogeneity of the seeds, flowering heads and siliquous

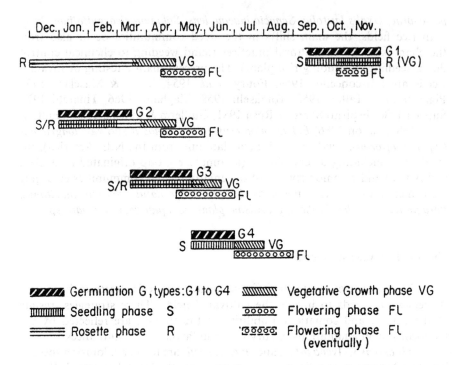

Figure 1. Simplified scheme of phenophases of different types of individuals of *Sinapis arvensis* L. according to their germinative aptitude.

have been pointed out by Jauzein (1980). This heterogeneity results in a large variability in seed germination capacity. In fact, four seed sorts (Montégut 1982) are related to the germination aptitude (Figure 1):
- the present seeds, or the seeds coming from the seed bank of the soil, able to germinate before winter: G1. This type of seed, particularly common in the Mediterranean region, evolves quickly, from the seedling stage to the rosette stage, going on to springtime. Flowering occurs from the middle of spring to before the autumn frosts.
- the seeds, coming from the seed bank, able to germinate with temperature above 0°C. during winter: G2, then flowering as G1.
- the seeds, coming from the seed bank, able to germinate at the beginning of spring, after the winter colds: G3, flowering as G1 and G2.
- the seeds, coming from the seed bank, able to germinate at the end of springtime and at the beginning of summer: G4. These individuals evolve directly, without rosette stage, and flower in summer, by long photoperiodic inductions.

Montégut (1984) regards *Sinapis arvensis* as annual species, winter prefering, with heterogeneous populations able to germinate at each season. It results for the global species an appearance of indifference to vernalisation.

Avena fatua and *Avena sterilis*

From three decades ago, the increasing cereal monocultivation, nitrogenous fertilisation, herbicide uses, and the late machine harvesting are factors favouring *Avena* spreading.

These species are common all over the Mediterranean region, but they occur within differentiate areas. *Avena fatua* is a more nordic species, present in Spain and Portugal only on the steppic plateaux, or in the northern areas of Greece and Yugoslavia. *Avena sterilis* ssp. *sterilis* is a southern species, and *Avena sterilis* ssp. *ludoviciana* a more atlantic one.

The diverse biotopes colonized by these species, hexaploïde in their originating area (Pamir, Minor Asia), is contrasting with the specific habitat of the diploïd populations (A.R.C. 1976). This phenomenon is similar to the distribution of the tetraploïd and diploïd *Dactylis glomerata* populations (Lumaret *et al.* 1987). Hexaploïdy favour feasible adaptation to new ecological constraints or competition relationships, and autofecondation results in steadiness of adapted taxonomic units.

The ability of *Avena fatua* and *Avena sterilis* to germinate in autumn or in spring in the Mediterranean region favours population selection, according to the herbicide effect. More is a differentiated dormancy according to the ecotypes, and to the seed position on the panicle or within the spikelets. This dormancy is a recessive character (hybridization experiments with *Avena sativa*). Its acquisition and its maintenance by dominant autofecondation is an advantage for the colonization of cultivated lands or disturbed sites. The early seedlings from big seeds are able to germinate in depth, and to develop a strong root system, with a high neat assimilation rate, particularly the nitrogenous assimilation rate, allowing an efficient competition with the cereal. The seed longevity is about only 9 years, but one individual can produce 500 seeds, a large strength for re-infestation.

Solanum nigrum

This species is an unsteady genetic complex, with a large variability in North America, and less variability in the Mediterranean region, due to its recent introduction. The most important European populations are hexaploïds (2n=72), with dominant autogamy (Gasquez *et al.* 1981), but it occasionally occurs with *Solanum americana* (2n=24), *Solanum sublobatum* (2n=24), *Solanum sarrachoides* (2n=24), and more frequently *Solanum luteum* (2n=48). Hybridization within these units are unknown. The genetic variability determines successful colonization in maizes and vineyards by the appearance of resistant populations to the herbicide applications.

This species, originating from tropical area, is adapted to the Mediterranean climate by a dormancy preventing any autumn or winter germination. Temperature above 15°C. and sufficiant moisture are necessary for germination.

72

This plant reaches its peak in May-June, continuing on to summer in irrigated fields. The seeds can't immediately germinate, according to an inhibition of the berry matter.

Although a C3 plant, *Solanum nigrum* increases rapidly in biomass, when faced with water. Irrigation favors its growing and its seed dissemination. Being a nitrophilous species, it spreads with nitrogenous fertilisers (Abdel Fattah 1982). The huge of seeds production, 30,000 seeds per individual, and the long-lasting of seeds, about 40 years (Keeley & Thullen 1983) make it a succesful invader.

Oxalis pes-caprae

This species, originating from the Cape province of South Africa, common in the Mediterranean region, is nowadays regarded as a naturalized species. Being a long time in gardens, hedges of nursery gardens, it is now spreading into cultivated lands. This rhizomatous geophyte is very aggressive, and eradicated with difficulty.

From an heterostyly system within the flower organization, sexual reproduction needs cross-fertilisation. For a long time, botanists considered it as a sterile plant, out of its originating area, according to the presence of only one type of flower. Chabrolin (1934) is probably the first to demonstrate, in Tunisia, that, if the sexual reproduction is exceptional, the late flowers (the flowering occurs from february to march) are able to produce seeds. These seeds germinate with summer rainfalls, producing sterile individuals, but regenerative individuals from bulbs. The sterility ratio is quite similar for individuals from South Africa or from the Mediterranean Basin (Meikle 1977). The life cycle of *Oxalis pes-caprae* is depicted on Figure 2.

Solanum eleaegnifolium

This aggressive species is known as a colonizing invader from various parts of the Mediterranean Basin, such as in Egypt (Tackholm 1956), in Israël (Dafni & Heller 1980, 1982), in Greece (Economidou & Yannitsaros 1975), in Yugoslavia (Gazi Baskova & Segula 1978, Pavletic *et al.* 1978), in Italy (Bianco 1976), in Spain (Carretero 1979), in Morocco (Tanji *et al.* 1974), in Algeria and in Tunisia (c.v.). Originating from South America, it is first reported in Morocco in 1950, and in Tunisia in 1986.

Solanum eleaegnifolium is a perennial species, with lignous lower stems, 100 cm high. The root system consists in a vertical root with subhorizontal rhizomes, giving shoots able to regenerate new individuals. The life cycle of the species is depicted on Figure 3.

For Morocco, germinations start in February and are very effective in March. For the remaining winter individuals, the growing, at the beginning of spring,

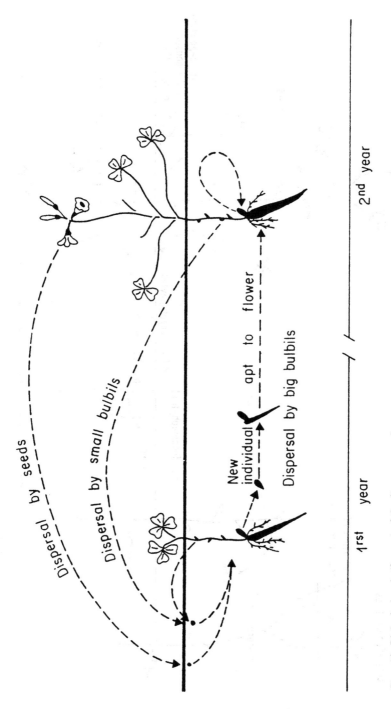

Figure 2. Biological cycle (2 years) of *Oxalis pes-caprae* L.

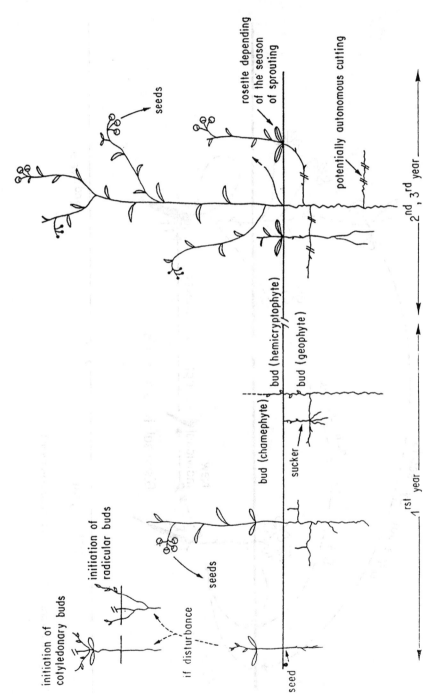

Figure 3. Biological cycle of *Solanum eleaegnifolium* Cav.

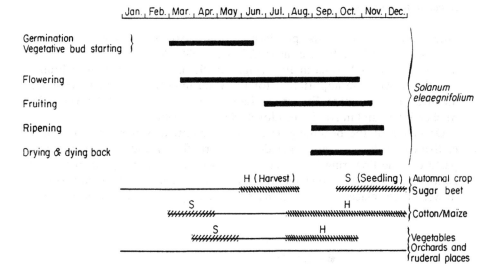

Figure 4. Phenophases of *Solanum eleaegnifolium* Cav. and life cycle duration of various crops in Morocco.

occurs from the regenerative buds of the remaining root stock (buds being flush with the soil, or just upper). Flowering starts in March-April, with an optimum in May, and go on to October. Fructification from June to November results in fruit ripeness from September to November. Fruits contain more or less 100 seeds. The plant disseminates by hydrochoric, zoochoric, or anthropochoric ways. The life cycle is quite similar in Greece, with a one month delay.

The ability of this species to survive agricultural disturbance is quite remarkable. It can occurs by:
- the sexual dissemination by seeds (germination rate is about 40% in Morocco). This is the therophytic strategy.
- the regenerative development in springtime
    (1) from aerial buds, located high on the remaining plant (chamephytic strategy),
    (2) from buds flush to the soil (hemicryptophytic strategy), or
    (3) from epiradicular buds (geophytic strategy).
- the regenerative development from root pieces, originated by the plough cutting, and disseminated by the agricultural machines.

The species is particularly aggressive in spring cultivations (cotton, maize), in Morocco, owing to the similarity of the life cycle with the cultivated plants (Figure 4).

Senecio inaequidens

Originating from the Cape province of South Africa, this species was first reported in north of France, near Calais in 1935 (Jovet & Bosserdet 1968), and in Italy, near Verona in 1947 (Pignatti 1982). Its introduction in France was linked with the importation for the wool industry. It is now becoming an invader in southern France, in Languedoc (Jovet & de Vilmorin 1975, Michez 1980), and in Aquitaine (Jovet 1971, Montégut 1982).

Occurring at the beginning as a ruderal plant, it is now spreading into cultivated fields (vineyards, meadows, lucern fields) and into fallows and pastures. The flowering heads have 80 to 100 flowers, giving 60 to 80% of germination rate after one month. The plant is cold resistant, and occurs in wet or dry places. It is an herbaceous chamephyte, with numerous stems

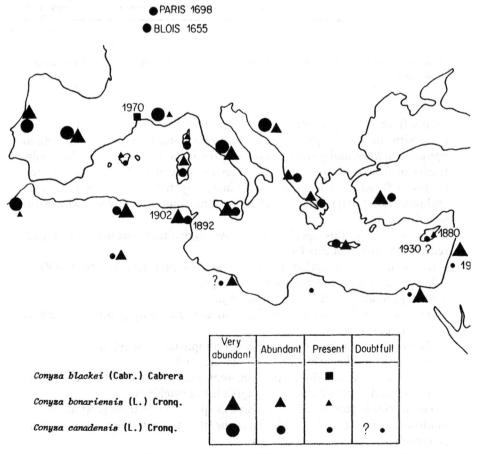

Figure 5. Spread of *Conyza blackei*, *C. bonariensis* and *C. canadensis* in the Circummediterranean area.

shotting from the bottom of the plant, able to regenerate adventive roots, from the base to the first floral ramifications, fastening the plant on to the soil. The main roots contain reserves available for regenerative bud growing. Spreading occurs by water or by wind. The flowering starts from the end of May until November-December. The germination periods take place after winter and spring. It sows itself in colonies, with an abundance of seeds. The species have not be found among annual cultivations.

The present spread of some weed invaders

The following maps (see Figures 5 to 9) illustrate, as examples, the spread of some invading weeds within the Mediterranean Basin. These examples are related to different periods of introduction.

The spatial distribution of *Conyza bonariensis* (from Central and South America), *Conyza canadensis* (from North America), and *Conyza blackei* (from South America) are depicted, with the indication of the first known data of registration of the species, on Figure 5.

Conyza bonariensis is more frequent within the eastern part of the Mediterranean Basin and *Conyza canadensis*. These two species have probably

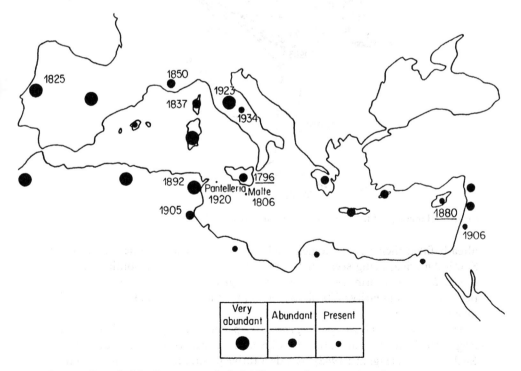

Figure 6. Spread of *Oxalis pes-caprae* L. in the Circummediterranean area (first record: 1796).

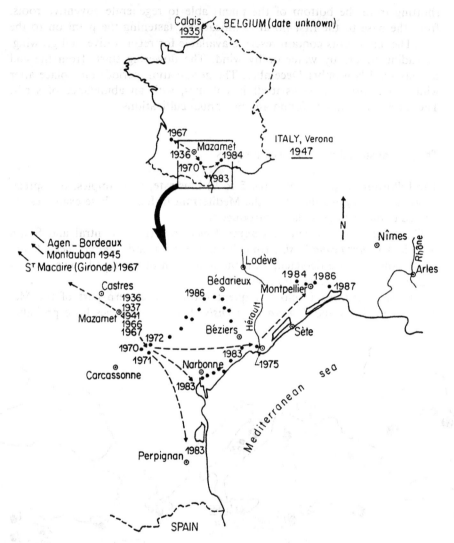

Figure 7. Chronology of the extension of *Senecio inaequidens* DC. in southern part of France.

already filled their new ecological niches. *Conyza blackei* is a newcomer from South America, being seen for the first time in 1970 in the South of France (Jovet & de Vilmorin 1975). This species grow in fallows and, casually in fields. But the similarity of its biology with the other *Conyza* makes it a potential invader.

The spread of *Oxalis pes-caprae* is depicted on Figure 6. This species originated from the Cape province of South Africa. Its first introduction into Sicilia, in 1796 (Pignatti 1982), as a decorative plant, seems to be the beginning of its spread in the Mediterranean Basin. It is today a dangerous weed in

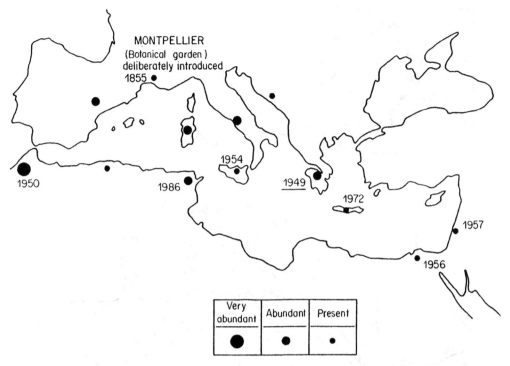

Figure 8. Spread of *Solanum eleaegnifolium* Cav. in the Circummediterranean area (first record: 1949).

irrigated fields (orchards, groves), gardens, market gardens.

The recent newcomer *Senecio inaequidens* is beginning to enlarge rapidly its range of distribution in the South of France and Italy (Figure 7).

Another newcomer, the *Solanum eleaegnifolium*, has a larger range than the precedent species (Figure 8). It is presently a very noxious weed in Morocco (Figure 9).

Discussion

Invasions in weed communities follow the more or less drastic changes in the agricultural practices. As the agricultural interference introduces a new disturbance within the agroecosystem, the level of the heterogeneity within the weed community is in a dynamic state, favourable for weed invasion processes.

From the antiquous agriculture interference within the Mediterranean Basin, weed populations were and are, from former times to nowadays, obliged to adapt, with their genetic and somatic variabilities, and their life cycle adjustments, their behaviour to survive these recurrent and fluctuated agricultural

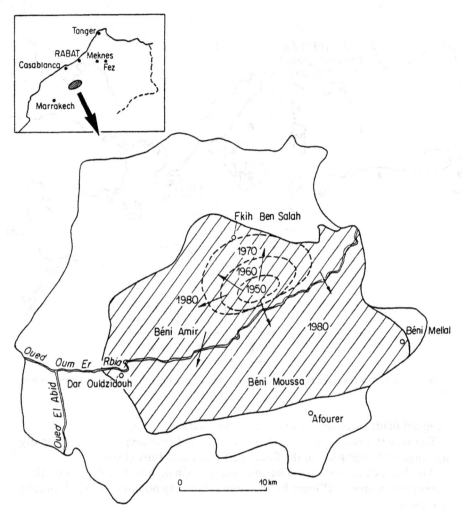

Figure 9. Chronology of the extension of *Solanum eleaegnifolium* Cav. in Morocco (in Tanji et al. 1985).

interferences, and the resulting new relationships between species. They have been able and are still able to differentiate, in space and time, weed populations adjusted to present cultivation, and by this way, promote feasible surge of invasion, resulting from climatic, edaphic, and agricultural shifting pressures.

The earliest weed communities resulted from the ability of the existing open-space species to colonize the new open areas created for cultivation. After the emergence and the differentiation of the weed flora, at the beginning of the agricultural era, the main shifts in weed communities occurred with species inputs from Eurasia, related to the first trades within these continents. The following inputs occured with the discovery and conquest of the New World,

81

and trades through the Tropical zones. The weed flux also spreads by intensified cultivation, including mechanisation and fertilisation. The latest drastic disturbances occured with the chemical weed control, selecting resistant weed populations to the weedkillers, the practice of new cultivations (particularly the summer cultivations), and the reclamation of new areas, which favours the spread of newcomers, from the surrounding or from distant countries.

The antiquity of the Mediterranean weed communities explains the present dominance of the Mediterranean weeds. This fact is according to the dynamic equilibrium model (Mac Arthur & Wilson 1967) where the common species of the continent (the autochtone or naturalized species of the Mediterranean zone, well adapted) extend their range of distribution into the nearly virgin islands (the new areas reclaimed for cultivation), setting against the exotic newcomers. Nevertheless, in some countries, as Crete in Greece, or Israël, the rapid agricultural extension favours, nowadays, spread of weed species originating from the Tropical zones or from the New World (Dafni & Heller 1980, 1982, Protopapadakis 1985). Weeds evolving as rapidly as cultivated plants, they will be feasible weed invasions as long as shifts interfere in agricultural managements and practices.

References

Abdel Fattah, H. 1982. Influence des techniques culturales sur les peuplements d'adventices des cultures de tomate dans la région du Vaucluse. *D.E.A. U.S.T.L.-E.N.S.A. Montpellier.* 46 pp.

A.R.C. (Agricultural research countil), 1976. *Wild oats in world agriculture.* D. Price Johnes ed. London. 296 pp.

Baker, H.G. 1974. The evolution of weeds. *Ann. Rev. Ecol. Syst.* 5: 1–24.

Barralis, G. 1966. Les adventices méditerranéennes dans la flore française. *Proc. 1st Intern. Symp. on Weed problems in the Mediterranean area. Oeiras (Portugal):* 5–16.

Bianco, P. 1976. Diffusione in Puglia del *Solanum eleaegnifolium* Cav. el suelo polimorfismo. *Pub. Univ. Bari. Ann. Fac. Agron. Italia.* 28: 229–241.

Bolos, O. de & Masclans, F. 1955. La vegetacion de los arrozales en la region mediterranea. *Collectanea Botanica (Barcelona).* 4. 3. 32: 415–434.

Carretero, J.L. 1979. *Solanum eleaegnifolium* Cav. y. *Cuscuta campestris* Yunker, nuevas especies para la flora espanola. *Collectanea Botanica.* 11: 143–145.

Chabrolin, Ch. 1934. Les graines d'*Oxalis cernua* Thung. en Tunisie. *Bull. Soc. Hist. Nat. Afr. du Nord* 25: 396–398.

Dafni, A. & Heller, D. 1980. The threat posed by alien weeds in Israël. *Weed Res.* 20. 5: 277–283.

Dafni, A. & Heller, D. 1982. Adventive flora of Israël. Phytogeographical, ecological and agricultural aspects. *Pl. Syst. Evol.* 149. 1: 1–18.

Economidou, E. & Yannitsaros, A. 1975. Recherche sur la flore adventice de Grèce. III: morphologie, développement et phénologie de *Solanum eleaegnifolium* Cav. *Rev. Ecol. Med.* 4: 29–44.

Ellenberg, H. 1968. Wege der Geobotanik zum Verständnis der Pflanzendecke. *Natuurwissensch.* 55: 462–470.

Espirito Santo, D. & Rosa, L. 1981. Contribuiçao para o conhecimiento da evoluçao da flora infestante dos arrozais. *Inst. Agron. Lisboa (Portugal).* 18 pp.

Foury, A., Perrin de Brichambaut, G. & Sauvage, Ch. 1954. Espèces adventices des rizières

82

du Rharb (Maroc). *La Terre Marocaine*. 300: 1-3.

Fox, M.D. & Fox, B.J. 1986. The susceptibility of natural communities to invasion. R.H. Groves & J.J. Burdon (eds.), *Ecology of biological invasions, an Australian perspective*. Austr. Acad. Sc. pp. 57-66.

Gasquez, J., Darmency, H. & Compoint, J.P. 1981. Etude de la transmission de la résistance chloroplastique aux triazines chez *Solanum nigrum* L. *C.R. Acad. Sc. Paris. D.* 292: 847-849.

Gazi Baskova, V. & Segulja, N. 1978. The appearance of dangerous weeds of the genus *Solanum* on the Kvarner Island of Plavnik. *Fragmenta Herbologica Jugoslavica.* 6: 55-59.

Greuter, W. 1971. L'apport de l'homme à la flore spontanée de Crète. *Boissiera.* 19: 329-337.

Guillerm, J.L. 1990. Weed invasions in agricultural areas. F. Di Castri & R.H. Groves (eds.), (in press): *Biogeography of Mediterranean invasions*. Cambridge Univ. Press.

Guillerm, J.L. & Maillet, J. 1982. Weeds of Western Mediterranean countries of Europe. W. Holzner & M. Numata (eds.), *Biology and ecology of weeds*. Junk Publishers, The Hague. pp. 227-243.

Guillerm, J.L. & Maillet, J. 1984. Influence de l'environment sur la flore des vignes désherbées chimiquement. Proc. 3rd Intern. Symp. E.W.R.S. on *Weed problems in the Mediterranean area*. Oeiras (Portugal). pp. 49-56.

Guillerm, J.L. & Sutisna, M. 1983. Caractérisation écologique des adventices des rizières en Camargue (France). C. Ferrari, S. Gentile, S. Pignatti & E. Poli (eds), *Le comunita vegetali come indicatori ambiantali.* Soc. Ital. Phytos. Bologna. (Italia). pp. 35-53.

Guillerm, J.L., Maillet, J., Sanon, M. & Barbier, J.M. 1989. Variabilité des communautés d'adventices des rizières en Camargue (France). Proc. 4th Intern. Symp. E.W.R.S. on *Weed problems in the Mediterranean area.* Valencia (Spain). 2: 312-320.

Holzner, W. 1978. Weed species and weed communities. *Vegetatio.* 38: 13-20.

Holzner, W. 1982. Concepts, categories and characteristics of weeds. W. Holzner & M. Numata (eds.), *Biology and ecology of weeds.* Junk Publishers. The Hague. pp. 3-20.

Jauzein, P. 1980. Caractéristiques physiologiques de la germination de *Sinapis arvensis* L. *6° Coll. Intern. Ecologie, biologie et systématique des mauvaises herbes. C.O.L.U.M.A. E.W.R.S. Montpellier.* 1: 63-72.

Jauzein, P. 1980. Héterogénéite des semences de *Sinapis arvensis* L. *6° Coll. Intern. Ecologie, biologie et systématique des mauvaises herbes. C.O.L.U.M.A. E.W.R.S. Montpellier.* 2: 327-335.

Jovet, P. 1971. Plantes adventices et naturalisées du Sud-Ouest de la France. *Boissiera.* 19: 329-337.

Jovet, P. & Bosserdet, P. 1968. *Senecio harveianus* Mac Orvan, relevé chronologique des observations en France. *Bull. Centr. Etud. Rech. Scient. de Biarritz. 7.* 2: 417-420.

Jovet, P. & de Vilmorin, R. 1975. Troisième supplément à *La flore descriptive et illustrée de la France* de l'abbé Coste.

Keeley, P.E. & Thullen, R.J. 1983. Influence of the planting date on the growth of *Solanum nigrum* L. *Weed Res.* 31. 2: 180-184.

Kornas, J. 1983. Man's impact upon the flora and vegetation in Central Europe. W. Holzner, M.J.A. Werger & I. Ikusima (eds.), *Man's impact on vegetation.* Junk Publishers, The Hague. pp. 277-286.

Kosinova, J. 1974. Studies on the weed flora of cultivated lands in Egypt. 4. Mediterranean and Tropical elements. *Candollea.* 29: 281-295.

Le Maignan, I. 1981. Contribution à l'étude des groupements de mauvaises herbes des cultures en France. Aspects synsystématiques et biologiques. *Th. 3° cycle. Univ. Paris-Sud Orsay.* 103 pp.

Lumaret, R., Guillerm, J.L., Delay, J., Loufti, A., Izco, J. & Jay, M. 1987. Polyploïdy and habitat differenciation in *Dactylis glomerata* L. from Galicia (Spain). *Oecologia (Berlin).* 73: 436-446.

Mac Arthur, R.H. & Wilson, E.O. 1967. *The theory of island biogeography.* Princeton Univ. Press. Princeton. New Jersey (U.S.A.).

Maghami, P. 1966. Biologie des adventices des rizières françaises. *Th. Dr Ing. Fac. Sc. Marseille.* 195 pp.

Maillet, J. 1981. Evolution de la flore adventice dans le Montpellierais sous la pression des techniques culturales. *Th. Dr Ing. U.S.T.L. Montpellier.* 200 pp.

Maillet, J. & Zaragoza-Larios, C. 1976. Flora adventicia de frutales en la provincia de Zaragoza. *8° Journées d'études A.I.D.A. Zaragoza (Espagne).* 15 pp.

Meikle, R.D. 1977. *Flora of Cyprus.* Vol. 1. Pub. Bentham-Noxon Trust, Royal Bot. Garden, Kew.

Michez, J.M. 1980. Biologie et écologie de *Senecio inaequidens* DC. *6° Coll. Intern. Ecologie, biologie et systématique des mauvaises herbes. C.O.L.U.M.A. E.W.R.S. Montpellier.* 1: 153–160.

Montégut, J. 1974. Mauvaises herbes des céréales méditerranéennes. Aspects géographiques et écologiques en France et en Espagne. *4° Journées Circum-méditerranéenes.* pp. 392–402.

Montégut, J. 1982. *Pérennes et vivaces nuisibles en agriculture.* S.E.C.N. Aubervilliers (France). 478 pp.

Montégut, J. 1984. Causalité de la répartition des mauvaises herbes, espèces indicatrices du biotope cultivé. *La Recherche Agronomique Suisse.* 1–2. 23: 15–46.

Moreira, I. 1976. Lignes de recherches actuelles sur la biologie et le contrôle des mauvaises herbes au Portugal. *8° Journées d'études A.I.D.A. Zaragoza (Espagne).* 16 pp.

Moreira, I., Guillerm, J.L., Caixinas, L., Espirito Santo, D. & Vasconcellos, T. 1986. *Mauvaises herbes des vignes et des vergers de l'Ouest du Bassin Méditerranéen.* F.A.O. – Inst. Agron. de Lisbonne (Portugal) ed. 185 pp. (version française et portugaise).

Pavletic, Z., Devetak, Z. & Trinajstic, I. 1978. A new remarkable habitat of the neophyte *Solanum eleaegnifolium* Cav. in flora of the Croatian Coast. *Fragmenta Herbologia Jugoslavica.* 6. 113: 69–72.

Pignatti, S. 1957. La vegetazione delle risaie pavesi. *Arch. Bot. Biog. Italia.* 33. 4–2: 3–67.

Pignatti, S. 1982. *Flora d'Italia.* Vol. 3. p. 130.

Pignatti, S. 1983. – Human impact on the vegetation of the Mediterranean Basin. W. Holzner, M.J.A. Werger & I. Ikusima, (eds.), *Man's impact on vegetation.* Junk Publishers The Hague. pp. 151–161.

Pinto Da Silva, A.R. 1971. Les plantes synanthropiques au Portugal et aux Açores. *Boissiera.* 19: 297–303.

Protopapadakis, E. 1985. Changement de la flore adventice des vergers d'agrumes en Crête sous la pression du désherbage chimique. *Agronomie.* 5. 9: 833–839.

Sanon, M. 1986. La flore adventice des rizières de Camargue: évolution récente et effets des techniques culturales. *D.A.T. Ecole Sup. Agron. Trop. I.N.R.A. Montpellier.* 143 pp.

Sutisna, M. 1980. Répartition actuelle et écologie des espèces spontanées des rizières de la région camarguaise, et cas particulier de *Leersia oryzoides* (L.) Sw. *Th. 3° cycle. U.S.T.L. Montpellier.* 172 pp.

Tackholm, V. 1956. *Student's flora of Egypt.* Cairo Univ.

Tallon, G. 1958. La flore des rizières de la région d'Arles et ses répercussions sur la culture du riz. *Vegetatio.* 7. 1: 20–42.

Tanji, A., Boulet, C. & Hammoumi, M. 1984. Contribution à l'étude de la biologie de *Solanum eleaegnifolium* Cav. (Solanacées), adventice des cultures dans le périmètre irrigué du Tadla (Maroc). *Weed Res.* 24. 6: 401–409.

Tanji, A., Boulet, C. & Hammoumi M. 1985. Etat actuel de l'infestation par *Solanum eleaegnifolium* Cav. pour les différentes cultures du périmètre du Tadla (Maroc). *Weed Res.* 25. 1: 1–9.

Tinarelli, A. 1973. *La coltivazione del riso.* Ed. Agric., Bologna (Italia). 425 pp.

Tomaselli, R. 1958. Aspetti della vegetazione in risaia de vicenda del Pavese e del Vercellese prima e dopo il diserbo. *Arch. Bot. Biog. Ital.* 34. 4. 3. 4: 217–253.

Vasconcellos, J. de C. 1959. Evoluçao da vegetaçao infestante dos arrozais no concelho da Azumbuja. *Min. Econ. Lisboa (Portugal).* 54 pp.

Yannitsaros, A. & Economidou, E. 1974. Studies on the adventice flora of Greece. *Candollea.* 29: 111–119.

Zohary, M. 1949–50. The segetal plant communities of Palestina. *Vegetatio*. 2: 387–411.
Zohary, M. 1983. Man and vegetation in the middle East. W. Holzner, M.J.A. Werger & I. Ikusima (eds.), *Man's impact on vegetation*. Junk Publishers. The Hague. pp. 287–295.

6. Widespread adventive plants in Catalonia

TERESA CASASAYAS FORNELL

Abstract.

About 450 alien vegetal species exist in Catalonia, most of which grow in humanized habitats, although a few are settled in natural vegetation. This chapter includes a small selection of adventive plants in Catalonia which present a well-defined distribution and information of their history it has been available. These exotic plants are presented in different groups according to their time of arrival: early historic times to the end of the 15th century, after the 15th century, 1850 to the early 1900's, and from 1950 onwards. Finally, there are general conclusions about the alien plants in Catalonia.

Introduction

Up till now the invasion of exotic species into more or less natural vegetation has presented no specific problem in Catalonia (northeast Spain), since none of the introduced species can be considered as particularly harmful. Most alien species occupy heavily disturbed habitats; only a very small number of them settle in natural vegetation. There are, of course, a few weeds that interfere with man's activities, and yet they are always isolated cases. Examples among these are *Galinsoga parviflora* and *Oxalis latifolia* that cause problems for farmers, and *Carpobrotus edulis*, a fleshy plant from South Africa, that possibly needs to be controlled, since it is spreading fast and dislodging the natural vegetation of the rocky slopes of the northeastern Catalan coast.

The important increase of introduced species that has been observed since the beginning of the present century can be traced to the trading rise. We shall have to wait a few years before we are able to report on the behaviour of these plants and tell whether they may cause damage to the environment.

In the next pages, the status of a series of alien plants showing a high spreading ability in Catalonia are discussed. I have included only the species of very well-defined distribution and on which sufficient information is available. The earliest introduced species, up to the end of the 15th century,

---

F. di Castri, A. J. Hansen and M. Debussche (eds.), Biological Invasions in Europe and the Mediterranean Basin.
85–104. © 1990, *Kluwer Academic Publishers, Dordrecht.*

86

(*Archaeophyta*) will be dealt with first. The ones whose introduction took place after the 15th century up to the first half of the 20th will be studied in the second part, the third part will contemplate the species that were introduced between the middle of the past century and the beginning of the present one. Finally, the last part will examine the newcomers introduced over the last thirty years.

It should be remarked that the information contained in this chapter is part of a comprehensive study of the Catalan exotic flora (northeast Spain) over a 40,000 km^2 area.

Early introduced species

The flora of exotic species introduced in early stages of history (*Archaeophyta*) has become greatly impoverished in the last few years. Evidence for this impoverishment can be found in the progressive reduction of the plants that used to occur in fields of winter cereals, particularly *Lolium temulentum* and *Agrostemma githago*. Others, however, also introduced a long time ago, such as *Centaurea cyanus* which comes up in crops as well as in more or less ruderal habitats (field borders and roadsides) are still commonly found. According to several authors (Kornas 1971, Holzner 1982, Hilbig 1982, Harlan 1982, Glauninger & Holzner 1982, Holzner & Immonen 1982, and Misiewicz 1985, among others) the retreat of these species is probably due to the continuous application of herbicides and to the fact that seed purification is becoming more and more effective. It is assumed that these segetals (plants growing on winter cereal crops) were transported as waste, 4,000 to 5,000 years ago, together with cereal grains (*Hordeum, Triticum, Avena,* etc.) from central Asia steppes and arid regions close to the Mediterranean Sea.

This group of segetals are specialists (Holzner 1982) that have grown and evolved with agricultural plants, their seeds being sown and harvested together with those of the crops. Because the recent agricultural revolution has altered environmental conditions faster than their adaptation mechanism could follow, they are dying out.

On the other hand, the distribution area of some other archaeophytes has remained unchanged. Among these are the Giant reed (*Arundo donax*) and the Walnut tree (*Juglans regia*) voluntarily introduced by man a long time ago.

Arundo donax is a grass from eastern Asia which has several agricultural uses: it is used as a windbreak, for settling disturbed soils and also, as stalks driven into the ground for propping field vegetables such as bean plants, tomato plants, etc. It occurs in the moist soils of field borders, roadsides and riverbanks in nearly all Catalonia excepting the Pyrenees. It forms quite a typical community described as *Arundini-Convolvuletum sepium* Association, which floristically is very poor in species. The dominant species is the Giant reed around which one or two native climbing plants, *Calystegia sepium* and *Cynanchum acutum* coil. It should be added that presently there exist introduced plants that also climb giant reeds. *Ipomoea indica, I. purpurea* and *Araujia sericifera* are probably the most abundant ones. In the vicinity of water, autochthonous communities of *Phragmition* are often replaced by communities of giant reeds, a species characterized by its vegetative reproduction system by rhizomes (it has never been found reproduced by seeds). It is precisely this feature that has led some authors to regard it as pseudo-naturalized.

Juglans regia was introduced especially for its timber and fruit, the walnut. It is indigenous to a vast ill-defined range stretching from eastern temperate Europe to the Himalayas and southeastern China. Its cultivation dates back to such remote times that it makes it difficult to know where it is spontaneous and where it is introduced (Lopez & Mielgo 1984). It is a naturalized plant, showing a preference for valley-wood communities, but becomes particularly abundant in alder-woods, ash-woods, etc. of hilly regions.

Species introduced after the 15th century

While many archaeophytes are being reduced more and more, and some of them are dying out, the number of recently introduced foreign plants (*Kenophyta sensu* Kornas 1968; *Neophyta sensu* Meusel 1943; *Xenophyta sensu* Greuter 1971) are clearly increasing in importance. According to Kornas (1971) 'synanthropic plants now enjoy almost unlimited possibilities of spreading as a result of the extraordinary degree of communication and trading development' (in French).

However, because not all alien newcomers are able to adapt and survive, many of them die out. Others, on the contrary, are successful in their new environment, although settling most of the time in disturbed communities such as cultivated fields, roadsides, waste ground, etc. These exotic plants, settled in humanized habitats, are named *Epoecophyta*. Very few succeed in natural vegetation; plants settled in natural habitats are named *Holoagriophyta*. An example among these is *Robinia pseudacacia*. It became extinct in Europe during the glaciations and was reimported from North America during the 17th century (1600 or 1601 depending on the authors) when Jean Robin, a botanist and herbalist to the king, first planted this species in Paris, at Place Dauphiné. In spite of its early introduction it seems that it did not start spreading through Europe until the 19th century (Moiroud & Capellano 1981). The oldest specimen in Europe is at Paris Jardin des Plantes, where it was planted in 1636 by J. Robins' nephew (López & Mielgo 1984). In the Iberian Peninsula it was already cultivated in the 18th century. In his flora J. Quer (1762) reports on its occurrence in Barcelona gardens. It is a *Leguminosae* of rapid growth and an excellent colonizer, since its ability to develop root suckers enables it to spread very quickly. As a result, it has often been used to stabilize disturbed soils: road taluses, railroads, etc. Because it usually reproduces by root saplings (sometimes arising 12 m away from the mother-plant), seed reproduction occurring only on woods, Flahault (1899) regards it as a not naturalized plant. Thellung (1912), however, would discard such conclusion for considering it unsufficiently reasoned. Boring & Swank (1984) report that in its native surroundings, southeastern United States (northern part of the Appalachian Mountains), it behaves like an opportunist plant, occupying the primary regeneration stages of decidous forests. In the United States this species is highly used as a protection against erosion and for restoring the surface of mines. In Catalonia it was abundantly introduced into natural and semi-natural ecosystems, the latter being understood as ecosystems slightly altered by man's activities or his animals. At present it is still used both as an ornamental species and a shade-tree, since besides the typical species, there exists a large number of garden varieties. It is found largely naturalized in most parts of the country, occupying above all the verges of communication routes (roads, paths, railway lines, etc.) and woodlands habitats: humid valley-bottoms and river valley-woods. In cool sites it forms thriving colonies along riverbanks, chiefly in alder-woods where it dislodges the original vegetation.

There exists a group of plants, originated from the American continent, that were voluntarily introduced during the centuries that followed its discovery. That transport was, of course, not a one-sided one. On the contrary, the human colonists took not only their livestock and crops with them, but also hundreds of plants colonists. These aggressive colonizing species competed with the American native vegetation (Holzner 1982).

As many species indigenous to America (e.g., *Agave americana, Opuntia ficus-indica, Helianthus tuberosus* and *Oenothera biennis s.l.*) have become part of our environment, it is difficult for us to imagine a Mediterranean landscape without them.

Agave americana, native to Mexico, was first introduced into Europe probably in Spain, in the 16th century. When, at the turn of the century, it flowered for the first time there, it caused sensation (Martins 1855). We know from Clusius that it did not occur spontaneously on the eastern coast of Spain in the middle of the 16th century, but that only cultivated specimens of it could be found in a few gardens. On the contrary, a hundred years later it was already fairly abundant along the Mediterranean coast. Martins (1862) reports that both Linné (*Prolepis Plantarum Amoenitates Academicae*) and Goethe (*Die Metamorphose der Pflanzen*) had already stated that the plant flowered when its vegetative growth was weakened and stopped.

Opuntia ficus-indica is also one of the first plants introduced into Europe by the Spaniards. This plant, so peculiar in its appearance, was imported from tropical America. From the 16th century it spread quickly throughout the Mediterranean region. Such rapid propagation was due, among other reasons, to the fact that in addition to exploiting its fruit, it was used for rearing a Coccus insect from which a dyeing material was obtained. As a result of the discovery of anilines, artificial colourings, at the beginning of the 20th century, it lost all its dyeing interest (Mathon 1981). However, cochineals are still reared in the Canary Isles nowadays.

Agave americana and *Opuntia ficus-indica* are abundant in the lowlands adjoining the Catalan coast in particular, and along the Mediterranean seaboard in general. In Catalonia both of them grow on dry and rocky hillsides (Figure 1). They often occur in shrubby vegetation and grassland communities with or without an upper layer of pines (*Pinus pinea, P. halepensis*), and beside native plants such as *Ulex parviflorus, Spartium junceum, Hyparrhenia hirta*, etc.

Helianthus tuberosus, a plant native to North America and introduced into Catalonia in the 17th century, has settled in a totally different habitat. Grown for its edible tubers as well as for ornament, it is widely naturalized by now and, at the end of summer and in autumn its flowers form large and impressive yellow masses along riverbanks, moist roadsides and the borders of vegetable fields. As it does not appear to reproduce by seeds in Europe (Hanf 1984), it is held to do so vegetatively by tubers. It is a highly invading species that, once introduced, is very difficult to eradicate because of the large number of tubers it develops. Indeed, a single one of those, even of small size, can produce a new shoot and, therefore, a new colony. It is widely distributed in the more humid areas of Catalonia, but restricted to moist sites such as riverbanks and the margins of irrigation ditches in the driest and most continental part of it. It grows from sea level up to 1100 m (1500 m in very sunny and warm places).

Figure 1. Coastal landscape of NE Catalonia with *Agave americana* L. and *Opuntia ficus-indica* (L.) Miller.

90

Nowadays there exist many other naturalized species widely distributed throughout Catalonia. Particularly worth mentioning among these is *Crepis sancta*. It is a *Compositae* original from the southeastern Mediterranean fringe, introduced into France at the end of the 18th century and growing in moderately dry and more or less stony soils such as grasslands, non irrigated cultures – vineyards, almond orchards, olive groves, etc., – fallows and the borders of pathways. It flowers generously in spring, covering the ground with an uninterrupted carpet of yellow flowers.

Ailanthus altissima is also very common in the vicinity of communication routes. It is indigenous to eastern Asia and at present ubiquitous throughout Catalonia. Once introduced somewhere it is very difficult to eradicate because of its great ability to develop root suckers. Its introduction into France dates back to the middle of the 18th century, its leaves being used there to feed certain silkworms (Pinto da Silva, pers. com.).

Oenothera biennis s.l. comprises a group of species of extreme taxonomic difficulty. There is quite a dispute among authors (Raven 1964, Cleland 1972, Rostanski & Glowacki 1977, etc.) with regard to its native range: for some of them it is indigenous to America, for others to Europe. *Oenothera* is an American genus held to be introduced into Europe in the 17th century. It can be assumed that introduced seeds were sown in botanical gardens from where they escaped to new areas. However, the most usual mode of seed introduction should probably be traced to the transport of *Oenothera* seeds together with the ballast of ships that was thrown away in the vicinity of the harbour after the crossing (Cleland 1972). It is a highly developing genus that has produced, in Europe, a large number of hybrids and races non-existent in America. According to Renner (1950, in Cleland 1972), *O. biennis* is the most extensive species in Europe and has yielded many hybrid races established in different areas of Europe. In fact, in Catalonia (particularly in the north) the *Oenothera biennis s.l.* species are naturalized in the sandy and stony soils of mainly riverbanks, roadsides, waste ground, beaches, etc.

Another problematical taxon is the genus *Xanthium*. How many species of *Xanthium* are there in Europe and how many in Catalonia? A careful study of this genus should be carried out to be able to answer this question. *X. spinosum*, a South American *Compositae* introduced into Europe in the 17th century, has been reported from Catalonia. It seems that it came to Spain and Portugal. From their journey through Spain and Portugal at the end of the 17th century, Tournefort and Salvador brought seeds of that plant already occurring in Portugal with them and planted them in the botanical gardens of Barcelona (Spain) and Montpellier (France) (Font i Quer 1976). At present it comes up in ruderal sites: verges of pathways, waste ground, etc., in nearly all Catalonia.

Xanthium italicum includes a group of species whose taxonomy is very problematical. In addition, its native range is in dispute. On the one hand, some authors assert that it is indigenous to America, and on the other, fossil infructescences of it dating from Roman times have been found in Europe

(Hanf 1984). In Catalonia it often occurs on the banks of rivers, borders of ponds, refuse heaps, waste ground, etc.

Species introduced from 1850 to the early 1900's

The spreading and invading ability of some of these species is considerable, since they occur throughout the Catalan territory. Most of them (*Amaranthus muricatus, Paspalum dilatatum, Sporobolus indicus*, etc.,) grow in heavily human disturbed communities: agricultural fields, borders of communication routes, waste ground, etc., whereas the ones that succeed also in natural or seminatural communities are rare. An example among these is *Aster squamatus* which is common in halophilous grasslands close to the seashore or in river valley communities.

The following section will be concerned with the distribution, modes of introduction, spreading mechanisms and ecology of some of the above-mentioned species. It should be remarked that such information is not always easy to obtain and that, consequently, our assertions must be regarded as

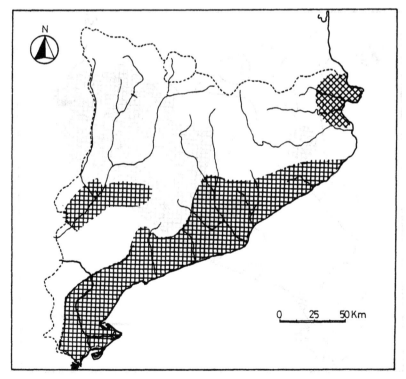

Figure 2. Distribution of *Amaranthus muricatus* (Moq.) Gillies ex Hicken.
* First occurrence.

simple hypotheses in almost every case.

Amaranthus muricatus (Figure 2) is a prostrate plant with narrow, elongated leaves. It flowers in summer as most species for this genus do. It is native to the Argentinian Pampa. In Spain it was first recorded by Sennen in Benicarló in 1908 (Sennen 1917). A year later it was found in Tarragona from where it spread northward along the coastal railway line and road up to Figueras (northeastern extremity of Catalonia). In less than ten years it traveled more than 300 km. With regard to its introduction, nothing can be asserted with certainty. The fact that its first occurrence was registered in Benicarló and not in Barcelona may indicate that seeds were probably unintentionally brought by sea to some harbour of the Valencia region together with goods or packing materials, from where they were transported northwards by the littoral goods trains and road. At present it is found also in the interior, nearly always in railway stations or their vicinity and even in town streets. It has a preference for euthropic soils and is resistant to dry conditions and low temperatures. It belongs to ruderal communities where it occurs beside other species of *Amaranthus: A. deflexus, A. blitoides*, etc. In Catalonia as well as in the warmest parts of Spain (south and east) and the Balearics (Carretero 1979) it is a

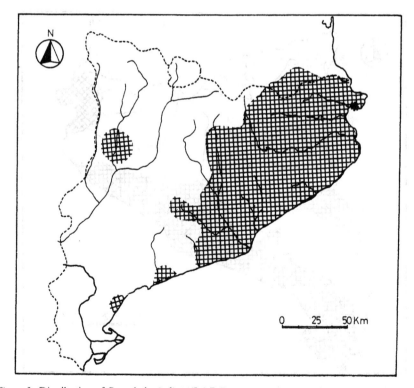

Figure 3. Distribution of *Sporobolus indicus* (L.) R.Br.
* First occurrence.

naturalized plant. It was reported from Port Juvenal (Montpellier, France) (Godron 1853) in the middle of the 19th century, whereas it was detected in Belgium (Visé 1958), Italy (Sortino 1970) and in much of the Mediterranean region (Greuter *et al.* 1984) in more recent years.

Sporobolus indicus (Figure 3) is a grass indigenous to the tropical regions of America. From the gathered data we can assume that it was first recorded in Europe in the outskirts of Roses (northeast of Catalonia) by Bubani in September 1853 (Bubani 1901). It is a most invading species (as already stated by the botanists of the beginning of the century: Codina 1908, Cadevall 1911, Sennen 1917, etc.) which spread along communication routes, being then abundant in the northeastern part of Catalonia. Its multiplication is ensured both by its high seed output and its ability to develop rhizomes. This is the reason why it is often used as lawn grass. It is at present extremely widespread in waste grounds, grasslands, railway stations, harbours and along riverbanks in both sandy and silty soils as well as strictly organic ones. In the northeastern part of the country it springs up also between road paving stones, whereas in the southern and inland regions its abundance decreases. In the rest of Europe it was reported from France (Thellung 1912), Portugal (Pinto da Silva

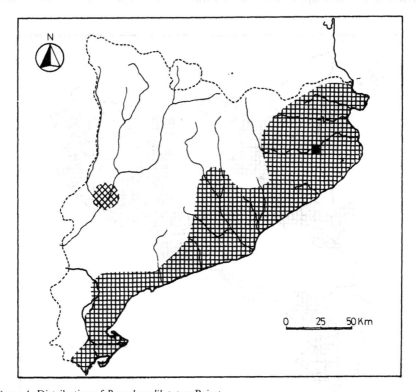

Figure 4. Distribution of *Paspalum dilatatum* Poiret.
* First occurrence.

et al. 1968), Açores, Italy and Bulgaria (Hansen in Tutin *et al.* 1980). In Argentina it occupies grasslands communities with different species of *Stipa* in more or less well-drained soils (Lewis *et al.* 1985).

Paspalum dilatatum (Figure 4) is a grass also indigenous to America. Its native range stretches from Brazil to Argentina (Pinto da Silva 1940). The oldest mention of it in Catalonia dates back to 1907 when Codina (1908) found it at La Cellera, along the river Ter. This is also the first record for Spain. A few years later it appeared in the sandy shores of the river Fluvià (Sennen 1912). In fifty years approximately it spread to a large part of the northeast fringe of Catalonia and at present its distribution is still widening. It prefers humid areas along riverbanks, irrigation ditches, margins of pathways, in grasslands and also in the vicinity of railway stations. In spite of this marked preference it can endure fairly long periods of drought. It is sometimes found as a weed in irrigated orchards (apple, peach orchards, etc.), where it forms large masses. Since it does not seem to be a competitor to the fruiter and affords a lower layer helping to keep a wetter microclimate, farmers aprove of it. It is sometimes used as lawn grass, in which case the gardener's repeated cuttings cause the plant to flower almost at ground surface. In the United States it is much appreciated as a forage plant, especially for dairy cattle,

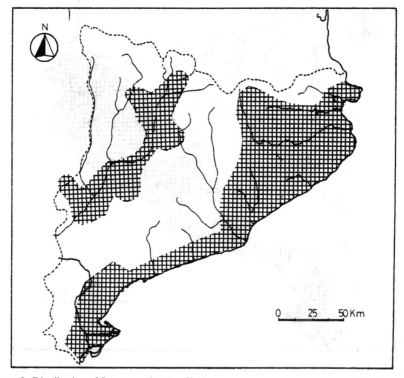

Figure 5. Distribution of *Bromus catharticus* Vahl.

either fresh or dried. In Catalonia it occupies at present an almost uninterrupted area stretching between the two extremes of the litoral fringe. In the interior it is nearly always confined to moist sites: margins of irrigation ditches, riverbanks, irrigated fields, etc. In Great Britain its introduction is traced to the wool industry (Ryves 1974), as it is presumably the case in Catalonia where introduced grains were carried away by water and consequently dispersed along the rivers. It is also assumed that trains and wheeled vehicles contribute likewise to its propagation. In the rest of Spain it is reported from the north (Paunero 1962, Bascones 1982, Aseginolaza *et al.* 1984), the center (Paunero 1962) and the east (Hernández 1978). It is also present in Portugal (Pinto da Silva 1940), France (Heckel 1906), Great Britain (Lousley 1961, Ryves 1974), Italy (Pignatti 1982), Kriti (Greuter *et al.* 1985) and in Asia, Africa and Australia (Häfliger & Scholz 1980).

Bromus catharticus (Figure 5) is a grass native to South America. Its distribution is almost identical to that of the species cited above, althoughh showing a greater tendency than the latter to penetrate inland by rivers. It was probably introduced into Catalonia at the beginning of the present century, even though we neither know where exactly nor in which year. The fact is that in 1929 it was very rare (Sennen 1929) and that it did not start spreading abundantly until the early 1970's. At present its distribution is still increasing in importance. It occurs mostly along riverbanks and roadsides, between rail sleepers, in waste grounds, grasslands and also as a weed in irrigated orchards, particularly in the northeastern part of Catalonia. However, benefiting from irrigated fields it becomes more and more widespread in the interior. It grows from sea level up to 1000m (Andorra). In spite of being a forage plant we do not think that it was introduced intentionally, since it has never been found cultivated. Its introduction should probably be traced to the textile industry, transport being perfomed by water (rivers), rail or road. It is abundant in Portugal (Pinto da Silva, pers. com.), irregularly widespread throughout Spain and also present in France (Godron 1853), Belgium (Visé 1958), Great Britain (Lousley 1961), Kriti (Greuter *et al.* 1985), Africa, Asia and Oceania (Häfliger & Scholz 1981). This species is known to be confused especially with regard to its correct name. From Pinto Escobar's research (1981) it seems that it must be called *B. catharticus* and not *B. willdenowii*.

Species introduced from 1950 onwards

I will discuss, finally, a few adventive species recently introduced into the country, most of them unintentionally, and showing a high colonizing ability.

Setaria geniculata is another highly polymorphous exotic grass. It apparently originates in America. According to Rominger (1962), it would be the most common species of that genus in North America. It was first reported from Spain in 1964 at Castelló de la Plana (Calduch 1968). Nearly ten years later

96

it was registered in two localities of the coastal part of the Tarragona region. In 1980 it was found in the central area of Catalonia (Casasayas & Masalles 1981) and it seems that from that date onwards it has become more and more abundant, mostly on the borders of the main road stretching along the Catalan coast. There is not doubt that the latter and the railroads as well were the means of propagation of that plant. At present it is basically coastal, having a preference for railway stations, roadsides and waste grounds. Sometimes it also springs up in large masses along rivers. Local specimens, up till now always found in the vicinity of railway stations have been registered in the interior. It is no less present in sandy, rather dry soils than in clay loam humid depressions. It grows from sea level up to approximately 800m (in Montseny). In Spain it is known from the north (Lainz 1967, Aseginolaza *et al.* 1984), the south (Rico 1984) and the east (Carretero & Esteras 1983), wheras in Portugal the first record of it dates back to 1961 (Paiva 1961). Its existence, however, had already been mentioned in the south of France – Port Juvenal – in 1846 (Thellung 1912) and in Great Britain (Lousley 1961), and more recently in Sweden (Hylander 1971), Denmark (Hansen 1974), Italy (Viegi *et al.* 1974b), Belgium, the Netherlands (Auquier 1979) and southeastern Europe (Häfliger & Scholz 1980). Its introduction into France and Great Britain can apparently be traced to the wool industry. As far as Spain is concerned, we do not know yet how it was introduced. Trading activities may be in part responsible for its introduction, but nothing positive can be asserted for the moment.

Chloris gayana an African introduction, was abudantly sown together with other grasses to stabilize the taluses of the littoral motorway. It is a stoloniferous plant of fast and vigorous growth. As it took it only a few years to grow subspontaneously, without the influence of man, it is found at present in localities fairly distant from the place where it was once introduced. We believe that it will be naturalized within a few years time.

Galinsoga parviflora is a North American plant occurring in Europe since the end of the 18th century. It was grown in Madrid and Paris botanical gardens (Jovet & Vergnet 1928). In the 19th century it was usual to find it escaped from cultivated plots throughout Europe. However, in Catalonia it seems that it was in Cambrils agricultural fields, not far from the city of Tarragona, that it was first collected in 1972 (Folch & Abellà 1974). Later is was recorded in other sites so distant from its original place of introduction that its presence there is assumed to answer to new introductions. One among these sites is located in the Gironès region into which it may have been introduced by cotton-waste, since this material was used as field manure (Girbal 1984). In Catalonia this species is of very local occurrence so far but always growing in high quantities. It has a preference for irrigated fields. Its profilic seed output and very short life cycle enabling it to produce numerous generations per annum and to grow throughout the year except during the coldest months make it a very difficult species to eradicate.

In 1985 we detected the first stands of *Galinsoga ciliata* in different sites of Catalonia. This species appears to be more invading than *G. parviflora*. However, it is only within a few years time that we shall be able to report on behaviour of both of them.

I include *Galinsoga parviflora* in the present section (although it occurred in the Iberian Peninsula in the 18th century). I believe that the field specimens have proceeded from new introductions.

There exist other plants, such as species of *Oxalis*, which are extremely invading field agrestal weeds. *O. latifolia* from tropical South America, is the species that causes the most trouble in our country. It is very difficult to eliminate since it reproduces vegetatively by means of its bulbs. However, at present there exist a few herbicides effective against both aerial and underground parts of the plant.

Aster pilosus (Figure 6) is perhaps the earliest introduced species of this last group, and the one most typically distributed. This *Compositae* native to America was confused until recently (Girbal 1984) with another related species, also originating in America, *A. lanceolatus*. It is mentioned for the first time in a publication dating from 1952 in which Bolòs reports it from Gerona under the name of *A. tradescantii*. Three years later, in 1955, Esteve reports it as subspontaneous from Alt Empordà (Gerona, northeast Catalonia)

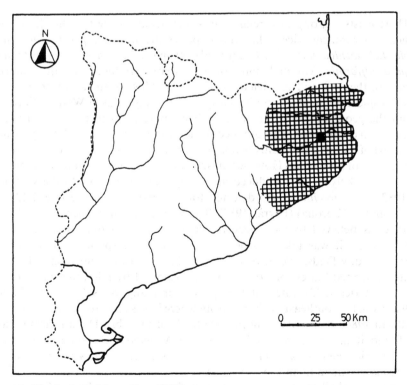

Figure 6. Distribution of *Aster pilosus* Willd.
* First occurrence.

98

(sub *A. lateriflorus*). Then it became extensive in abandoned fields, along roadsides, riverbanks, etc., throughout northeastern Catalonia. Its introduction could be related to the fact that it is grown in gardens for the great ornamental quality of its abundant flowers at the end of summer and in autumn. It should be noted that it is a plant of high morphological and phenological variety.

Rhagodia nutans is an Australian species of the family *Chenopodiaceae*. It is a policarpous, prostrate plant with small triangular leaves that become fleshy in highly saline environments. Very characteristic of this plant is its small, bacciform red fruit containing a black seed germinating extremely easily. It has been detected nowhere else in Europe. In Catalonia its origin could be traced to Blanes Botanical Garden 'Mar i Murtra', since it is in its vicinity that it was first recorded in 1971 (Bolòs & Vigo 1979). For the time being we do not know whether this plant was cultivated in the Garden (at present it is not) or introduced there together with earth or seeds of any Australian ornamental plant. Nowaways it is extremely abundant not only in the Botanical Garden surroundings but also along the margins of paths, in waste grounds and especially on the seashore cliffs and rocks, as a fleshy plant, of the southern part of the Costa Brava, from Blanes to Tossa de Mar. As mentioned earlier, it will be possible to report on its behaviour only within a few years time.

There exists a group of foreign plants that always grow together in a very typical environment: rice fields. Such species are *Bergia capensis, Lindernia dubia, Echinochloa oryzoides, E. oryzicola, Ammannia coccinea, A. robusta* and *Cyperus difformis*. Although some of these were introduced into the country ages ago, they are still restricted to rice fields. Most of them are of tropical or subtropical origin and occur in nearly all the paddy fields of Western Europe. From the geographical point of view their interest lies in this most typical location. In comparison with other countries, the agrestal flora (plants growing in crops) of Catalan rice fields is not very rich, partly because of the excessive application of herbicides. However, in spite of that we think that new adventive species are likely to be introduced mostly by migrating birds. In the summer of 1982 *Najas gracillima* was detected for the first time in the rice fields of Alt Empordà (Gerona) (Farràs 1984). This delicate plant indigenous to North America is believed to have occurred here for quite a long time, but that until recently it was taken to be *N. minor*, a native species widespread in Catalan paddy fields. We also think that this species did not come directly from the United States, but that it was transported by migrating birds from the Italian rice fields where it has grown for many years (Cook 1973), as well as from the southeast of France from where it was reported lately (Molinier 1980). In 1984, while surveying the rice fields of the Ebro Delta (Tarragona), we found it in abundance and along with *N. minor*. The other rice fields of the Iberian Peninsula should be surveyed in order to know whether *Najas gracillima* occurs in them or is able to do so.

Finally we shall mention *Ehrharta longiflora*, a South African *Gramineae* collected for the first time in Catalonia near Blanes Botanical Garden 'Mar

99

i Murtra' in 1981. It had never been recorded in the Iberian Peninsula before. It was kindly identified for me by Professor J. Malato-Beliz from Elvas (Portugal). As for its European occurrence, it is reported from Czechoslovakia (Dvorak & Kühn 1966), Great Britain (Ryves 1974) and the Netherlands (Auquier 1980). It is an annual plant that apparently grows very well and reproduces easily by seeds. At present it is important in town ruderal sites. Its origin could be traced to the introduction of seeds into the Botanical Garden.

Conclusion

Most of the Catalan alien species occur in segetal and ruderal communities, that is to say heavily human-disturbed habitats. Very few of them succeed in natural vegetation. In recent years both the composition and distribution of exotic species have been subjected to considerable change. New agricultural technologies such as the use of herbicides and the cleaning of seeds have resulted in the retreat of a large number of *Archaeophyta*. On the other hand, there exists a series of recently introduced plants quickly increasing in importance (*Kenophyta sensu* Kornas 1968, *Neophyta sensu* Meusel 1943, *Xenophyta sensu* Greuter 1971). Most archaeophytes are specialized weeds ('specialists' *sensu* Holzner 1982) that is to say, species that evolved in the course of agricultural history beside certain cultivated plants and developed an adaptation to a specific type of crop. On the contrary, non-specialized weeds ('colonizers' *sensu* Holzner 1982) are becoming more and more important. They are not adapted to any crop in particular but are capable of succeeding in a wide range of environmental conditions.

Human disturbance in Catalan flora dates back to remote ages. According to Guillerm & Trabaud (1980), man started to frequent the Mediterranean region 500,000 years ago. There is no doubt that the alteration produced then is insignificant in comparison with modern interferences.

Catalan exotic flora has not always originated in the same place, nor has it presented the same composition throughout the ages. The majority of *Archaeophyta* are natives of the Middle East and central and western Asia, whereas at present the more recently introduced species come mostly from the American continent. Commercial exchanges are most likely responsible for this change of origin. Catalonia imports many American products, especially feed. The delivered feed grains are not always well cleaned and contain weed seeds and other impurities.

In Catalonia there exist around 450 exotic species. Though most of them were introduced voluntarily by man for cultivation (as crop plants, ornamentals, etc.), they are but subspontaneous species.

With regard to their frequency of occurrence, the further you move away from the coast towards the interior, the fewer they become. The same pattern is displayed in altitude: the lowland areas are richer in synanthropic species

100

than the hilly regions. The highest frequency occurs, as it is expected, in the surroundings of the largest towns as well as in the regions under irrigated fields, where in the ones where natural communities are extensive the frequency percentage is very low.

References

Ahti, T. & Hamet-Ahti, L. 1971. Hemerophilous flora of the Kunsamo district, northeast Finland, and the adjacent part of Karelia, and its origin. *Ann. Bot. Fenn.* 8: 1–91.
Aseginolaza, C. & Gómez, D. *et al.* 1984. Catálogo florístico de Alava, Vizcaya y Guipúzcoa. Gobierno Vasco. Vitoria.
Auquier, P. 1979. Le genre *Setaria* Beauv. en Belgique et au Grand-Duché de Luxembourg. *Lejeunia* 97: 1–13.
Auquier, P. 1980. Graminées adventices nouvelles ou intéressantes pour la flore belge, II. *Bull. Soc. Roy. Bot. Belgique* 113(1): 3–13.
Bascones, J.C. 1982. Flora vascular de la Navarra húmeda. I. Poaceae. *Publ. Biol. Univ. Navarra. Ser. Bot.* 1: 33.
Bolòs, O. de 1952. Notas florísticas. III. *Collect. Bot., Barcelona* 3(2): 185–197.
Bolòs, O. de & Vigo, J. 1979. Observacions sobre la Flora dels països catalans. *Collect. Bot., Barcelona* 11: 25–89.
Boring, L.R. & Swank, W.T. 1984. The role of Black Locust (*Robinia pseudacacia*) in forest succession. *Journal of Ecology* 72: 749–766.
Bubani, P. 1897–1901. Flora Pyrenaea. 4 Vol.
Buckley, R. 1981. Alien plants in central Australia. *Bot. Journ. Lin. Soc.* 82: 369–379.
Cadevall, J. 1911. Notas fitogeográficas criticas. *Bot. Soc. Esp. Hist. Nat.* 11(5): 225–256.
Cadevall, J. *et al.* 1913–1937. Flora de Catalunya. Vol. I–VI. Barcelona.
Calduch, M. 1968. Plantas de mi herbario. Nota sobre el género *Setaria* P. Beauvois. *Collect. Bot., Barcelona* 7(1): 151–163.
Carretero, J.L. 1979. El género *Amaranthus* en España. *Collect. Bot., Barcelona* 11: 105–142.
Carretero, J.L. & Esteras, F.J. 1983. Algunas gramíneas de interés corológico para la provincia de Valencia. *Collect. Bot., Barcelona* 14: 215–219.
Casasayas, T. & Masalles, R.M. 1981. Notes sobre flora al·lòctona. *Butll. Inst. Catalana Hist. Nat.* 46 (*Sec. Bot.* 4): 111–115.
Chevalier, A. 1944–1945. Plantes adventices introduites par la guerre actuelle à Paris et aux environs. *Bull. Soc. Bot. France* 91: 102–106.
Cleland, R.E. 1972. *Oenothera*: Cytogenetics and Evolution. Experimental Botany: An Intern. Series of Monograph. Vol. 5. Academic Press. London.
Codina, J. 1908. Apuntes para la Flora de la Sellera y su comarca. Col. Méd. de la Prov. Gerona. Gerona.
Cook, C. 1973. New and noteworthy plants from the northern Italian ricefields. *B.E.R. Schweiz. Bot. Ges.* 83(1): 54–65.
Cosson, E. 1860. Appendix Florulae Juvenalis ou liste des plantes étrangères récemment observées au Port-Juvénal près de Montpellier. *Bull. Soc. Bot. France* 6: 605–615.
Cosson, E. 1864. Appendix Florulae Juvenalis Altera, ou deuxième liste des plantes étrangères récemment observées par M. Touchy au Port-Juvénal près Montpellier. *Bull. Soc. Bot. France* 11: 159–164.
Dvorak, J. & Kühn, F. 1966. Eingeschleppe Pflanzen im Areal der Wollspinnerei 'Mosilana' in Brno. (Brünn). *Preslia* 38: 327–332.
Esteve, F. 1955. Reseña de una excursión Botánica al Alto Ampurdán: Vegetación de la sierra de Roda y Plana de Castelló (Prov. de Gerona). *Anales Jard. Bot. Madrid* 14: 555–596.
Farràs, A. 1984. *Najas gracillima* (A. Braun et Englem.) Magnus a Catalunya. In Notes Breus

101

sobre la flora dels Països Catalans. *Butll. Inst. Catalana Hist. Nat.* 51 (*Sec. Bot.* 5) 178.

Flahault, C. 1899. La naturalisation et les plantes naturalisées en France. *Bull. Soc. Bot. France* 46: XCI–CVIII.

Folch, R. & Abellà, C. 1974. *Galinsoga parviflora* Cav. i *Guizotia abyssinica* (L.) Cass. dos adventicias nuevas para la flora catalana. *Collect. Bot., Barcelona* 9: 183–189.

Font i Quer, P. 1976. Plantas medicinales. El Dioscórides renovado. Editorial Labor. Barcelona.

Franchet, A. 1872. Sur une flore adventice observée dans le Département de Loir-et-Cher en 1871 et 1872. *Bull. Soc. Bot. France* 19: 195–202.

Gaudefroy, M. 1872. La florule obsidionale des environs de Paris en 1872. *Bull. Soc. Bot. France* 19: 266–277.

Girbal, J. 1984. Flora i vegetació del Gironès. Tesi doctoral inèdita.

Glauninger, J. & Holzner, W. 1982. Interference between weeds and crops: A review of literature. In: W. Holzner & M. Numata (eds.), *Biology and ecology of weeds*, Junk Publishers. The Hague. pp. 149–159.

Godron, A. 1853. Florula Juvenalis seu Enumeratio et Descriptio Plantarum, é seminibus exoticis inter lanas allatis enatarum in campestribus Portûs Juvenalis propé Monspelium. *Mém. Acad. Scien. et Lettr. de Montp., sec. Médic.* 1(4): 409–456.

Greuter, W. 1971. L'apport de l'homme à la flore spontanée de la Crète. *Boissiera* 19: 329–337.

Greuter, W. *et al.* 1984. Med-Checklist. 1. Pteridophyta (ed. 2). *Gymnospermae. Dicotyledones (Acanthacaeae-Cneoraceae)*. Ed. Conserv. & J. Bot. Genève.

Greuter, W. *et al.* 1985. Additions to the flore of Crete. 1973–1983(1984). III. *Willdenowia* 15: 23–60.

Guillerm, J.L. & Trabaud, L. 1980. Les interventions récentes de l'homme sur la végétation au nord de la méditerranée et plus particulièrement dans le sud de la France. *Naturalia Monspeliensia, n° Hors Série*: 157–171.

Häfliger, E. & Scholz, H. 1980. Grass Weeds 1. Documenta CIBA-GEIGY. Basel.

Häfliger, E. & Scholz, H. 1980. Grass Weeds 2. Documenta CIBA-GEIGY. Basel.

Hanf, M. 1984. Ackerunkräuter Europas mit ihrem Keimlingen und Samen. BASF. Ludwigshafen.

Hansen, A. 1974. Gramineernes udbredelse i Danmark. IKKE-naturaliserede arter. *Bot. Tidsskr.* 68: 345–357.

Harlan, J.R. 1982. Relationships between weeds and crops. In: W. Holzner & M. Numata (eds.), *Biology and ecology of weeds*. Junk Publishers. The Hague. pp. 91–96.

Heckel, E. 1906. Sur *l'Ambrosia artemisiaefolia* L. et sa naturalisation en France. *Bull. Soc. Bot. France* 53: 600–620.

Hernández, A.M. 1978. Notas sobre la flora murciana. I. Plantas de los alrededores de Murcia. *Anales Univ. Murcia* 32(1,2,3,4): 71–83.

Hilbig, W. 1982. Preservation of agrestal weeds. In: W. Holzner & M. Numata (eds.), *Biology and ecology of weeds*. Junk Publishers. The Hague. pp. 57–59.

Holub, J. 1971. Notes on the terminology and classification of the synanthropic plants, with examples from Czechslovak flora. *Saussurea* 2: 5–18.

Holzner, W. 1982. Concepts, categories and characteristics of weeds. In: W. Holzner & M. Numata (eds.), *Biology and ecology of weeds*. Junk Publishers. The Hague. pp. 3–20.

Holzner, W. & Immonen, R. 1982. Europe: An overview. In: W. Holzner & M. Numata (eds.), *Biology and ecology of weeds*. Junk Publishers. The Hague. pp. 203–226.

Hylander, N. 1971. *Prima loca plantarum vascularium Sueciae. Plantae subspontaneae vel in tempore recentiore adventitiae. Svensk. Bot. Tidsskr.* 64(1970), suppl. 332 pp.

Jalas, J. 1955. Hemerobe und hemerochore Pflanzenarten. Ein terminologischer Reformversuch. *Acta Soc. Fauna et Flora Fenn.* 72(11): 1–15.

Jorgensen, P.M. & Ouren, T. 1969. Contributions to the Norwegian Grain Flora. *Nytt. Mag. Bot.* 166: 123–137.

Jovet, P. & Vergnet, J. 1928. Notes sur deux adventices. *Galinsoga parviflora* Cav. et *Artemisia annua* L. *Bull. Soc. Bot. France* 75: 930–945.

Kornas, J. 1966. Influence of Man and his economic activities on the vegetation of Poland.

102

Synanthropic flora. In: W. Szafer (ed.), *The Vegetation of Poland*. Pergamon Press. Oxford. pp. 97-137.

Kornas, J. 1968a. A geographical-historical classification of synanthropic plants. *Mater. Zakladu Fitosocjol. Stosowanej U.W.* 25: 33-41.

Kornas, J. 1968b. A tentative list of recently introduced synanthropic plants (Kenophytes) established in Poland. *Mater. Zakladu Fitosocjol. Stosowanej U. W.* 25: 43-53.

Kornas, J. 1971. Changements récents de la Flore polonaise. *Biological Conservation* 4(1): 43-47.

Kornas, J. 1976. Decline of the European Flora; facts, comments and forecasts. *Phytocoenosis* 5:(3-4): 173-185.

Kornas, J. 1978. Remarks on the analysis of a synanthropic flora. *Acta Bot. Slovaca Acad. Sci. Slovacae.* Ser. A 3: 385-393.

Kornas, J. 1982. Man's impact upon the flora: processes and effects. *Memorabilia Zool.* 37: 11-30.

Kornas, J. 1983. Man impact upon the flora and vegetation in Central Europe. In: W. Holzner & M.J.A. Werger & I. Ikusima (eds.), *Man's impact on vegetation.* Junk Publishers. The Hague. pp. 277-286.

Krippelová, T. 1978. Beitrag zur Klassifikation synanthroper Pflanzengesellschaften. *Acta Bot. Slovaca Acad. Sci. Slovacae.* Ser. A 3: 395-399.

Krippelová, T. 1981. Synanthrope Vegetation des Beckens Kosická Kotliná. *Vegetácia CSSR, B4. pp. 216. Veda Verlag Slowakisch. Akad. Wissensch. Bratislava.*

Kurtto, A. 1982. Plants introduced with sunflower seeds from the USA into Helsinki in 1980. *Memoranda Soc. Fauna Flora Fenn.* 58(1): 7-15.

Kuzmanov, B.A. & Kozuharov, S.I. 1971. Aliens in the Bulgarian flora. *Boissiera* 19: 319-327.

Lainz, M. 1967. Aportaciones al conocimiento de la flora gallega. V. *Anales Inst. For. Inv. Exper.* 52 pp.

Lewis, J.P. & Collantes, M.B. *et al.* 1985. Floristic groups and plant communities of southeastern Santa Fe, Argentina. *Vegetatio* 60: 67-90.

Linkola, K. 1916. Studien über den Einfluss der Kultur auf die Flora in den Gegenden nordlich vom Ladogasee. *Acta Soc. Fauna et Flora Fenn.* 45(1): 1-432.

López, A. & Mielgo, M. 1984. Arboles de Madrid. Comunidad Autónoma de Madrid.

Lorenzoni, G.G. 1977. Considerazioni sulla flora sinantropica. *Informat. Bot. Ital.* 9(3): 262-269.

Lousley, J.E. 1961. A census list of wool aliens found in Britain 1946-1960. *Proc. Bot. Soc. Br. Isl.* 4: 221-247.

MacDonald, A.W. 1983. Alien trees, shrubs and creepers invading indigenous vegetation in the Hluhluwe-Umfolozi Game Reserve Complex in Natal. *Bothalia* 14(3-4): 949-959.

MacDonald, I.A.W. 1984. Is the Fynbos Biome Especially Susceptible to Invasion by Alien Plants? A Re-analysis of Available Data. *S. African Journal of Science* 80(8): 369-377.

MacDonald, I.A.W. & Jarman, M.L. (ed.) 1984. Invasive Alien Organisms in the terrestrial Ecosystems of the Fynbos Biome, South Africa. *S. African Nat. Science. Progr.* Report n°85.

Martins, C. 1855. De l'introduction en Europe, de la naturalization et de la floraison de *l'Agave americana. Bull. Soc. Bot. France* 2: 6-15.

Martins, C. 1862. Lettre à M.J. Gay, sur la floraison simultanée de 1500 *Agave americana* dans les plaines de Mustapha près Alger. *Bull. Soc. Bot. Fr.* 9: 146-147.

Mathon, C.C. 1981. L'origine des plantes cultivées. Phytogéographie appliquée. *Ecologie appliquée et sciences de l'environment* 5.

Mennema, J. en & Ooststroom, S.J. van 1974. Nieuwe vondsten van zeldzame planten in Nederland, hoofdzakelijk in 1973. *Gorteria* 7(5): 65-83.

Mennema, J. en & Ooststroom, S.J. van 1975. Nieuwe vondsten van zeldzame planten in Nederland, hoofdzakelijk in 1974. *Gorteria* 7(12): 185-206.

Meusel, H. 1943. Vergleichende Arealkunde. I. Berlin-Zehlendorf.

Michael, P.W. 1981. Alien Plants. In: R.H. Groves (ed.), Australian Vegetation. Cambridge University Press. pp. 44-64.

103

Misiewicz, J. 1985. Investigations on the synanthropic flora of Polish sea harbours. *Monographiae botanicae* 67: 5–68.

Moiroud, A. & Capellano, A. 1981. Fixation d'azote chez les espèces ligneuses symbiotiques. II. Reprise de l'activité fixatrice (reduction de C_2H_2 chez *Robinia pseudacacia* L. au printemps). *Bull. Soc. Bot. France (Let. Bot.)* 4,5: 239–248.

Molinier, R. 1980. Catalogue des plantes vasculaires des Bouches- du Rhone. *Bull. Mus. Hist. Natur. Marseille* Tome IL.

Mühlenbach, V. 1979. Contributions to the Synanthropic (Adventive) Flora of the Railroads in St. Louis, Missouri, U.S.A. *Ann. Missouri Bot. Gard.* 66(1): 1–108.

Mühlenbach, V. 1983. Supplement to the contributions to the synanthropic (adventive) flora of the railroads in St. Louis, Missouri, U.S.A. *Ann. Missouri Bot. Gard.* 70: 170–178.

Paiva, J.A.R. 1961. Subsidios para o conhecimiento da flora portuguesa. I. *Ann. Doc. Brot.* 27: 17–35.

Paunero, E. 1962. Las Paníceas españolas. *Anales Jard. Bot. Madrid* 20: 51–90.

Pignatti, S. 1982. Flora d'Italia. 3 Vol. Edagricole.

Pinto da Silva, A.R. 1940. O Genero *Paspalum* em Portugal. *Agron. Lusit.* II: 5–23.

Pinto da Silva, A.R. 1942. Algumas considerações sôbre as plantas vasculares subespontaneas em Portugal. *Agron. Lusit.* 4: 213–221.

Pinto da Silva, A.R. 1971. Les plantes synanthropiques au Portugal continental et aux Açores. *Boissiera* 19: 297–303.

Pinto da Silva, A.R. & Teles, A.N. 1962. *Setaria geniculata* (Lam.) P. Beauv. In Plantas novas e novas areas para a flora de Portugal VIII. *Agron. Lusit.* 29(1,2): 179.

Pinto da Silva, A.R., Teles, A.N. & Rainha, B.V. 1968. *Sporobolus indicus* (L.) R.Br. In Plantas novas e novas areas para a flora de Portugal IX. *Agron. Lusit.* 29(1,2): 5–6.

Pinto Escobar, P. 1981. The genus *Bromus* in northern S. America. *Bot. Jahrb. Syst.* 102: 445–457.

Quer, J. 1762. Flora española o historia de las plantas que se crían en España. I Vol. Madrid.

Raven, P.H. 1964. The generic subdivision of *Onagraceae*, tribe *Onagreae*. *Brittonia* 16(3): 276–288.

Rico, E. 1981. Algunas plantas del NE cacereño. *Anales Jard. Bot. Madrid* 38(1): 181–186.

Rominger, J.M. 1962. Taxonomy of *Setaria* (Gramineae) in North America. *Illinois Biological Monographs* 29.

Rostanski, K. & Glowacki, Z. 1977. The distribution of the Species of the Genus *Oenothera* L. in the Central Part of Eastern Poland. *Fragm. Flor. et Geob.* 23(3,4): 309–316.

Ryves, T.B. 1974. An interim list of the woll-alien grasses from Blackmoor, North Wales, 1962–1972. *Watsonia* 10: 35–48.

Schroeder, F.G. 1969. Zur Kassifizierung der Anthropochoren. *Vegetatio* 16(5–6): 225–238.

Sennen, Fr. 1917. Flore de Catalogne. Additions et commentaires. *Treball de la Inst. Catalana Hist. Nat.* 3: 55–266.

Sortino, M. 1970. *Amaranthus muricatus* Gillies nuova avventizia della flora italiana. *Lav. Ist. Bot. e Giard. Col. di Palermo* 24: 203–208.

Suominen, J. 1979. The grain immigrant flora of Finland. *Acta Bot. Fenn.* 111: 1–108.

Tanji, A. & Boulet, C. 1986. Diversité floristique et biologie des adventices de la région du Tadla (Maroc). *Weed Research* 26: 159–166.

Terpo, A. & Egyedne-Balint, K. 1985. Subspontaneous woody plants of the Hungarian flora. *Public. Universit. Horticulturae* 15: 117–126.

Thellung, A. 1910. La flore adventice de Montpellier. Résumé d'un mémoire inédit sur le même sujet. *Bull. Soc. Languedoc. Géogr.* 33: 1–32.

Thellung, A. 1912. La Flore adventice de Montpellier. *Mém. Soc. Sc. Nat. Cherbourg.* 38: 57–728.

Thellung, A. 1915. Pflanzenwanderungen unter dem Einfluss des Menschen. *Englers Bot. Jahrb.* 53(116): 37–66.

Tutin, T.G. *et al.* 1964–1980. Flora Europaea. Vol. I–V. Cambridge.

Viegi, L. 1974. Deffinizione e nomenclature delle specie esotiche della Flora Italiana. *Informat.*

104

Bot. Ital. 6: 136–137.

Viegi, L., Cela Renzoni, G. & Garbari, F. 1974a. Flora esotica d'Italia. *Lavori Soc. Ital. Biogeogr.* 5: 125–220.

Viegi, L., Garbari, F. & Cela Renzoni, G. 1974b. Le esotiche avventizie della Flora Italiana. *Informat. Bot. Ital.* 6(3): 274–280.

Viegi, L. & Cela Renzoni, G. 1981. Flora esotica d'Italia: 1e specie presenti in Toscana. Consiglio Nazionale delle Ricerche. Pavia.

Visé, A. 1958. Florule adventice de la vallée de la Vesdre. *Bull. Soc. Roy. Bot. Belgique* 90: 287–305.

Yannitsaros, A. 1982. The adventive Flora of Greece. A Review. *Bot. Chron.* 2(2): 159–166.

Yannitsaros, A. Economidou, E. 1974. Studies on the adventive flora of Greece I. General remarks on some recently introduced taxa. *Candollea* 29: 111–119.

7. History and patterns of plant invasion in Northern Africa

E. LE FLOC'H, H.N. LE HOUEROU and J. MATHEZ

Abstract

The history of plant invasion in Northern Africa is mainly concerned with two different aspects: naturalization and expansion of exotic species on the one hand, and the colonization of new land by local species on the other hand. Both processes generally result from the activities of man and his livestock: deliberate or inadvertent introduction of exotics that may become invasive, expansion of weeds initially linked to cultivation (native and exotics), encroachment of piospheric* vegetation types of nitratophytes linked to high and long standing density of people and/or stock, or of species which are ignored by herbivores.

The patterns of plant invasion in time and space depend on three major groups of factors: anthropozoic, climatic and edaphic. The intimate mechanism of invasion, however, remains insatisfactorily explained in most cases; for instance the reasons why an exotic may litteraly 'burst' and then regress. In this paper we provide a number of examples of invasion and of control mechanisms (climatic, edaphic, biological, human). Some 'mysterious' patterns of distribution and some no less mysterious disappearances are addressed.

Plant invasion in Northern Africa is concerned with a relatively small number of species, less than 200 (i.e., less than 4% of the flora). From the viewpoint of the geographic origin of the invading species, it is a surprising fact that nearly 50% come from the New World. Most of expanding species belong to commensal taxa linked to human activities: cultigenic, range invaders and piospheric nitratophytes.

Dangerous exotic pests are very rare, contrary to what happened on other continents.

Rangelands are progressively infested by unpalatable weed invaders that, almost invariably, belong to the native flora; they generally have a very large area of distribution in the Mediterranean Basin throughout North Africa and the Near East.

* See glossary of terms used at the end.

F. di Castri, A. J. Hansen and M. Debussche (eds.), Biological Invasions in Europe and the Mediterranean Basin.
105–133. © 1990, *Kluwer Academic Publishers, Dordrecht.*

106

Most of the few exotic invaders tend to occupy ecological niches in which, for various reasons, competition from native species is moderate, such as fresh sediments consecutive to floodings, urban and suburban zones, etc. Most of those belong to the 'r' selection type which allocate a large proportion of energy invested in the production of diaspores such as *Xanthium, Conyza, Dittrichia, Heliotropium, Blackiella, Echinochloa, Cyperus*, etc.

Introduction

Africa north of the Sahara occupies a huge area of some 6 million km^2. It has extremely diverse ecological conditions, despite being entirely subjected to a mediterranean climate with various proportions of humid and desert subclimates.

Altitude for instance may vary from 4,200 m a.s.l. in the Jbel Toubkal, Morocco, to –133 m in the Qattara depression (Egypt). Mean annual rainfall may vary from 2,000 mm near Collo, Algeria, to virtually zero in the Libyan Desert. Mean minimum winter temperatures may vary from –2°C or less (S.W. Algeria) to +10°C (S.W. Morocco and N.E. Libya). All sorts of geological substrates and soil types may be encountered in this huge area.

Northern Africa has been inhabited by man since at least 0.5 million years B.P., with a very strong impact in the Neolithic and early historical times from ca. 5000 B.P. to present. As a consequence, vegetation types have been submitted to anthropozoic pressure for a longer period of time than most other geographic zones of the world, the Near East excepted. In many cases, vegetation may come back to pristine conditions when the anthropozoic pressure is discontinued or released, as happened in several occasions in history as a result of warfare, disease and socio-economic regression. In many cases, however, vegetation would undergo more or less deep and permanent changes in structure and composition whenever human pressure is so heavy and long lasting as to prevent any recovery towards the primeval flora. In such cases a true phytogeographic mutation may occur, particularly when pristine vegetation has been cleared for cultivation over very large areas, whether cultivation is then abandoned or not. A number of such cases have been documented in Northern Africa for instance when forest vegetation evolved into shrubland or into steppic communities under anthropogenic pressure (Le Houérou 1959, 1969, 1981a, 1981b).

A brief history of crop expansion in Northern Africa

Crop history in the Maghrib countries is very old, it dates back to the Neolithic some 7000 BP, particularly in the Lower Nile Valley and somewhat later (3000–3500 BP) further west. At the beginning of historical times, some 2500–3000 years ago, agriculture was already well developed in Northern Africa as witnessed in the writings of Greek historians such as Xenophon, Theophrastus, Plato, Herodotus and others, who lived between 500 and 200 BC, and later from Carthaginian and Roman writers. Mago, a Carthaginian agronomist, wrote a treatise of agronomy in 28 volumes some 200 years BC, extracts of which reached us through Latin translations. Latin authors published similar works dealing partly with agriculture in Northern Africa (then called Ifriqia) among those Cato the Elder (234–149 BC), Plinius the Elder, Columella, Varro and Palladius are the best known. For a detailed account of early agriculture in Northern Africa see: Nisard 1877, Tissot 1884–1886, Gsell 1911, 1913, 1928, Cagnat 1912, 1916, Camps-Faber 1953, Camps 1960, 1974. Throughout history there have been periods of intense human activities and agriculture expansion followed by periods of regression and abandonment. These flows and ebbs are of paramount importance to the understanding of the evolution of flora and vegetation in Africa north of the Sahara; they are briefly recalled below.

Barley (*Hordeum vulgare* L.) is most probably the oldest cereal cropped in the Mediterranean Basin; some prehistorians believe it was already known to the epipaleothic men (Iberomaurusian) some 5000 years BC, a fact which seems quite plausible since it was cultivated in Greece, the Near East, Egypt and Cyrenaica as early as 5000–7000 BC (Chevalier 1949, Clark 1971, Zohary 1973, Wendorff *et al.* 1979). *Hordeum vulgare* seems to have evolved from *H. spontaneum* which is native and still common in the semi-arid zones of the Near East, Greece and southern Italy, under ecological conditions very close to those of Northern Africa.

It is widely accepted that the neolithic civilizations (including the post-Capsian cultures that prevailed in North Africa) have been a landmark in the way in which prehistoric man used to deal with his environment. He then used new techniques: terra cotta, livestock husbandry, cereal cultivation, cultivated fibers and fruits. Neolithic man had thus to burn for grazing and to clear for cultivation. It is likely, however, that the North African rainfed 'farmer' used more rough techniques than those that had evolved under irrigated farming in 'Fertile Crescent' of the Near East and the Nile Valley. It would seem that perfection in farming and animal husbandry came from the East.

New techniques (and crops?) came during the Chalcolithic and Iron ages until the Phoenician times, starting ca. 800 BC. Great progress occurred during the Phoenician times (814–146 BC); by 200 BC farming was not substantially different from what it is at present: the main crops were the same and the techniques were not fundamentally different judging from the writings of Mago, confirmed by innumerable archaeological findings. It is Cato the Elder who, showing the Roman Senate a handful of figs from North Africa as a symbol of Carthaginian prosperity, precipitated the declaration of the 3rd Punic War.

The Roman period (146 BC–650 AD) is fairly well known as North Africa then became one of the main granaries of Rome, with a large state owned grain trade organization the Annona, and the source of one of its main supplies of wine, fruits, olive oil, livestock and wild beasts for the Circus games. It

108

was also its closest source of food supply (Sicily excepted). After this period of nearly one thousand years of continuous agricultural expansion for ca. 500 BC to 500 AD where Northern Africa had probably a population as large as present (Le Houérou 1969), came the decline with some early up and downs between 500 and 650 AD (Vandal & Byzantine dominations).

During the next 1,200 years prevailed the arab civilization and the domination of beduin shepherd tribes with a sharp regression in cultivation and probably a progressive evolution towards a return to pristine vegetation conditions due to low human pressure. The human population from the 10th to the 19th century was probably not much more than 1/5 to 1/10 of that of the 6th to 7th centuries; it remained about stable until the 20th cerntury. During the Arab times (7th through 19th centuries) many introductions were made from the East along the caravan roads, the Silk Road from China to the Mediterranean shores then to the Atlantic (mulberries, citruses, sugar cane, banana, pulses, fibers and spices, nutfruits (apricot, almond, pistachio)).

From the 15–16th centuries American species, introduced to Spain since Colombus' expeditions, were also introduced to North Africa when the Moors, expelled from the Iberic Peninsula in 1610 and the Andalusians settled down in Northern Africa, bringing, among other species, the cacti, *Agave,* etc...

The 19th century AD was a time of great change due to the systematic introduction of a large number of plant species (with the procession of inadvertently accompanying diaspores). There was a kind of mysticism among early scientists (1850–1940) for 'acclimatization', a kind of quest for miracle plants. For a hundred years probably more than one thousand plant species of potential economic interest were deliberately introduced from various arid and semi-arid zones of the world. Among this wealth of introduction one must cite, Australian *Eucalyptus, Acacia, Atriplex, Casuarina*, etc. and many cultivars of cereals as well as many fruit tree species and cultivars in which wine and olive played a major role as they did some 2,500 years earlier.

The evolution of vegetation and ecosystems is particularly obvious during the present century owing the exponential increase in human and livestock populations. This exponential growth increased suddenly toward 1930. The demographic increase in human population for the region as a whole has been 3% per annuum for the past 30 years (i.e., a doubling period of 23 years). Under such heavy pressure, vegetation undergoes drastic and probably irreversible changes towards an ever increasing degree of 'artificialization' due to wood cutting, clearing, burning, overstocking, overcultivation, urbanization, expansion of irrigated schemes (reservoirs, artesian aquifers), soil erosion, floddings, sedimentation, decreasing soil fertility and productivity on the watersheds, drainage and drying up of marshes, ponds and lakes. All these human activities result in a rarefaction of many species and the expansion of others: segetal, cultigenic, ruderal, nitrophilous. Table 1 is an attempt to sum up the information available on the early cultivation of plant species in the area under study. There are many interrogation marks as some species may have been cultivated long before reference to their existence has been

Table 1. Synoptic table of plant crops introduced in Northern Africa.

| Period | Name | | Origin | Sources |
	Scientific name	Common name		
Paleolithic	*Hordeum vulgare* L.	barley/orge	Syria, Cyrenaïca	Wendorf *et al.* 1979
Neolithic	*Triticum* sp.	wheat/blé		Gsell 1928
	Phoenix dactylifera L.	date palm/palmier	Sahara	
	Ziziphus sp.	date palm/jujubier		Chevalier 1932
	Lens culinaris Medicus	lentil/lentille		
	Vicia faba L.	broad bean/féve	Europe, Asia	Chevalier 1932
	Lawsonia inermis L.	henna/henné	Iran, India, Ethiopia	
	Pisum sativum L.	common/pea/pois		van Zeist 1980
	Vicia ervilia (L.) Willd.	ers/vesce		van Zeist 1980
?	*Cicer arietinum* L.	chick-pea/pois chiche		van Zeist 1980
Phenician	*Olea europaea* L.	olivier/olive	Mediterranean region	Herodotus/van Zeit
	Vitis vinifera L.	grape vine/vigne	Armenia	Herodotus/van Zeit
	Ficus carica L.	tongue fig/figue	North Africa (?), Caucasus	Van Zeit 1980
	Cucurbita moschata Duch.	cushaw/courge	?	
	Punica granatum L.	pomegranate/grenade	Asia minor, Armenia	Bonnet & Baratte 1896
	Medicago sativa L.	alfalfa/luzerne	Mediterranean region	
	Linum usitatissimum L.	flax/lin	Oriental Europe	Chevalier 1932, 1939
	Trigonella foenum-graecum L.	fenugreek/fenugrec	Asia minor	
Roman	*Cucumis sativus* L.	cucumber/concombre		Plinius
	Allium sativum L.	garlic/ail		Plinius
	Allium cepa L.	onion/oignon		Plinius
	Cynara cardunculus L.	cardoon/cardon	Mediterranean	Plinius
	Cuminum cyminum L.	cumin/cumin	Mediterranean	Plinius
	Cicer arietinum L.	chick-pea/pois chiche	Europe, Orient	Columella (in Nisard)
	Malus domestica Borkh.	quince/cognassier	Mediterranean, Caucasus	Précheur-Canonge
	Pennisetum miliaceum L.	millet/mil	Tropics	
	Prunus dulcis (Miller) Webb	almond tree/amandier	North Africa, Central Asia	

Table 1. (Continued).

Period	Name		Origin	Sources
	Scientific name	Common name		
	Pyrus malus L.	apple/pommier	Europe, Asia	Précheur-Canonge
	Pyrus communis L.	pear/poire	Europe, Northern Asia	Précheur-Canonge
(700..	*Prunus armeniaca* L.	apricot/abricot	Central Asia, Arabia	Plinius
	Prunus persica (L.) Batsch	peach/pêche	Orient	
?	*Morus alba* L.	mulberry/murier	China-Japan	
?	*Morus nigra* L.	mulberry/murier	Temperate Asia	
?	*Cyperus esculentus* L.	earth almond/souchet	South Europe/North Africa	
	Saccharum robustum L.	sugar cane/canne à sucre	New Guinea (via China)	
	Ceratonia siliqua L.	carob tree/caroubier	Spain, Syria	Chevalier 1939
	Cannabis sativa L.	fimble/chanvre	Central Asia	
	Solanum melongena L.	eggplant/aubergine	Tropical Asia	
(1000..	*Citrus aurantium* L.	orange/orange	Tropical Asia	Jean Léon L'Africain
	Citrus medica L.	citron/citron	Tropical Asia	Jean Léon L'Africain
	Carum carvi L.	caraway/carvi	Europe, Northern Asia	Edrissi
16 and 17th	*Agave americana* L.	american aloe/agave	Tropical America	
	Arachis hypogaea L.	peanut/arachide	Tropical America	
	Opuntia ficus-barbarica Berg.	prickly pear/figue de Barbarie	America C. & S.	Monjauze, Le Houérou
	Zea mays L.	corn/maïs	Tropical America	Chevalier 1932
?	*Datura metel* L.	downy thornapple/datura	Tropical America	
?	*Nicotiana rustica* L.	tobacco/tabac	Mexico	Kress
	Capsicum annuum L.	Cayenne pepper/piment de Cayenne	Tropical America	Kress
	Capsicum frutescens L.	Chili pepper/piment		Kress
	Cucumis melo L.	melon/melon	Old world Tropics	Jean Léon L'Africain
	Ipomea batatas (L.) Lam.	sweet potato/patate douce	Tropical America	Hedrick
	Lycopersicon esculentum Mil.	tomato/tomate	Tropical America	Kress
	Morus alba L./*M. nigra* L.	mulberry/murier	Chine-Japon/Asia	Jean Léon L'Africain
	Phaseolus sp.	bean/haricot	America	Chevalier 1932

Table 1. (Continued).

Period	Name		Origin	Sources
	Scientific name	Common name		
19th century	*Gossypium barbadense* L.	egyptian cotton/coton	South America	Jean Léon L'Africain
around 1830	*Avena sativa* L.	oats/avoine	Europe	Chevalier 1932, 1939
	Beta vulgaris L.	beet/betterave	Europe, North Africa	Chevalier 1932, 1939
	Elaeagnus angustifolia L.	russian olive/olivier de Bohème	Europe, North Asia	
	Eriobotrya japonica lindl.	loquat/Nèfle du Japon	Japon, China	
	Casuarina sp.	casuarine/filao	Australia	Chevalier 1932
?	*Diospyros kaki* L.	kaki/plaqueminier		
around 1850	*Pelargonium radula* L'Herit.	kaki/géranium rosat	Réunion, Comores	Chevalier 1927
around 1860	*eucalyptus* sp.	gum trees/Eucalyptus	Australia	Picot/Trabut 1922
	(ex. *E. globulus* L.)			Chevalier 1932
	Acacia sp.	acacia/acacia	Australia	Battandier
around 1890	*Schinus molle* L.	false pepper/faux poivrier	South America	Battandier
	Atriplex sp.	saltbushes/atriplex	Australia	Battandier
	(ex. *A. semibaccata* R. Br.)			
	Blackiella inflata Aellen	saltbush	Australia	Battandier
	Ailanthus altissima Swingle	japan varnish tree/vernis	Asia	
	Melia azedarach L.	persian lilac/lilas d'Inde	India	Chevalier 1932
around 1899	*Tipuana speciosa* Benth.		South America	Chevalier 1932
	Parkinsonia aculeata L.	Jerusalem thorn	North America	Chevalier 1932
	Myoporum parviflorum R. Br.	coolaba	Australia	Chevalier 1932
	Bougainvillea sp.	bougainvilla/bougainvillier	South America	Chevalier 1932

112

made by historians, geographers or travellers. Furthermore, some plants may undergo sequential expansions and regressions over time according to socio-economic and socio-cultural circumstances.

Table 1 warrants the following comments:

(i) A first phase is concerned mostly with pan-Mediterranean species of recent domestication, mainly cereals, pulses, vegetables and fruits from the Eastern Mediterranean (Near East), Europe, and occasionally central Asia and the Far East. This phase occured during the Neolithic which also witnessed the early stages of livestock husbandry in the Maghreb (not necessarily nomadic in nature, contrary to a common belief).

(ii) The Arab period saw further introductions of crops, mostly fruit crops from the Middle and Far East.

(iii) The 16th century corresponds to the beginning of the American phase of plant introduction through Spain and the Moors when the latter were expelled therefrom, as mentioned above.

(iv) The 19th century and the first half of the 20th, the colonial times was a period of systematic introduction of all sorts of useful plants from various tropical and subtropical zones from all over the world. Poletaeff (1953) states that about 1,000 plant species were introduced annually in Tunisia mainly as seeds. This author also mentions that the most successful introductions came from regions located between 30 and 40° of latitude in both hemispheres, i.e. southern USA, northern Mexico, China, Japan south and southwest Australia, New Zealand, South America (Argentina, Uruguay, Chile, southern Brazil) South Africa and naturally from the Mediterranean Basin itself. This is a kind of 'isolatitudinal' introduction law, quite logical in terms of bioclimatology (photoperiodism and temperatures) and biogeography.

Naturalized and acclimated species (deliberately or inadvertently introduced)

The introduction of cultivated species (cereals, fodder, pulses, fiber, dying, fruits, vegetables, etc.) and pastoral nomadism have obviously strongly influenced the history of flora and vegetation in Northern Africa:
– through the seeds of wild species inevitably mixed with crop seeds.
– through the zoochorous species disseminated by livestock, particularly (but not only) exozoochorous in sheep fleeces and endozoochorous in ruminants, birds, hogs, etc.

Many environments were disturbed for cultivation first in the plains and valleys with deep soils but also by overstocking and overgrazing, by fallowing. Xenophon (4th century BC) states, for instance, that clear fallowing was the up-to-date technique of cultivation and soil preparation in his time. In spite of a few attempts (Bonnet 1894, Labbe 1950, 1955) the history of deliberate and inadvertent introductions still remains to be described, for the most part.

In Table 2 we provide a list of species which are regarded as naturalized

Table 2. Plant taxa naturalized in Northern Africa.

	Date of introduction (or 1rst citation)	Way of introduction	Origin	Biological type	Bioclimate type
Abutilon theophrasti Medicus		deliberately	Far East	T	SA-H
Agave americana L.		deliberately	North America	H	A-H
Agrostemma githago L.	1854	inadvertently	Europe, Near East	T	SA-H
Ailanthus altissima (Miller) Swingle		deliberately	Europe, Near East	Ph	SH-H
Aloe vera (L.) Burm. fil.			Tropical Africa	H	A-SH littoral
Alopecurus myosuroides Hudson	1896 (T)	inadvertently	North America	T	SA
Amaranthus albus L.	1854 (T)	inadvertently	North America	T	A-H
Amaranthus blitoides S. Watson	*	inadvertently	North America	T	SA-H
Amaranthus cruentus L.	1854 (T)	inadvertently	North America	T	SA
Amaranthus deflexus L.	1850	inadvertently	South America	T	A-SH
Amaranthus hybridus L.	*	inadvertently	Tropical America	T	A-SH
Amaranthus muricatus (Moq.) Gillies		inadvertently	Argentina	T	A
Amaranthus retroflexus L.	* 1854 (T)	inadvertently	North America	T	A-SH
Amaranthus viridis L.		inadvertently	Tropical America	T	SA
Arundo donax L.		deliberately	Europe, Near East	NPh/G	wet soils
Asclepias curassavica L.		deliberately	Tropical America	Ch/NPh	A
Asclepias fruticosa L.		deliberately	South Africa	Ch/NPh	SA-SH
Asparagus officinalis L.	1896 (T)	deliberately	Europe, Asia	G	oases
Aster squamatus (Sprengel) Hieron.	1952 1916 (T)	inadvertently	South America	T (1-2)	A-H wet soils
Atriplex semibaccata R. Br.	1895 (T)	deliberately	Australia	Ch	A-SA salted soils
Azolla filiculoides Lam.	1971	inadvertently	Neotropic (S. America)	Hyd.	SA
Bidens aurea (Ait.) Sherff	1952	inadvertently	Central America	T	A
Bidens pilosa L.		inadvertently	South America	T	SA
Blackiella inflata (F.V. Muell.) Aellen	1895 (A)	deliberately	Australia	Ch	SA-H
Brachiaria eruciformis (Sm.) Griseb.	1951	inadvertently	Paleo-subtropical	T	A-SA
Bromus willdenowii Kunth		inadvertently	South America	T	SH-H
Callitriche deflexa Hegelm.		inadvertently	S. Hemisphere	T	SA-SH wet soils
Cannabis sativa L.		deliberately	Near & Middle East	T	SH-H oases
Carpobrotus edulis (L.)N.E.Br.		deliberately	South Africa	Ch	SA-SH littoral

Table 2. (Continued).

	Date of introduction (or 1rst citation)		Way of introduction	Origin	Biological type	Bioclimate type
Chenopodium ambrosioides L.		XVIe (T)	deliberately	North America	T	A-SH
Chenopodium giganteum D. Don	XVIe		deliberately	Far East (Nepal)	T	A-SH
Chloris gayana Kunth			deliberately	Tropical Africa	H/G	rivers banks
Consolida ambigua (L.) P.W. Ball & Heywood		1927 (T)	deliberately	Europe, Asia	T	SA-H
Consolida orientalis (Gay) Schrödinger	préhist ?		inadvertently	Europe, Asia	T	A-SA
Conyza bonariensis (L.) Cronq.		1892 (T)	inadvertently	South America	T	A-H
Conyza canadensis (L.) Cronq.		1892 (T)	inadvertently	North America	T	A-H
Coronopus didymus(L.) Sm.		1900	inadvertently	South America	T	SA-H
Cotula coronopifolia L.		1883 (T)	inadvertently	South Africa	T	SH
Cuscuta campestris Yuncker	1955		inadvertently	North America	T	SA
Cuscuta epithymum (L.) L.	1912	XIIIe (T)	inadvertently	South/West Asia	T	SA
Cuscuta australis R. Br.	1941		inadvertently	Europ. med. (Italy, Yougoslavia)	T	SA
Cymbalaria muralis Gaert., Meyer & Scherb.				Italy	H	SH-H
Cyperus difformis L.	1963 (M)		inadvertently	?	G	rice-fields
Datura metel L.	1907 (A)	1883 (T)	deliberately	Tropical America	T	SA
Datura stramonium L.		1892 (T)	deliberately	America	T	A-H
Echinochloa colonum (L.) Link			inadvertently	Java - Malaisya	T	A-H wet soils
Echinochloa oryzoides (Ard.) Fritsch	1950 (A)		inadvertently	South Europe	T	rice-fields
Elaeagnus angustifolia L.			deliberately	Europe, Asia, Mediterranean	Ph	oases
Eleocharis flavescens (Poiret) Urban	1971		inadvertently	Central America	T/Hyd	free water
Eragrostis curvula (Schrad.) Nees			deliberately	Africa, Tropical America	H	SA
Euphorbia lathyris L.			deliberately	North Mediterranean	T (2)	SA-SH
Heliotropium curassavicum L.		1854 (T)	deliberately	North America	Ch	S-SA wet, salted soils
Iris belouini Bois & Cornuault			deliberately	?	G	SA-SH

Table 2. (Continued).

	Date of introduction (or 1rst citation)	Way of introduction	Origin	Biological type	Bioclimate type
Iris germanica L.		deliberately	Oriental Mediterranean	G	H
Lantana camara L.		deliberately	Tropical America	Ph	SA-H
Lepidium sativum L.	1854 (T)	deliberately	Europe, Asia	T	SA-H
Lippia canescens Kunth	1896 (T)	deliberately	South America	C	wet soils
Lycium barbarum L.		deliberately	China	NPh	A-H
Mentha spicata L.		deliberately	Oriental Mediterranean	H	SH-H
Moluccella laevis L.		inadvertently	Oriental Mediterranean	T	SA
Nicotiana glauca R.C. Graham	1909 (T)	deliberately	South America	C	wet soils
Nigella sativa L.	1896 (T)	deliberately	South America, Near East	T	SA-H
Oenothera biennis L.	*	?	North America	T (2)	SH-H
Opuntia ficus-barbarica A. Berger	XVIe (M.A.T.)	deliberately	North America (Mexico)	P	A-SH
Oxalis compressa Jacq.		deliberately	Austral Africa	G	SH
Oxalis pes-caprae L.	1892 (T)	inadvertently	Austral Africa	G	A-H
Oxalis purpurea L.		deliberately	Austral Africa	G	H
Paspalum dilatatum Poiret		deliberately	Tropical America & Africa	G/H	rivers banks
Paspalum paspalodes (Michx) Scribner		deliberately	Tropical America & Africa	G	rivers banks
Pennisetum villosum R. Br. ex Fresen	1934	deliberately	Ethiopia, Arabia	H	SH-SA
Phalaris canariensis L. ssp. *canariensis*		deliberately	Canaria	T	SH-SA
Phytolacca americana L.	1896 (T)	deliberately	North America	H	SA-H
Plumbago europaea L.		deliberately	South Africa	Ch	SH-H
Ricinus communis L.	*	deliberately	Tropical Africa, India	NPh/Ph	A-SH
Rubia tinctorum L.	1896 (T)	deliberately	Europe, Near East	H/G	oases
Salpichroa origanifolia (Lam.) Baillon	1940 ?	inadvertently	South America	Ch Liana	SA-H
Solanum elaeagnifolium Cav.	1944 (M)	inadvertently	South America	G/H	A-SA (irrigated fields)
Solanum laciniatum Ait.	1943 (M)	deliberately	Australia, New Zealand	Ch	SA
Spartina densiflora Brongn.		inadvertently	South America	H	SH littoral, wet soils

Table 2. (Continued).

	Date of introduction (or 1rst citation)	Way of introduction	Origin	Biological type	Bioclimate type
Spinacia oleracea L.		deliberately	Near East	T	SH-H
Tropaeolum majus L.		deliberately	South America	T	SH-H
Veronica persica Poiret		inadvertently	Orient	T	SH
Withania somnifera (L.) Dunal		deliberately	Europe , Mediter., Africa	C/NPh	A-H
Xanthium x brasilicum		inadvertently	South America	T	A-SH
Xanthium spinosum L.	1960	inadvertently	South America	T	A-SH
Xanthium strumarium L.	1960	inadvertently	South America	T	A-SH
Zantedeschia aethiopica (L.) Sprengel		deliberately	South Africa	G	H

Particular symbols

* = several centuries
M = Morocco
A = Algeria
T = Tunisia

Biological type (Raunkiaer's system)

Ph = phanerophyte
NPh = nanophanerophyte
Ch = chamaephyte
H = hemicryptophyte

G = géophyte
T = therophyte
2 = bisannual
Hyd = hydrophyte

Bioclimate type

H = Humid ecoclimatic zone (P 800 mm)
SH = Subhumid ecoclimatic zone (600 P 800 mm)
SA = Semi arid ecoclimatic zone (400 P 600 mm)
A = Arid ecoclimatic zone (100 P 400 mm)
S = Saharan ecoclimatic zone (Hyperarid) (P 100 mm)

(subspontaneous) in Northern Africa. These have been introduced deliberately or inadvertently and they sometimes come from floristically very remote regions. They no longer depend on man for their propagation and survival in the region under consideration. We have removed from this table some species the 'naturalization' of which may be contended by some botanists and malherbologists such as, for instance, some species which have remained restricted to limited areas or that may be regarded as intermittently naturalized or having transitorial remanence. Some such species may undergo pulsations which are brief and unpredictable. Their temporary abundance seems linked to temporarily favorable ecological conditions (such as extended floodings for hygrophytes which would then disappear whenever a sequence of drought years comes in succession until the next succession of flodding events). A few cases of such temporary explosions have been documented. Draz (1954) has explained the case of *Kochia indica* which, deliberately introduced by seeds from Australia in 1945 in the region of Mersa-Matruh, Egypt, became invase in this part of northern Egypt in the early 1950's but almost disappeared later.

Table 2 has been somewhat abridged purposedly in order to eliminate some contended cases, but there are perhaps as many as 100 to 200 species that should be considered naturalized in North African countries (i.e., between 2 and 4% of the flora); we are far from the 13% of naturalized species reported in the Californian flora (Raven 1977).

A brief analysis of the information

Date of introduction
Data on this subject are unfortunately too fragmentary to warrant much valid conclusion. The introduction of adventitious species as well as of cropped species in the course of centuries often occurred at an unknown time. Rather rare are the cases when a precise date may be ascribed to their introduction. *Atriplex semibaccata* and *Blackiella inflata* were introduced about the same time, for sure before 1904. *Acacia saligna* (*A. cyanophylla*) and a number of other phyllodineous *Acacias* from Australia were introduced in Algeria in the early 1870's (Bert 1886). *Eucalyptus* species were introduced from 1850 onwards (Picot 1928). Some new hybrids between *Eucalyptus globulus* and *E. rudis* (*E. algeriensis*) and between *E. globulus* and *E. botryoides* (*E. trabutii*) were indentified in 1904 and 1920 respectively (Picot 1928). Rice cultivation expanded in the Lower Cheliff Valley of western Algeria in the early 1950's was first free of specialized adventitious weeds: *Echinochloa oryzoides, Scirpus mucronatus* and *Cyperus difformis* (Dubuis & Simonneau 1956). In the late 1950's, however, Dubuis and Faurel (1957) noted the frequence of *E. oryzoides* in the Lower Cheliff and in the Mitidja valleys (Algeria) and *Cyperus difformis* was recorded in rice fields of Morocco in 1963.

Information on the progression of a particular species is occasionally

118

available. For example, *Heliotropium curassavicum* was first recorded in Tunisia by Desfontaines in 1874 near Sousse. In 1921, Burollet and Boitel (1921) found it abundant in that location, apparently not aware of Desfontaines' record. One year later, the same species was found again in a small site near Monastir (Boitel 1923). In 1955, Labbe found five new sites in Tunisia. It is now common on saline soils around most oases in this country. *Aster squamatus* migrated since early 1940's from Tunis southwards at an average rate of some 10 km per annuum (Labbe 1955). In Morocco, (Mathez & Sauvage 1970) *A. squamatus* was first recorded by Perrin de Brichambaut and Sauvage in 1952 in the Rharb plains. In 1969, this species was abundant in this part of Morocco in ricefields, others irrigated fields and along river banks. *A. squamatus* is at present considered a common ruderal species in those parts. *Solanum elaeagnifolium* recorded in irrigated zones of Morocco since 1944 and recorded for Algeria by Marlierre (in Tanji *et al.* 1985) had not been found elsewhere in northern Africa until it was found in an irrigated field at El Alem near Kairouan, Tunisia in 1986.

Means of introduction
Table 2, column 2 shows that nearly 50% of the native or cultivated species were purposely introduced for cultivation initially. If some of those are now considered as weeds some others are still deliberately propagated for economic reasons. Such is the case of *Opuntia ficus-barbarica (O. ficus-indica)* which incidently does not constitute an invading pest in Northern Africa as it does in other subtropics, in part because of high grazing pressure, in part for bioclimatic reasons (winter rains and low winter temperatures).

Origin
Table 2 shows that 45% of the introducedd species come from the New World (17% from North America and 28% from South America) while 20% originate in the Near and Middle East, 17% from southern Africa, 13% from Eurosiberia and only 3% from the Far East.

Species expanding in Northern Africa

One should first remark how difficult it is to label a species as 'invading' in such a large territory as Northern Africa, since a given species may have a different status in this respect in various parts of the region at a given time, and in a given location at different times (Guillerm *et al.* 1986). Such or such species, to day stabilized in such and such place, may have been an invader there not long ago, a status that it may still have in other places. Such other species may not expand geographically anymore but become more frequent and increase in density as well; it is then labelled 'infesting'.

Table 3 shows that the proportion of expanding introduced species is relatively small: 18 among the species included in Table 2 (i.e., 20%). Out

Table 3. Native or introduced plant taxa undergoing a strong expansion in Northern Africa (personal data).

	Status	Bioclimate type	Biological type
I. Exotics (= Introduced)			
Ailanthus altissima (Miller) Swingle	rud.	A-H	Ph
Amaranthus deflexus L.	rud.	A-SH	T
Amaranthus retroflexus L.	comm./cult.	A-SH	T
Amaranthus viridis L.	comm./rud.	SA	T
Aster squamatus (Sprengel) Hieron.	rud./hyg.	A-H	T (1-2)
Atriplex semibaccata R. Br.	rud./nitr./hal.	A-SA	Ch
Blackiella inflata (F.V. Muell.) Aellen	rud./nitr.	A-SA	Ch
Conyza bonariensis (L.) Cronq.	rud./fal.	A-H	T
Cuscuta epithymum (L.) L.	par.	SA	T parasite
Echinochloa oryzoides (Ard.) Fritsch	irr. (rice fields)	rice fields	T
Heliotropium curassavicum L.	rud./comm./ hyg./hal.	S-H (wet soils)	Ch
Nicotiana glauca R.C. Graham	rud.	SA-H (wet soils)	NPh/Ph
Oenothera biennis L.	rud.	SH-H (wet soils)	T (2)
Oxalis pes-caprae L.	rud./comm.	A-H	G
Ricinus communis L.	rud.	A-SH	NPh/Ph
Rubia tinctorum L.	rud./comm.	A oases	H/G
Solanum elaeagnifolium Cav.	comm./irr.	A-SA	H/Ch
Xanthium spinosum L.	rud.	A-SH	Ch
II. Natives (= Spontaneous)			
Aizoon canariense L.	waste/nit.	S-SN	T
Aizoon hispanicum L.	waste/nit.	A-H	T
Ammi majus L.	comm./rud./cult.	A-H	T
Ammi visnaga L. (Lam.)	comm./rud.	SH-H	T
Anagallis foemina Miller	comm./cult.	S-H	T
Anthemis arvensis L.	comm./cult.	A-SA	T
Arisarum vulgare Targ.-Tozz.	comm./cult.	SA-H	G
Arrhenatherum album (Vahl) W.D. Clayton	rud./past.	SH-H	H
Artemisia campestris L.	post.past/rud./ tox.	S-SH	Ch
Asphodelus fistulosus L.	post.past/tox.	A-H	T
Asphodelus microcarpus Salzm. et Viv.	post.past/tox.	A-SA	G
Asphodelus tenuifolius cav.	post.past	S-SA	T
Astragalus armatus Willd.	post.past/spiny	S-SA	Ch
Atractylis serratuloïdes Sieb. ex Cass.	post.past/spiny	S-SA	Ch
Avena sterilis L.	rud./comm.	A-H	T
Ballota hirsuta Benth.	rud./comm./ unpal.	A-SH	Ch
Beta vulgaris ssp. *maritima* (L.) Arcangeli	rud./hyg./hal.	SA-H (littoral)	G
Biarum bovei Blume	nit.	SA-SH	G
Brachypodium distachyon (L.) Beauv.	post.past.	S-H	T
Brassica tournefortii Gouan	comm./cult.	A-SH	T
Bromus madritensis L.	rud.	A-H	T

120

Table 3. (Continued).

	Status	Bioclimate type	Biological type
Bromus sterilis L.	rud.	SH-H	T
Calendula arvensis L.	rud./nit.	SA-H	T
Calotropis procera (Ait.) Ait.	post.past./tox.	S-A	NPh
Calycotome villosa (Poiret) Link	post.past./spiny	S-SH	Ch/NPh
Calystegia sepium (L.) R.Br.	rud.	SH-H	G/H
Cardaria draba (L.) Desv.	rud./comm.	A-SH	H
Carduncellus choulettianus (Pomel) Batt.	post.cult./comm.	SA-SH	Ch
Carduncellus eriocephalus Boiss.	post.cult.	S-A	T
Carduncellus mareoticus (Del.) Hamelt	post.cult.	A	Ch
Carduncellus pinnatus (Desf.) DC.	post.cult.	SA-SH	H
Carduus getulus Pomel	post.cult.	S-SA	T
Carduus involucrata Poiret	post.cult.	A-H	H
Carduus lanata L.	post.cult.	SH-H	T
Carthamus lanatus L.	post.cult.	A-H	T
Centaurea dimorpha Viv.	post.past/spiny	A-SH	H/G
Centaurea calcitrapa L.	post.past	SA	T
Chenopodium murale L.	rud./nit.	S-H	T
Chrozophora obliqua (Vahl) A. Juss. ex Spreng.	comm./rud.	A-SH	T
Chrysanthemum coronarium L.	comm./cult.	S-H	T
Chrysanthemum trifurcatum Desf.	comm./cult.	S-SA	H
Cirsium arvense (L.) Scop.	post.past./spiny	SH-H	G
Cistus monspeliensis L.	pyro.	SH-H	Ch
Cleome amblyocarpa Barr. & Murb.	post.past./unpal.	S-SA	T
Convolvulus althaeoides L.	comm./cult.	A-H	H
Convolvulus arvensis L.	comm./cult.	A-H	T
Convolvulus tricolor L.	comm./cult.	SA-H	T
Coronilla scorpioides (L.) Koch	comm./cult.	A-H	T
Cuscuta epithymum (L.) L.	comm./cult.	A-H	T parasite
Cynanchum acutum L.	oases	S-SH (hygro)	G
Cynara cardunculus L.	post.cult./ post.past./spiny	A-H	H
Cynodon dactylon (L.) Pers.	comm./nit.	S-H	G
Cyperus esculentus L.	oases	SA-H	G
Cyperus rotondus L.	oases	A-SA (irrigated fields)	G
Diplotaxis harra (Forsk.) Boiss.	post.cult./ post.past./unpal.	A-SA	T/H
Diplotaxis simplex Spreng.	comm.	A-SA	T
Dittrichia viscosa (L.) W. Greuter	rud.	SA-H	Ch
Dittrichia graveolens (L.) W. Greuter	rud.	SA-H	Ch
Ecballium elaterium (L.) A. Richard	rud./nit.	SA-H	T
Echium italicum L.	post.past./rud.	SH-H	H/T
Echium sabulicola Pomel	post.past./rud.	Emex spinosa	comm.
Eremobium aegyptiacum (Spreng.) Asch. & Schw.	rud./post.past./ comm.	S-A	T
Eryngium campestre L.	post.past./spiny	A-H	H
Eryngium dichotomum Desf.	post.past./spiny	SA-SH	H

Table 3. (Continued).

	Status	Bioclimate type	Biological type
Eryngium tricuspidatum L.	post.past./spiny	SA-H	H
Eryngium triquetrum Vahl	post.past./spiny	SA-H	H
Foeniculum vulgare Miller	rud.	A-H	H
Fumaria densiflora D.C.	comm.	A-SH	T
Galium verrucosum Hudson	comm.	A-SH	T
Gladiolus italicus Miller	comm.	A-H	G
Glaucium corniculatum (L.) J.H. Rudolph	comm.	A-H	T
Gypsophila pilosa Hudson	comm./cult.	A-SA	T
Haloxylon scoparium Somel	post.past./unpal.	A-SA	Ch
Haplophyllum tuberculatum (Forsk.) Juss.	comm./nit.	A-SA	T/H
Heliotropium europaeum L.	rud.	A-H (hydro)	T
Hirschfeldia incana (L.) Lagrèze-Fossat	comm./rud.	SA-SH	T
Hertia cheirifolia (L.) O.K.	post.past./unpal.	A-SH	H/Ch
Hertia maroccana Batt.	post.past./unpal.	A-SA	NPh
Hordeum murinum L.	rud./nit.	A-H	T
Hyoscyamus albus L.	rud./nit.	A-H	T
Hyoscyamus niger L.	rud./nit.	SH	T
Hypericum triquetrifolium Turra	post.past./ post.cult./tox.	SA-H	H/Ch
Ifloga spicata (Forsk.) Schultz Bip.	post.past.	S-SA-H	T
Imperata cylindrica (L.) Raeuschel	rud./hygr.	A-SA (phreato)	G
Launaea nudicaulis (L.) Hooker fil.	comm.	A-SA	T/H
Launaea resedifolia (L.) O. Kuntze	comm.	S-SH	T
Lolium multiflorum Lam.	rud.	SA-H	T/H
Lolium perenne L.	rud./comm.	A-H	H
Lolium rigidum Gaudin	comm.	SA-H	T/H
Lycium shawii Roem. & Sch.	rud.	S-SH	Ch/NPh
Malva parviflora L.	rud./nit.	S-H	T
Matthiola longipetala (Vent.) DC.	post.past./comm.	S-SA	T
Melilotus segetalis (Brot.) Ser.	comm./cult.	SH-H	T
Medicago polymorpha L.	comm./cult.	SA-SH	T
Mercurialis annua L.	rud./waste	SA-H	T
Mesembryanthemum crystallinum L.	rud./pion./waste	A-SH	T
Mesembryanthemum nodiflorum L.	rud./nit./hal.	A-H	T
Onopordum arenarium (Desf.) Pomel	comm./rud.	S-SA	H
Onopordum confusum Pamp.	comm./rud.	A-SH	T (2)
Onopordum espinae Cosson & Bonnier	comm./rud.	A-SA	H
Otospermum glabrum (Lag.) Willk.	rud./hyd.	SH-H	T
Peganum harmala L.	rud./post.past./ nit./tox.	A-SH	H
Petroselinum crispum (Miller) A.W. Hill	rud.	SH-H	H
Phragmites australis (Cav.) Trin. ex Steud.	rud./hyg.	S-H (phréato.)	H
Raphanus raphanistrum L.	rud./comm.	SA-H	T
Rapistrum rugosum (L.) All.	rud./comm.	A-SH	T
Ridolfia segetum Moris	comm./hyg.	A-H	H
Rorippa sylvestris (L.) Besser	rud./hyg.	A-H (hydro.)	H

122

Table 3. (Continued).

	Status	Bioclimate type	Biological type
Salsola kali L.	post.cult./rud.	SA-H (littoral)	T
Scolymus grandiflorum Desf.	post.past./spiny	SA-H	H
Scolymus hispanicus. L.	post.past./spiny	SA-H	H/T
Setaria verrticillata (L.) Beauv.	comm./oases	A-SH	T
Silene colorata Poiret	post.past.	A-H	T
Silybum eburneum Coss. & Dur.	rud./post.past./ spiny	A-H	T/H
Silybum marianum (L.) Gaertner	rud./post.past./ spiny	SA-H	T
Sinapis arvensis L.	comm./cult.	A-H	T
Sisymbrium irio L.	rud./comm./ post.cult.	A-H	T
Solanum nigrum L.	post.past./tox.	A-SA	T
Sonchus oleraceus L.	cult./oases/nit.	A-SH	T
Spergularia marina (L.) Griseb.	rud./hyg./hal.	A-SH	H
Stipa capensis Thunb.	post.past./unpal.	S-SA	T
Stipagrostis pungens (Desf.) de Winter	post.past./unpal.	S-SA	G
Thapsia garganica L.	post.past./tox.	A-H	G
Thesium humile Vahl	cult./par.	S-H	Ch/parasite
Thymelaea hirsuta (L.) Endl.	post.past./unpal.	A-SH	Ch/NPh
Torilis nodosa (L.) Gaertner	rud./comm.	A-H	T
Vaccaria pyramidata Medicus	comm./cult.	A-H	T
Verbascum sinuatum L.	rud.	A-H	H
Withania somnifera (L.) Dunal	rud./nit.	A-SA	T

Status

comm.	= commensal (here of crops)
cult.	= cultigenic
fal.	= fallows
hal.	= halophyte
hyd.	= hydrophyte (free water)
hyg.	= hygrophyte (wet soils)
unpal.	= unpalatable
phreato.	= phreatophyte
irr.	= irrigated fields
nit.	= nitrophyte, nitrophilous
oases	= oases
par.	= parasite
litt.	= littoral (sea-side)
pion.	= pioneering species
post.cult.	= post cultural
post.past.	= post pastoral (= range invader)
pyr.	= pyrophytic (tolerant to fire)
rud.	= ruderal
seg.	= segetal
spiny	= spiny
tox.	= toxic
waste	= wastelands

Biological type (Raunkiaer's system)

Ph	= phanerophyte
NPh	= nanophanerophyte
Ch	= chamaephyte
H	= hemicryptophyte
G	= géophyte
T	= therophyte
Hyd	= hydrophyte

Bioclimate type

H = Humid ecoclimatic zone (800 mm <P)
SH = Subhumid ecoclimatic zone (600 mm <P< 800 mm)
SA = Semi arid ecoclimatic zone (400 mm <P< 600 mm)
A = Arid ecoclimatic zone (100 mm <P< 400 mm)
S = Saharan ecoclimatic zone (Hyperarid) P< 100 mm)

Table 4. Main native invading cultigenic species (crops and fallows) (amount to 50 species).

Ammi majus	*Coronilla scorpioides*	*Hirscheldia incana*
Ammi visnaga	*Cynanchum acutum*	*Hypericum triquetrifolium*
Anagallis arvensis	*Cynara cardunculus*	*Launaea nudicaulis*
Anagallis foemina	*Cynodon dactylon*	*Lolium rigidum*
Anthemis arvensis	*Cyperus esculentus*	*Matthiola longipetala*
Avena sterilis	*Cyperus rotondus*	*Medicago polymorpha*
Beta macrocarpa	*Diplotaxis harra*	*Melilotus segetalis*
Beta vulgaris	*Diplotaxis simplex*	*Rapistrum rugosum*
Brassica arvensis	*Ecballium elaterium*	*Raphanus raphanistrum*
Calendula arvensis	*Echium italicum*	*Ridolfia segetum*
Centaurea dimorpha	*Fumaria densiflora*	*Salsola kali*
Chrysanthemum coronarium	*Galium verrucosum*	*Setaria verticillata*
Chrozophora obliqua	*Gladiolus segetum*	*Sinapis arvensis*
Cirsium arvense	*Glaucium corniculatum*	*Sonchus oleraceus*
Convolvulus althaeoides	*Gypsophila pilosa*	*Thesium humile*
Convolvulus arvense	*Haplophyllum tuberculatum*	*Vaccaria pyramidata*
Convolvulus tricolor	*Heliotropium europaeum*	

Table 5. Main native invading nitratophytes (amount about 25 species).

Aizoon canariensis	*Halogeton sativus*	*Peganum harmala*
Aizoon hispanicum	*Hyoscyamus albus*	*Salsola longifolia*
Arisarum vulgare	*Hyoscyamus niger*	*Scolymus maculatus*
Ballota hirsuta	*Malva parviflora*	*Silybum eburneum*
Biarum bovei	*Marrubium alysson*	*Sisymbrium irio*
Calotropis procera	*Mercurialis annua*	*Sisymbrium marianum*
Chenopodium murale	*Mesembryanthemum crystallinum*	*Verbascum sinuatum*
Emex spinosus	*Mesembryanthemum nodiflorum*	

Table 6. Main range native invaders (amount about 50 species).

Arisarum vulgare	*Carlina involucrata*	*Imperata cylindrica*
Artemisia campestris	*Carlina lanata*	*Lycium arabicum*
Asphodelus cerasifer	*Carthamus lanatus*	*Mesembryanthemum*
Asphodelus fistulosus	*Centaurea calcitrapa*	*nodiflorum*
Asphodelus microcarpus	*Cistus monspeliensis*	*Onopordon arenarium*
Asphodelus tenuifolius	*Cleome amblyocarpa*	*Onopordon nervosum*
Astragalus armatus	*Diplotaxis harra*	*subsp. platylepis*
Atractylis serratuloides	*Dittrichia graveolens*	*Onopordon espinae*
Biarum bovei	*Dittrichia viscosa*	*Peganum harmala*
Brachypodium distachyon	*Eryngium campestre*	*Silene colorata*
Bromus rubens	*Eryngium dichotomum*	*Silybum eburneum*
Calycotome villosa	*Eryngium tricuspidatum*	*Stipa capensis*
Carduncellus choulettianus	*Eryngium triquetrum*	*Thapsia garganica*
Carduncellus eriocephalus	*Foeniculum vulgare*	*Thymelaea hirsuta*
Carduncellus mareoticus	*Haloxylon salicornicum*	*Waronia saharae*
Carduncellus pinnatus	*Haloxylon scoparium*	*Withania sommifera*
Carduus gaetulus	*Hertia cheirifolia*	*Zilla spinosa*
Carlina corymbosa	*Hordeum murinum*	

124

of those, only a few are considered very 'dangerous weeds' at the present time: *Cuscuta epithymum, Oxalis pes-caprae, Echinochloa oryzoides.*

A few others may be labelled 'agressive invaders': *Heliotropium curassavicum, Aster squamatus, Nicotiana glauca, Blackiella inflata, Conyza bonariensis, Xanthium spinosum.* All those latter have American origin, except *Blackiella* which is Australian.

The native 'invading' species are far more numerous: about 133 (3% of the flora). Most of those species are either:

(i)   Cultigenic and segetal species (Table 4)
(ii) Range invaders (unpalatable) whose expansion results from overstocking (Table 5)
(iii) Nitratophytes linked to the piosphere (Table 6)

Some of those species, ignored by livestock, are more or less toxic or have anti-herbivore chemical repellents or are spiny hence difficult to tackle (many thistles).

The ecological status of the expanding species

In each case, we have retained the most frequent situation, but this may substantially vary for a given species that may alternately behave as ruderal, cultigenic or else. Information is cross-checked in Table 7.

The figures in Table 7 are rather surprising even when is considering the relative subjectivity of such concepts as ruderal or cultigenic. Introduced species (exotics) preferentially settle in environments where they undergo weak or moderate competition from species favoured by man (along roadsides, wasteland, etc.). Although present on farmland, these introduced species are virtually absent on rangelands.

Table 7. Ecological status of invading species in Northern Africa (%).

	Ruderal (wasteland)	Cultigenic (cropland)	Range invaders	Miscellaneous
Introduced species	72	22	–	6
Native species	30	38	28	4

Bioclimate and expanding species

Many species are present in several types of Mediterranean bioclimate. Table 8 shows the percentage of species from Table 3 in the main bioclimatic zones of the Maghrib; 75% of the expanding species are found in the Subhumid, Semi-Arid and Arid bioclimatic zones undergo the heaviest human interference: expansion of cultivation at the expense of forest, shrublands and steppe, intense

Table 8. Bioclimate types and expanding species (%).

	Humid	Sub-humid	Semi-Arid	Arid	Saharan Hyper Arid	Azonal
Introduced species	13	23	31	29	2	4
Native species	17	23	21	22	8	2

overstocking, irrigation, greenhouse cultivation, fast growing urbanization, and so forth.

Biological types of expanding species

We have retained the commonest biological type for any given species; but this does not go without ambiguity as a given species may exist under more than one biological type. Such information is shown in Table 9.

Again, the above figures are rather surprising since nearly 45% of the exotics are shrubs and trees (Ch, NPh, Ph), against only 11% among the natives. On the other hand, the proportion of annual species (Therophytes) is high in both cases (45 and 50%).

Table 9. Biological types (Raunkiaer's system) of expanding species (%) (signification of Biological types symbols at the end of Table 2).

Biological	Ph	NPh	Ch	H	G	T	Hyd.	Miscel.
Introduced species	5	11	29	5	5	45	–	–
Native species	–	1	10	11	10	50	1	7

Discussion and conclusion

The analysis of a given parameter, however important it may be, cannot provide a fully satisfactory explanation of the facts recorded, without reference to the other factors. As a matter of fact, the relatively high number of expanding species in the Semi-Arid bioclimatic zone cannot be explained in the same way for the exotics and the natives, although in both cases therophytic biotypes are dominant. The exotics actually colonize ruderal environments where they play a pioneering role. Table 10 for instance examines the contingency between biotypes and ecological status.

This table shows that most exotics are either ruderal or cultigenic species. In the first case, most species are shrubs (Ph, NPh, Ch) while cultigenic species are primarily annual (Therophytes). Conversely, the native ruderals are rarely

126

Table 10. Distribution (%) of expanding species according to biological types and ecological status.

		Ph/NPh	Ch	H	G	T	Hyd.	Misc.
Ruderal	exotics	17	28	11	5	22		
	natives		2	7	2	19		1
Cultigenic	exotics					17		
	natives		1	6	4	25		2
Range invaders	exotics							
	natives	1		6	8	2	8		2
Miscellaneous	exotics							
	natives		1		1			2

shrubs (2%) but essentially herbs (Therophytes). On the other hand, shrubs are rare among expanding natives. Most of the native expanding shrubs are range invaders; those are usually unpalatable dwarf shrubs such as *Haloxylon salicornicum, H. scoparium* or spiny bushes (*Calycotome villosa*) or species poisonous to livestock (*Hypericum triquetrifolium, Thapsia, Ferula...*) whose density increases with range overstocking. Environments subject to invasion are chiefly wasteland (ruderal), cropland (cultigenic) and recently cleared land. In other environments, infestation results from the expansion of recognized noxious species such as *Oxalis pes-caprae*. Some wastelands have soils rich in nitrates due to an abundance of dung, a fact that favours the selection of some species which then may show a very strong vigour and vegetative development (*Scolymus, Silybum*, etc.). Cultivated land also offers a number of particular attributes. First tilling, particularly when often repeated, tends to increase soil and microclimatic homogeneity. It is, thus, evident that annual psammophytes may also colonize silty soils in southern Tunisia when these are frequently tilled (Telahigue *et al.* 1987). These expanding annual species are thus protected from competition by perennials because of the frequent tilling. On the other hand, segetal species also benefit from a selective advantage whenever their biological cycle is similar to that of the crop with which they share water and nutrients and are, therefore, more difficult for man to eradicate. They also produce large amounts of seeds (up to 200,000 seeds per plant of *Chrysanthemum coronarium*, according to Laumont) (Chabrolin 1934). These seeds may remain alive in the soil for many years, thus ensuring reinfestation over time. Agricultural tilling practices may also favour some species of weeds in some environments:

(i)   The absence of fertilizing in traditional cropping systems inevitably results in decreasing overall soil fertility and productivity down to a metastable low status. This low fertility corresponds in turn to a low rate of cover of the crops (cereals), to a proliferation of species indicative of low fertility status such as *Papaver rhoeas, P. dubium, Rumex bucephalophorus, Hypericum triquetrifolium...*, and to an increased sensitivity of crops to parasitic or hemi-parasitric species such as *Thesium humile.*

127

(ii) The high rate of weed seeds in the traditional systems of cereal seed production, processing and trading (no cleaning, no testing, no certification) ensures a permanent reinfestation of weeds in cropland. Ducellier (1920) counted 500 to over 10,000 weed seeds per kilogram of locally traded cereal seeds in Algeria. The situation in this respect may not be very different nowaways particularly in small farming operations, where seed production is carried out on the farm, most of the time without any cleaning, seed stocks being renewed from certified commercial seeds only at intervals of several years (5–10 or more).

(iii) The burning up of stubble favours the multiplication of weed species whose germination is enhanced by fire or simply fire-tolerant by reducing competition from fire sensitive species.

(iv) The utilization of herbicides favours tolerant species and populations that may have evolved towards tolerant types.

(v) The tillage system used (deep tilling versus chiselling), or non tillage systems by burying seeds or not, influence species which germinate in the dark or those that need light.

The arid steppic rangelands of Northern Africa, which are very ecologically and floristically very close to those of the Near and Middle East, must have undergone early introduction of invaders from the East (central and middle Asian steppes). This may have occurred since the Neolithic. As a matter of fact, a number of range invaders are common throughout the arid steppes of the Old World from Mongolia to Morocco and Spain (e.g., *Peganum harmala, Reaumuria, Artemisia, Cynanchum, Nitraria, Lycium, Tamarix, Arthrophytum, Haloxylon, Aeluropus*, etc.) with a rather large number of vicariant species from East to West (Le Houérou 1985). This is perhaps the reason why so few deliberately introduced range species have become naturalized (*Atriplex semibaccata* being a prominent exception together with *Blackiella inflata* introduced for range improvement by Trabut in 1895 (Trabut 1904), which turned out to be a noxious range invader). In fact, *Atriplex semibaccata* behaves as a ruderal species. Comparable facts are recorded for other tropical forage species: e.g., many species of *Stylosanthes* from South America introduced for pasture improvement in the Paleotropics and Australia actually behave as ruderal species in their Neotropic native habitat (Reid, pers. comm., 1976).

Herd and range management, continuous grazing, and permanent overstocking result in overgrazing and hence the prevalence of annual species over perennials. Establishment of annuals is much easier under such conditions, due to their fast early stages of development. They contrast in this respect with perennials whose early development stages are characterized by slow growth, hence high sensitivity to grazing, at least for the palatable species. Fast early growth also allows for annuals to escape various ecological stresses (water, cold); a well documented case being that of the Cheat Grass (*Bromus tectorum*) which thus eliminated Bluebunch wheat Grass (*Agropyrum spicatum*) from large tracts of lands in the northwestern U.S. (Harris 1967). The effects of some ecological factors (such as drought) may be enhanced by man thus

128

giving a selective advantage to species that are better adapted to this factor. This is usually the case for therophytes which are drought evading species.

In response to various kinds of agression by man, plant communities are sometimes able to regenerate to close to pristine conditions when the disturbing factor is discontinued, but in most cases regeneration is prevented by more or less continuous human interference and as a consequence native species are not able to reconquer the ground lost in their initial geographic area of distribution. The most tolerant or aggressive natives and/or exotics take advantage of the lessened competition resulting from anthropozoïc interference. They are thus made able to expand in environments which have been disturbed by man and stock. Flora and vegetation are then progressively and deeply altered.

Glossary of the technical terms used

Adventitious	Inadvertently, fortuitiously introduced.
Commensal	Sharing a same given habitat e.g. commensal to man = ruderal, commensal to cereals = segetal etc.
Cultigenic	Species common in and linked to cultivated fields and fallows; includes segetal species but not restricted to them.
Exotic	Originating from abroad, usually overseas.
Infesting	Species expanding at a fast rate 'in terms of density and frequency' tending to become an agricultural pest.
Introduced species	Not belonging to the native flora.
Invading	Species having an expanding status either in terms of geographical area or in terms of increasing frequence and density.
Native	Beloging to the natural flora = spontaneous.
Naturalized	Introduced species which reproduces naturally without the deliberate assistance of man and stock and hence are or tends to become a part of the local flora.
Nitratophilous and *Nitrophilous*	Preferring or linked to soils enriched in nitrogen particularly nitrites and nitrates (dung & waste).
Overgrazing	Grazing pressure in excess of the carrying capacity over prolonged periods.
Overstocking	Stocking rate above the carrying capacity over prolonged periods.
Piosphere	Waste land with abundance of livestock dung around e.g. watering points, villages, etc.
Ruderal	Plant species commensal to man usually frequent and abundant in and around human settlements.
Segetal	Plant species common in cereal crops and tied to this habitat = weeds in cereal fields.
Spontaneous	Native, belonging to the local flora.

References

Abdul Wahab H.H. 1954. Les steppes tunisiennes (région de Gammouda) pendant le Moyen-Age. *Cahiers de Tunisie* 2: 5–16.
Arnoulet F. 1965. Notes sur l'histoire de l'agriculture en Tunisie. *I.B.L.A.* Tunis 28 (112): 401–418.

Babou, 1905. Projet de fixation de dunes de Bizerte. *Revue Horticole Tunisienne*: 93-136.

Baraban L. 1887. A travers la Tunisie. Etude sur les oasis, les dunes, les forêts, la flore et la géologie. Rothschild, édit. Paris, 227 p.

Battandier A. 1904. Modifications de la flore atlantique, acquisitions, extinctions, plantes intermittentes. *Bull. Soc. Bot. Fr.* 51: 345-350.

Bert J. 1886. Note sur les essences australiennes introduites en Algérie. *Bull. Min. Agric.* V, 6: 436-442.

Berte A. 1892. Flora invernale nei dintorni della citta di Tunisi. *Riv. Ital. Sc. Nat.* 12: 21-22; 39-40; 55-56.

Boitel, 1923. Présence de l'*Heliotropium curassavicum* L. à Monastir (Tunisie). *Bull. Soc. Hist. Nat. Afr. du Nord*, 14 (2), 65 p.

Bonnet G. & Baratte G. 1896. Cataloque raisonné des plantes vasculaires de la Tunisie. Imprimerie Nat. Paris, 519 p.

Boulos L. 1967. On the weed flora of Aswan, Egypt. *Botiniska Noticer* 120: 368-372.

Boulos L. & El-Hadidi M.N. 1968. Common weeds in Egypt. Dar El Maaref Press, Cairo, 160 p.

Boulos L. & El-Hadidi M.N. 1984. The weed flora of Egypt. *American Univ. Press.* Cairo, 178 p.

Brullo S. 1985. L'ordre *Brometalia rubenti-tectori* en Cyrénaïque septentrionale. Colloques Phytosociologiques. XII. Les végétations nitrophiles et anthropogènes. Bailleul, 1983: 269-281.

Brun A. 1979. Recherches palynologiques sur les sédiments du Golfe de Gabès: résultats préliminaires. *Géologie Méditerranéenne.* La Mer Pélagienne 6 (1): 247-264.

Brun A. 1983. Etude palynologique des sédiments marins holocènes de 5 000 BP à l'actuel dans le Golfe de Gabès (Mer Pélagienne). *Pollen et Spores* 25 (3-4): 437-460.

Burollet P.A. 1927. Le Sahel de Sousse. Monographie phytogéographique. *Ann. Serv. Bot. Agron. de Tunisie* 4, 2: 1-270.

Burollet P.A. & Boitel, 1921. Présence de l'*Heliotropium curassavicum* sur un point de la côte orientale tunisienne. *Bull. Soc. Hist. Nat. Afr. du Nord*, 12: 178-179.

Cagnat R. 1896-1912. L'Armée romaine d'Afrique et l'occupation militaire de l'Afrique sous les empereurs. Leroux Edit., Paris, 1 vol., 432 p.

Cagnat R. 1914. la frontière militaire de la Tripolitaine à l'époque romaine. *M.A.T.* XXXIV: 77-109.

Cagnat R. 1916. L'annone d'Afrique. *M.A.T.* XI, 247 p.

Camps G. 1961. Aux origines de la Berbérie, Massinissa ou les débuts de l'histoire. *Libyca* VIII: 1-320.

Camps G. 1974. Les civilisations préhistoriques de l'Afrique du Nord. Doin édit., Paris.

Camps-Faber N. 1953. L'olivier et l'huile dans l'Afrique romaine. *Publi. Servic. des Antiquités*, Alger. 95 p.

Chabrolin Ch. 1934. Les mauvaises herbes. *Ann. Serv. Bot. et Agr. de Tunisie* 11: 5-64, 12 pl. photos.

Chabrolin Ch. 1934. Monographie d'une santalacée: le *Thesium humile. Ann. Serv. Bot. et Agron. de Tunisie* 11: 65-192, 18 pl..

Chabrolin Ch. 1934. Les graines d'*Oxalis cernua* Thung. en Tunisie. *Bull. Soc. Hist. Nat. Afr. du Nord* 25: 396-398.

Chevalier A. 1927. Le geranium rosat en Afrique du Nord. *Rev. Bot. Appl. et d'Agric. Colon.* 7 (68): 273-275.

Chevalier A. 1932. Les productions végétales du Sahara et de ses confins Nord et Sud. Passé, présent, avenir. *Rev. Bot. Appl. et Agron. Colon.* 12 (133-134): 669-919.

Chevalier A. 1938. Le Sahara, centre d'origine de plantes cultivées *Société de Biogeographie* 6. La vie dans la région désertique nordtropicale de l'Ancien Monde: 307-322.

Chevalier A. 1939. Les origines et l'évolution de l'agriculture méditerranéenne. *Rev. Bot. Appl. et Agric. Trop.* 19 (217-218): 613-662.

Chevalier A. 1947. Nouvelles remarques sur les Acacias africains du groupe *A. seyal. Rev. Internat.*

Bot. Appli. et Agric. Trop. 27 (301-302): 470-483.

Chevalier A. 1947. Les jujubiers ou *Ziziphus* de l'Ancien Monde et utilisation de leurs fruits. *Rev. Internat. Bot. Appl. et Agric. Trop.* 27 (301-302): 470-483.

Chevalier A. 1947. Dossier sur les cactus (Opuntias). Espèces fruitières et fourragères. Espèces nuisibles. *Rev. Internat. Bot. Appl. et Agric. Trop.* 27 (301-302): 444-457.

Chevalier A. 1949. L'origine des plantes cultivées dans l'Afrique du Nord et le Sahara. Trav. Bot. dédiés à R. Maire, *Bull. Soc. Sc. Nat. d'Afrique du Nord.* Mém. h.s., Alger: 51-56.

Clark J.D. 1971. A re-examination of the evidence for agricultural origins in the Nile Valley. *Proceedings of the Prehist. Soc.* 37: 34-79.

Corippus J. 6th Century. De Bellis Libycis. Libri VII in Monumenta Germaniae Historia. Trad. Alix (1898-1902), *Revue Tunisienne.*

Cuénod A. 1913. Discours d'ouverture de la session de botanique: Notes sur la flore tunisienne. Congrès de l'A.F.A.S. Tunis 42e session: 282-293.

Cuénod A. Pottier-Alapetite G. & Labbe A., 1954. Flore analytique et synoptique de la Tunisie. S.E.F.A.N. Tunis, 1 vol., 287 p.

Desfontaines R.L. 1798-1800. Flora Atlantica. Paris, 4 vol.

Despois J. 1940. La Tunisie Orientale: Sahel et Basses Steppes. *P.U.F.* Paris (2e édition 1955), 550 p.

Draz O. 1954. Some desert plants and their uses in animal feeding (*Kochia indica* Wight; *Prosopis juliflora* D.C.). *Public. de l'Institut du Désert d'Egypte* n° 2, 95 p.

Dubuis A. & Faurel L. 1956. Notes sur les *Kochia* observés dans le Nord de l'Afrique. *Bull. Soc. Hist. Nat. de l'Afr. du Nord*, 48: 471-493.

Dubuis A. & Faurel L. 1957.- Notes de floristique nord-africaine: I. *Bull. Soc. Hist. Nat. de l'Afr. du Nord*, 48: 471-493.

Dubuis A. & Simonneau P. 1956. La végétation des rizières en Oranie. Gouv. Général de l'Algerie. Dir. Serv. Colon. et Hydraulique, 75 p., 1 carte.

Ducellier L. 1914. Note sur la végétation de l'*Oxalis cernua* Thung. en Algérie. *Rev. Gen. de Botan.* 25: 217-227.

Ducellier L. 1923. Le *Ridolfia* des moissons ou faux fenouil en Algérie. *Rev. Agric. de l'Afrique du Nord* 9: 538-540.

Ducellier L. & Maire R. 1923. Végétaux adventices observés dans l'Afrique du Nord. *Bull. Soc. Hist. Nat. Afr. du Nord* 14: 304-325.

Ducellier L. & Maire R. 1925. Végétaux adventices observés dans l'Afrique du Nord. (2e note). *Bull. Soc. Hist. Nat. Afr. du Nord* 16: 126-131.

Durand G. & Baratte G. 1910. Cataloque raisonné des plantes de Tripolitaine. Genève: 330 p., 20 pl.

El Bekri A.O. 1068. Description de l'Afrique du Nord. Traduct. De Slane. Jourdan édit. Alger (1913): 405 p.

El-Hadidi N. & Kosinova J. 1971. Studies on the weed flora of cultivated land in Egypt. 1. Preliminary studies. *Mitt. Bot. Staatssaml. München* 10: 354-367.

Erroux J. 1962. Les blés des oasis sahariennes. *Inst. Rech. Sahar., Univ. d'Alger*, Mém n° 7: 180 P.

Evreinoff V.A. 1947.- Le caroubier ou *Ceratonia siliqua* L. *Rev. Internat. Bot. Appl. et d'Agric. Trop.* 27 (299-300): 389-401.

Fauconnier R. & Bassereau D. 1970. La canne à sucre. G.P. Maisonneuve et Larose Editeurs, Paris, 468 p.

Franclet A. & Le Houérou H.N. 1971. Les *Atriplex* en Tunisie et en Afrique du Nord. Doc. Techn. n° 7. FO: SF/TUN 11, FAO-Rome: 249 p.

Francois L. 1928-1929. Les semences de plantes adventives dans les céréales. *Ann. Agron.* 45: 543-467; 46: 176-193.

Godard F. 1928. El Hamra (*Hypericum crispum*). *La Tunisie Agricole* 29: 1-3.

Gounot M. 1959. Contribution à l'étude des groupements messicoles et rudéraux de la Tunisie. *Ann. Serv. Bot. et Agron. de la Tunisie* 31: 282 p.

Gsell S. 1911. Le climat de l'Afrique du Nord dans l'antiquité. Alger.

131

Gsell S. 1913-1928. Histoire ancienne de l'Afrique du Nord, Hachette, Paris, 8 vol.

Gsell S. 1926. La Tripolitaine et le Sahara au 2e siècle de notre ère. Extr. de *Mém. de l'Acad. Inscrip. et Belles Lettres*, Paris.

Guillerm J.L. Le Floc'h E. Maillet J. & Boulet C. 1987.– The invading weeds within the western mediterranean basin (this volume).

Harris G.A. 1967. Some competitive relationships between *Agropyron spicatum* and *Bromus tectorum. Ecol. Monogr.* 37: 89–111.

Hendrik U.P. (edit.) 1972. Sturtevant's Edible plants of the World. Dover Public., Inc., New York, 686 p.

Hejny S. & Kosinova J. 1977. Contribution to synanthropic vegetation of Cairo. Caire Univ. *Herbarium Publ.* (7 & 8): 273–286.

Herodote, 470 B.C. Histoires. Traduct. P.E. Legrand. Les Belles Lettres édit., Paris (1968-1970), 10 Voi.

Ibn El Awam, 1158. Le livre de l'Agriculture (Kitab el Felaha). Bouslama Tunis (2e edit. 1977), 3 vol.: 657, 293 et 460 p.

Ibn Hauqal, Xth Cent. Description de l'Afrique du Nord, Paris (Trad. de Slane), 1 Vol.

Ibn Khaldoun, XIth Cent. Histoire des berbères et des dynasties musulmanes de l'Afrique septentrionale. Alger 1852-1856 (Trad. De Slane), 4 vol.

Ibn Khaldoun, XIth Cent. Prolégomènes. Paris 1863-1868 (Trad. De Slane), 3 vol.

Imam M. and Kosinova J. 1972. Studies on the weed flora of cultivated land in Egypt. 2. Weeds of Rice fields. *Bot. Jahrbücher für Systematik* 92 (1): 90–107.

Jahandiez E. Maire E. et Emberger L. 19331-1941.– Cataloque des plantes du Maroc. Imp. Minerva. Alger, 4 vol., 1179 p.

Jean Léon L'Africain, 1550 – Description de l'Afrique. (Nelle édition traduite de l'italien par A. Epaulard et annotée par Epaulard, A., Monod, Th., Lhote, H. et Mauny, R. 1959). Librairie d'Amérique et de l'Orient. Adrien Maisonneuve – Paris 2 tomes 630 p.

Keith H.G. 1970. A preliminary check list of the Libyan Flora. Minist. of Agric. Tripoli–Libya, 2 vol., 1047 p.

Kolendo J. 1984. Le rôle économique des Iles Kerkena au premier siècle avant notre ère (A propos du Bell. Afr. VIII et XXXIV). *Bull. archeol. du CTHS nouv. sér.*, 17 B: 241–249.

Kosinova J. 1972. On the weed flora. *Egyptian Bot. Soc.* Yearbook 2, 1–6.

Kosinova J. 1974. Studies on the weed flora of cultivated land in Egypt. 3. Distributional types. *Bot. Jahrb. Syst.* 94 (4): 449–458.

Kosinova J. 1974. Studies on the weed flora of cultivated land in Egypt. 4. Mediterranean and Tropical elements. *Candollea* 29: 281–295.

Kosinova J. 1975. Weed communities of winter crop in Egypt. *Preslia*, Praha 47: 58–74.

Kress H.J. 1977. Eléments structuraux 'andalous' dans la génèse de la Géographie culturelle de la Tunisie. Traduit de: *Marburger Geographishe Schriften* – Marburg (7-3), 32 p. ronéo.

Labbe L. 1950. Plantes spontanées de Tunisie à floraison estivale. *Bull. Serv. Bot. et Agr. Tunisie* (20): 3–26.

Labbe L. 1955. Contributions à la connaissance de la flore phanérogamique de la Tunisie. 5. Espèces subspontanées et naturalisées. *Bull. Soc. Sc. Nat. Tun.* 8 (1–2): 97–117.

Lagierre R. 1966. Le coton. G.P. Maisonneuve et Larose Editeurs, Paris, 306 p.

Laumont P. 1931. Les mauvaises herbes et leurs graines en Algérie. *Rev. Agric. de l'Afrique du Nord*: 262–270.

Laumont P. et Gueit M. 1953. Comportement en Algérie de *Kochia indica* Wight. *Bull. Soc. Agric. d'Agérie* (578): 1–16.

Lavauden L. 1928. La fixation et le reboisement des dunes de Bizerte. *Rev. des Eaux et Forêts*, LXVI (6): 351–361.

Le Floc'h E. 1983. Contribution à une étude ethnobotanique de la flore tunisienne. Imp. Officielle Tunis, 402 p.

Le Houérou H.N. 1959. Recherches écologiques et floristiques sur la végétation de la Tunisie méridionale. *Mém H.S. Inst. Rech. Sah. Alger*, 2 parties 510 p., pochette tabl. et cartes.

Le Houérou H.N. 1968. La désertisation du Sahara septentrional et des steppes limitrophes.

132

Ann. Alg. de Geogr. 3, 6: 2–27.

Le Houérou H.N. 1969. La végétation de la Tunisie steppique (avec références aux végétations analogues au Maroc, en Algérie et en Libye). *Ann. Inst. Nat. Rech. Agron. Tunis,* 1 vol. texte 624 p., 1 pochette tabl. et cartes.

Le Houérou H.N. 1981a. Long term dynamics in Arid Land vegetation and ecosystems of North Africa. In D.W. Goodall and R.A. Perry (eds.), *Arid Land Ecosystems: their structure, functioning and management.* Cambridge Univ. Press. London. *Internat. Biol. Progr.* Vol. 17: 375–384.

Le Houérou H.N. 1981b. Impact of man and his animals on mediterranean vegetation. In: F. di Castri, D.W. Goodall & R.D.L. Specht (eds.), *Mediterranean type shrublands.* Ecosystems of the world. Vol. 11, chapt. 25: pp. 479–521.

Le Houérou H.N. 1986. The desert and arid zones of Northern Africa. In: M. Evenari an D.W. Goodall (eds.), *Hot deserts and Arid shrublands.* Ecosystems of the World, Elsevier, Amsterdam. Vol. 12b, Chap. 4: pp. 101–147.

Le Houérou H.N. 1987. Aperçu écologique des déserts chinois. *Soc. Biogéogr.* Paris, 63 (2): 35–69.

Leone g. 1924. Consolidamento e imboschimento della zone dunose della Tripolitania. *L'Afrique Colon.* XVIII, 9: 299–308.

Main F. 1927. Culture du Ricin en Algérie. *Rev. Bot. Appl. et Agric. Colon.* 7 (68): 287.

Maistre J. 1964. Les plantes à épices. G.D. Maisonneuve et Larose Editeurs, Paris, 289 p.

Martinier M. 1928. Note sur les assolements. *La Tunisie Agricole,* 29 (6): 1–11.

Mathez J. et Sauvage Ch. 1970.– Nouveaux matériaux pour la flore du Maroc. 1er fascicule. *Bull. Soc. Sc. Nat. et Phys. du Maroc* 49: 81–107.

Mollard G. 1950. L'évolution de la culture et de la production du blé en Algérie de 1830 à 1939. Editions Larose, Paris, 127 p.

Monjauze A. et Le Houérou H.N. 1965. Le rôle des Opuntia dans l'économie agricole Nord-africaine. *Bull. Ec. Nat. Sup. Agr. Alger,* 8–9: 1–85.

Nisard M. 1877. Les agronomes latins: Caton Columelle, Palladius, Varon. Firmin Didot edit., Paris, 650 p.

Perrot E. 1927. Principales plantes adventices nuisibles aux céréales dans la région de Batna. *Rev. Agric. de l'Afrique du Nord:* 456–461.

Picot G. 1928. La culture et l'utilisation des eucalyptus dans l'Afrique du Nord (d'après le Dr. L. Trabut). *Rev. Appli. et Agron. Colon.* 8 (86): 715.

Platon, 380 BC. Oeuvres complètes. Timée-Critias. (Trad. A. Rivaud). Les Belles Lettres édit. Paris (1970): 274 p.

Pline l'Ancien, 60 BC. Histoire Naturelle (Trad. Littré). Collect. Nisard. Paris (1877), 2 vol.

Poletaeff N. 1953. L'introduction de plantes nouvelles par le Service Botanique et Agronomique de Tunisie. *Rev. de la Soc. Hort. Tun.* 50, nelle sér. (2): 18.

Pottier-Alapetite G. 1952. Comment les plantes voyagent-elles? *Bull. Soc. Sc. Nat. de Tunisie* 5: 65–72, 2 pl.

Pottier-Alapetite G. 1979–1981. Flore de la Tunisie. Imprim. Offic. Tunis, 2 vol., 1190 p.

Précheur-Canonge T. s.d. La vie rurale en Afrique romaine d'après les mosaïques, P.U.F./ *Publications de l'Univ. Tunis, Fac. Lettres,* 1ère série, Vol., 6, 97 p.

Quézel P. et Santa S. 1962–1963. Nouvelle flore de l'Algérie. CNRS édit. Paris, 2 vol., 1170 p.

Raven P.H. 1977. The Californian Flora. In: M.G. Barbour & J. Major. *Terrestrial vegetation of California.* J. Wiley and Sons, N.Y. pp. 109–138.

Reed C.A. 1969. The pattern of animal domestication in the prehistoric Near East. In: P.J. Ucko and G.W. Dimbley (eds.), *The domestication and exploitation of plants and animals.* Duckworth, London. pp. 361–380.

Reille M. Triat-Laval H. & Vernet J.L. 1980 Les témoignages des structures actuelles de végétation méditerranéenne durant le passé contemporain de l'action de l'homme. Colloque Fondation L. Emberger, Montpellier 9–10 avril 1980: La mise en place de l'évolution et la caractérisation de la flore et de la végétation circumméditerranéenne 6b, 1 à 11.

Romains J. 1906. L'Horticulture en Tunisie. Notes sur quelques plantes nouvelles ou peu connues

133

introduites récemment en Tunisie. *Bull. Soc. Hort. Tunisie* 5 (19): 63:65.

Sallustius, 40 BC. *Bellum Jugurthinum* (Trad. A. Ernoux) Edit. Belles Lettres Paris (4e edit. 1960).

Saarisalo A. 1968. The weed vegetation of flower beds and irrigated vegetations in Cairo and some adjacents Province. *Ann Bot. Fennici.* 5: 1906–1968.

Sauvaigo E. 1899 – Ennumération des plantes cultivées dans les jardins de la Provence et de la Ligurie. Flora Mediterranea Exotica, 1ère partie 316 p., 2ème partie 96 p.

Schnell R. 1957. Plantes alimentaires et vie agricole de l'Afrique Noire. Editions Larose, Paris, 223 p.

Simpson N.D. 1932. A report on the weed flora of irrigated channels in Egypt Cairo. Goverment Press.

Tackholm V. 1974. Student's flora of Egypt. Coop. Printing. Cairo Univ. Cairo (2nd edit.), 888 p.

Tadros T.M. & Atta B.A.M. 1958. The plant communities of barley fields and uncultivated desert area of *Mareotis* (Egypt). *Vegetatio* 8: 161–175.

Tanji A. Boulet C. & Hammoumi M. 1985. Etat actuel de l'infestation par *Solanum elaeagnifolium* Cav. pour les différentes cultures du périmètre de Tadla (Maroc). *Weed Research* 25: 1–9.

Telahigue T. Floret Ch. & Le Floc'h E. 1987. Successions post- culturales en zone aride tunisienne. *Acta Oecologica, Oecol. Plant.*, 8, 22, 1: 45–58.

Thellung A. 1922. Encore *Erigeron kazwinshianus* DC. var. *mucronatus* (DC) Aschers. Le Monde des plantes (20) – 135 (3e série): 6–7.

Tissot P. 1939. Assolement du cotonnier au Maroc. *Rev. Bot. Appl. et Agric. Trop.* 19 (217–218): 704–706.

Trabut L. 1904. Naturalisation de deux *Atriplex* australiens dans le Nord de l'Afrique (*Atriplex halimoides* Lindl. *A. semibaccata* R. Br.). *Bull. Soc. Bot. Fr.* 51: 105–106.

Trabut L. 1922. Naturalisation d'un Eucalyptus en Algérie *Eucalyptus algeriensis* Trab. *Bull. Soc. Hort. Tun.* 20 (164): 42–43.

Vanden Berghen C. 1979. Quelques groupements végétaux nitrophiles reconnus à Djerba (Tunisie méridionale). *Doc. Phytosociologiques*, Lille, N.S., Vol. IV: 923–928.

Vanden Berghen C. 1980. Observations sur la végétation de l'Ile de Djerba (Tunisie méridionale). Note 4: La végétation adventice des moissons. *Bull. Soc. Roy. Bot. Belg.* 113: 33–44.

van Zeist W. 1980. Aperçu sur la diffusion des végetaux cultivés dans la région méditerranéenne. *Colloque Fondation L. Emberger* Montpellier 9–10 avril 1980: La mise en place, l'évolution et la caractérisation de la flore et de la Végétation circumméditerranéenne 9, 1 à 19.

Vavilov M.I. 1951. The origin variation. Immunities and breeding of cultivated plants. *Chron. Bot.* 13: 1–366.

Wendorf F. Schild R. El-Hadidi N. Close A.E. Kobusiewicz M. Wieckowska H. Issawi B. & Haas H. 1979. Use of barley in the Egyptian late paleolithic. *Science* 205 (4413): 1341–1347.

Whyte R.O. 1961. Evolution de l'utilisation des terres en Asie du Sud-Ouest. In: D. Stamp edit. History of Land-use in Arid zones. *Arid zone Research* 18: 65–132. UNESCO-Paris.

Xenophon, 400 BC. Economiques (Trad. P. Chantraigne) Les Belles Lettres édit. Paris (1971), 120 p.

Zeuner F.E. 1963. A history of domesticated animals. Hutchinson-London, 560 p.

Zohary D.D. 1973. The origin of cultivated cereals and pulses in the Near East. In: J. Wahrman and K.R. Lewis (eds.), *Chromosomes to-day 307–320.* J. Wiley and Sons, N.Y.

Zohary D.D. & Spiegel-Roy P. 1975. Beginning of fruit growing in the Old World. *Science* 187: 319–327.

8. Invasions of adventive plants in Israel

AMOTS DAFNI and DAVID HELLER

Abstract.

One hundred twenty-three adventive wild species occur in the flora of Israel: 42% are of tropical origin, 22.7% are North American. The geographical distribution suggests that about two thirds of the species reached Israel through neighbouring countries, while only one third arrived directly from their countries of origin. Four patterns of population dynamics can be recognized: accidental species, species with limited distribution, colonizing species, and species penetrating into natural habitats. There is insufficient information about the remaining species. About 30 species are widespread noxious weeds which now have economic significance in Israel; 20 others are known to have similar ecological tendencies in other countries. These species soon may become aggressive weeds in Israel.

Introduction

Almost every country contains at least some adventive species, mostly due to human activities which cause the importation of alien species (Good 1964).

Disturbed habitats generally supply favourable conditions for colonizing species (Harper 1977). Agricultural ecosystems are therefore ideal for the establishment of migrating alien species (Baker 1972). Many of these species are herbaceous annuals, many of which are regarded as weeds (Mulligan 1965).

In recent decades the most serious source of noxious weeds (especially in developing countries) is through accidental introductions from other floristic regions, especially North America, Australia, and South Africa (Parker 1977). A survey of the most important weeds of the world (Holm *et al.* 1977) reveals that most are alien species which were introduced mainly by activities of man.

The rapid development of new settlements in Israel in areas that were never before cultivated, opened habitats for colonization. The shift from extensive to intensive cultivation accompanied by heavy use of herbicides has caused considerable changes in the weed flora. Local plants decreased and new niches

---

F. di Castri, A. J. Hansen and M. Debussche (eds.), Biological Invasions in Europe and the Mediterranean Basin. 135–160. © 1990, Kluwer Academic Publishers, Dordrecht.

136

were opened for adventive plants. The phenomenon of replacing local weeds by cosmopolitan ones is well documented (Holm *et al.* 1977, Parker 1977, King 1966). In Israel this shift was found to be very rapid due to the high acceleration of agricultural development in a relatively short period.

The introduction of new crops (cotton has been grown for only 30 years), highly sophisticated irrigation systems (e.g. trickle pipes), and the massive enlargement of greenhouses and plastic cover tunnels offer new opportunities for adventive weeds.

The intensive exchange of agricultural material especially seed imports from the United States (e.g. cotton, sorghum and forage plants), has caused direct introduction of weeds, probably by means of seed impurities. In addition, exchange of agricultural products with the neighbouring countries has caused an import of adventive weeds which were known in these countries long before they reached Israel.

Adventive species (of which many are colonizers) often provide natural experiments on the structure and dynamics of populations. Basic problems of evolutionary processes, such as dispersal, establishment, r- and k-selection, weediness, colonization and competitive ability can be exemplified by adventive species. The adventive flora offers rewarding and undesigned opportunities to examine concepts of population biology under natural and seminatural conditions.

Another aspect of the dramatic invasions by alien plants in Israel (especially weeds) can be inferred through the botanical records. Eig (1927) in his milestone work 'On the vegetation of Palestine' mentioned only four adventive species as common weeds (*Amaranthus graecizans, Setaria verticillata, Oxalis cernua* and *Datural metel (= D. innoxia)*. Zohary (1941) in his survey on the weeds of Israel enumerated less than a dozen adventive species. Even in his later publications (Zohary 1962, 1981), adventive species are hardly mentioned. Today the most serious weeds, especially in irrigated fields, and the most common roadside plants are adventive species. Local plants are mostly minor weeds and show an impressive regression along roadsides and waste places. The worldwide trend of the replacement of local weedy flora by cosmopolitan aggressive plants (Holm *et al.* 1979) is well documented and Israel is not a unique case in this respect. The accelerated changes, especially during the last four decades are a manifestation of the rapid development of highly sophisticated agriculture which creates new open habitats for colonizing species.

The present study us an up-to-date revision of the adventive flora of Israel. In the last list (Dafni & Heller 1980), we included 73 species. The completion of Flora Palaestina Vol. IV (Feinbrun-Dothan 1986) and an extensive survey by the authors have yielded a list of 122 adventive species, which comprise 4.3% of the local flora. Special attention has been devoted to changes in population dynamics during the last decade.

137

Materials and methods

The identification of a plant species as an adventive species has been based primarily on local floras (Boissier 1867–1888, Post 1933, Eig *et al.* 1948, Zohary 1966, 1972, Feinbrun-Dothan 1978, 1986). For each species the relevant data about regional origin, world distribution, and biology were compiled from literature. The evidence for the first appearance of a species in Israel was collected mainly from the Hebrew University Herbarium in Jerusalem (HUJ). It is recognized that a certain period has to be allowed between the first invasion and the first collection. Nevertheless, this approach represents perhaps the most accurate documentation. Data on the population dynamics and species autecology were collected on field excursions between 1970 and 1988.

The inventory of the adventive flora

The Adventive Flora of Israel with extensive new observations and additional data is given in Table 1. Some general aspects are discussed in the following paragraphs.

Geographical origin and Immigration routes

Phytogeographical analyses (first column in Table 1) show that 10.9% of the species are South American, 42% tropical (19.3% neotropical, 22.7% paleotropical), 22.7% North American, 5.9% Australian and 10.9% of other origins.

The high frequency (53%) of tropical neophytes is obviously due to the enlargement of irrigated habitats (fields, gardens, greenhouses, plastic tunnels etc.). The combination of the local warm climate and the high soil humidity brings about a simulation of tropical habitats as pointed out by Baker (1972). This does not sustain the view (Yannitsaros & Economidou 1974) that climate is the determining factor in the establishment of adventives. Only three species (*Antirrhinum siculum, Anacyclus radiatus* and *Artemisia arborescens*) are of W. Mediterranean origin. In Israel the artificial ecological conditions have been more important than the climate for the establishment of adventive species. The North American neophytes are living evidence for the close agricultural connections between the United States and the Levant, (mainly through the import of cotton, corn, sorghum and forage plants); many of these species have appeared only during the past three decades. All the Australian neophytes were deliberately introduced as forage plants, have escaped cultivation, and now mainly appear along roadsides.

In order to reconstruct the immigration routes of adventive species, the steps of their arrival have to be dated. When it can be shown that a neophyte appeared first in a neighbouring country and more recently in Israel, it is likely that immigration took place in that sequence, e.g. through the exchange

Table 1. The adventive flora of Israel. Data on distribution, origin and ecological status. The origin and geographical ditribution data have been compiled from the following floras and sources: Americanos (1972), Boissier (1867-1888), Davis (1965-1985), Edgecombe (1970), Eig *et al.* (1948), Feinbrun (1978, 1986), Grunberg-Fertig (1966), Holm *et al.* (1977, 1979), Kosinova (1975), Meikle (1977, 1982), Parker (1973), Rechinger (1963), Täckholm (1974), Zohary (1966, 1972).

Species	Geograph-ical origin	First record from Israel	Habitat and crops	Distribution and expansion	Status as a weed	Presence in the Middle East	World distribution
Abutilon theophrastii Medik.	P	1942	F, Co	NC, E	M	T, Ir	NA, S. & E.Eu, Af, A
Achyranthes aspera L.	P	1900	F, R, O, G	N	m	S, L, E, A	M, A, SAf
Alternathera pungens Kunth	SA	1949	–	N	–	E	E.Af
Amaranthus albus L.	P	1879	F, V	C	m	All	T, ST
Amaranthus blitoides S. Wats.	NA	1949	F, G	C, E	M	All	Cos
Amaranthus chlorostachys Willd.	NA	1906	F, G, L	C	m	All	Cos
Amaranthus graecizans L.	NA	1906	F, V, Co	C	M	All	Cos
Amaranthus muricatus Gillies ex Moq.	SA	1964	W	Ca		All	C. & W.Eu
Amaranthus palmeri S. Wats.	NA	1957	F, S, V, I	C, Ce	M		NA, Eu
Amaranthus retroflexus L.	NA	1906	F, V	C	M	All	Cos
Amaranthus rudis Sauer	NA	1970	FP, H, C	NC, Ex	m		NA
Amaranthus spinosus L.	N	1960	?	Ca			W. & S.Af, SE. & E.As
Amaranthus viridis L.	?	1906	F, V	C	M	All	Cos
Anacyclus radiatus Loisel.	WM	1965	R, C	R, Ex	m		S.Eu
Anoda cristata (L.) Schlecht.	NA	1981	FP, F, Co	NC, Ex	m		NA
Antirrhinum siculum Mill.	WM	1889	Wa	NC	m		M.S.Eu
Araujia sericifera Brot.	SA	1976	O	NC, Ex	m, Ex		SA
Argemone mexicana L.	N	1953	Ca	R	–	S, L, E, A	Cos
Artemisia arborescens L.	WM	1875	Wa	R	–	T, S, J	WM
Aster subulatus Michx.	NA	1960	W, F, N, G	C, Ex	m	A, Iq	NA
Atriplex canescens (Pursh) Nutt.	NA	1974?	G	R	–	–	NA
Atriplex holocarpa F. Muell. (= Senniella spongiosa (F. Muell.) Aellen)	A	1972	W	R	–	E	A

Table 1. (Continued).

Species	Geographical origin	First record from Israel	Habitat and crops	Distribution and expansion	Status as a weed	Presence in the Middle East	World distribution
Atriplex nummularia Lindley	A	1975	G	R	–	Iq	A
Atriplex semibaccata R.Br.	A	1933	W	N	–	–	A
Atriplex suberecta Verdoorn (= *A. muelleri* Benth.)	A	1974	W	N	–	–	A
Bassia indica (Wight) A.J. Scott (= *Kochia indica* Wight)	A	1917	W	N	–	E, A$_2$	A, In
Bassia scoparia (L.) A.J. Scott (= *Kochia scoparia* (L.) Schrader)	A	1933	W	R		T, S, T$_2$	NA
Bidens pilosa L.	N	1954	F, G, N	C, Ex	m	E	Cos
Borago officinalis L.	M	1987	W	R	–	E, C, T	M
Bromus catharticus Vahl (= *B. unioloides* Kunth)	SA	1939	F, G	N	m	E	SA, NA, N.Af, E.As
Cassia obtusifolia L.	P	1985	FP	R	–	–	SA, NA, N
Cenchrus echinatus L.	N	1970	W, O, V	GEX	m		A, E, Af, E, As
Cenchrus incertus M.A. Curtis	NA	1953	W, F, V	Ex	m		SA, NA
Chenopodium ambrosioides L.	N	1906	F, W	C		All	Cos
Chenopodium ficifolium Sm.	E	1988	R	R		E, T, Ir, Iq	Eu
Chloris gayana Kunth	P	1933	F, G, N, R	C, Ex	m	E, A, Ir	Cos
Chloris virgata Swartz	P	1959	F, G	W, Ex	m	Iq, Ir, E, A	Cos
Conyza albida Willd. ex Spreng.	SA	1957	F, W, G, R	C, Ex	m	L	Cos
Conyza bonariensis (L.) Cronq.	SA	1896	F, W, I, G, N	C, Ex	M	All	Cos
Conyza canadensis (L.) Cronq.	NA	1939	F, W, I, G, N	C, Ex	M	All	Cos
Cotula anthemoides L.	P	1960	?	Ca	–	E, Ar	S.W.As
Cuscuta campestris Yuncker	NA	1951	F, H	Ca	m	–	Eu, M
Dalbergia sissoo Roxb.	In	1985	F	R	–	Iq, Ir	T, ST
Datura ferox L.	TA	1952	F, CO	R, Ex	M		SAf, A, In, Eu
Datura innoxia Mill.	N	1896	W, CO	C, Ex	M	All	Cos
Datura stramonium L.	NA	1912	F, W	C, Ex	–	All	TA, Af, Am
Dinebra retroflexa (Vahl) Panz.	P	1922	F, H	N	–	Iq, E, A	W.Ind, C.Af, SAf

Table 1. (Continued).

Species	Geograph-ical origin	First record from Israel	Habitat and crops	Distribution and expansion	Status as a weed	Presence in the Middle East	World distribution
Diplotaxis tenuifolia (L.) DC.	E	1941	G	R	–	L, T	Eu
Dodonaea viscosa L.	P	1938	R	R	–	E, A, Iq, Ir	T, ST
Eclipta prostrata (L.) L.	NA	1917	H	Ca	–	T, Iq, Ir	Cos
Eleusine indica (L.) Gaertn.	P	1935	F, G, I	C.Ex	m	Ir, E, Ar	Cos
Eragrostis echinochloidea Stapf	SAf	1980		R	–	–	SAf
Eragrostis japonica (Thunb.) Trin.	P	1927	H	R.Ex	–	A, E, Iq	P, ST, In, As
Eragrostis palmeri S.Wats.	N	1970	R	R.Ex	–		N, NA
Eragrostis prolifera (Swartz) Steud.	P	1953	?	R	–	–	
Eragrostis sarmentosa (Thunb.) Trin.	P	1923	H	N	–		T, As, SAf
Euphorbia geniculata Ortega	N	1941	F, G, Co	C, Ex	M	L, E, A, C, Ar	NA
Euphorbia hirta L.	N	1924	G, I	C, Ex	M	Sy, L, Iq, Ar	T, A, As
Euphorbia lasiocarpa Klotzsch	N	1972	G	N, E	m		N
Euphorbia serpens Kunth	NA	1940	G, C	N, Ex	m		NA
Euphorbia nutans Lag.	NA	1940	G	Ca	–		S. & C., Eu, S. & N., Am
Euphorbia prostrata Ait.	N	1948	W, F, G, I,	N, Ex	M		W, M, E, Af, NA
Fallopia convolvulus (L.) A. Löve (= *Polygonum convolvulus L.*)	Eu	1982	F, N	R, Ex		±r.E, S, L, T	P, SAf, A
Galenia secunda (L.f.) Sonder	SAf S.A	1970	R	R	–	–	SAf, SA
Galinsoga parviflora Cav.	N	1950	F, G, V, N	N, Ex	m	–	Cos
Gisekia pharnacioides L.	P	?	H	R	–	Ir, E, Ar	E.As, Af, W.Eu
Gnaphalium luteo-album L.	?	1966	R.H.	R, Ex	–	T, Ir, Iq, E, Ar	N. & W.Eu, C.Am, NA
Hedysarum coronarium L.	S.Eu	1986	R	R		E, L	S.Eu, C. & W.M
Helianthus annus L.	NA	1977	W, C, V, R, FP	N, Ex	m	All	NA
Heterotheca subaxillaris (Lam.) Britt. & Rusby	NA	1977	W, G, R, NC	C, Ex	M	–	NA
Ipomoea cairica (L.) Sweet	P	1875	H	R	–	S, L, A, E	Tr, P, As, Af
Ipomoea hederacea Jacq.	N	1984	F	R	–	–	NA

Table 1. (Continued).

Species	Geographical origin	First record from Israel	Habitat and crops	Distribution and expansion	Status as a weed	Presence in the Middle East	World distribution
Ipomoea indica (Burm.) Merr.	P.N.	1962	S	R	–	E	
Ipomoea pes-caprae (L.) R.Br.	P	1962	C	R	–	E, A	Cos
Ipomoea triloba L.	N	1984	FP, CO, H	R	–	–	NA, Au
Lantana camara L.	N	?	G	R	–	All	T, ST, Eu
Maireana brevifolia (R.Br.) P.G. Wilson (= Kochia brevifolia R.Br.) P.G. Wilson	A	1948	W	NC, E	–		Au
Maireana sedifolia (F.Muell.) P.G. Wilson	Au	1980?	G	R			Au
Mantisalca salmantica (L.) Brig. & Cavill.	M	1966	W	R	–	E, C, T	M
Momordica balsamina L.	P	1884	F	N	–	E, A	N, D
Nicandra physaloides (L.) Gaertn.							
Nicotiana glauca Graham	SA	1898	W, Wa	C, E	–	T, S, L, E	M, Au, NA, SA
Oenothera drummondii Hook.	NA	1912	S	C, NC	–	–	NA, WM, S. & E.As, Au
Oenothera laciniata Hill	NA	1985	G	R			Eu
Oenothera rosea L'Her.	NA	1890	N, F	R, Ex			NA
Opuntia ficus-indica (L.) Mill.	SA	1894	W, R	C	–	All	Cos
Oxalis corniculata L.	P	1906	G, N, V, O, Co	C, E	M	All	Cos
Oxalis europaea Jordan	Eu	1983	G, N	R, Ex	m	All	NA, E.As, Eu
Oxalis pes-caprae L.	SAf	1906	G, F, W, O, N	C, E	M	T, L, S, C, E	M, S.Eu, NA
Panicum antidotale Retz.	In	1948	W, Ir	R	m	Iq	NA
Panicum capillare L.	NA	1930	W, L, G, N	R, Ex	m		NA, E, Eu
Panicum maximum Jacq.	P	1970	F	R	m	–	Cos
Panicum miliaceum L.	In	1926	F	Nc, N	–	T, L, Ir, C	C.As, N.Af, In, Eu, Au, SAf, NA
Parkinsonia aculeata L.	N		R	E			Af, ST
Parthenium hysterophorus L.	N	1980	FP, Co, S	R. Ex	m	–	NA, In, Au, SAf
Paspalum dilatatum Poir.	N	1970	F, W, G, I, O	N, Ex	M	T, Iq, E	Cos

Table 1. (Continued).

Species	Geograph-ical origin	First record from Israel	Habitat and crops	Distribution and expansion	Status as a weed	Presence in the Middle East	World distribution
Paspalum paspaloides (Michx.) Scribn.	NA	1939	F, G, N, I, L	C	M	Iq, E	S. & W.Eu, E.As
Passiflora morifolia Masters	N	1984	O	R	–	–	NA, Au, SAf
Phyllanthus rotundifolius Klein ex Willd.	P	1978	G	R	–	E, Ar	P
Physalis angulata L.	N	1977	F, G	R, Ex	m	E	NA, N, S. & E.Af
Phytolacca americana L.	NA	1898	W	R	–	T, S, L	NA
Ricinus communis L.	P	1894	W.R.	N	–	All	S.Eu
Rorippa prostrata (J.P. Bergeret) Schinz et Thell.	Eu	1984	G	R			C. & E.Eu
Sesbania sesban (L.) Merill	P	?	F.R.	R		A.E, Iq	T, ST
Setaria glauca (L.) Beauv.	P	1929	Ir, G, F, Co	N	m	All	Cos, ST
Setaria verticillata (L.) Beauv.	Eu	1911	G, F, Ir	N	m	All	Cos, ST
Setaria viridis (L.) Beauv.	Eu	1922	F	C	m	All	Cos
Sida acuta Brum.f.	P	1980	H, FP	R		E	Af, NA
Solanum capsicoides All.	NA	1977	G	R		–	NA
Solanum cornutum Lam.	N	1953	F, V	N	m	–	E. & C.Eu, N, NA
Solanum dulcamara L.	Eu	1983	H	R		S, L, T, Ar	Eu
Solanum elaeagnifolium Cav.	NA	1957	F, W, G, N, I	C, E	M	E, Ir, Ar	Au, SAf, In, SA
Solanum laciniatum Aiton	AN	1985				–	Eu
Solanum sarrachoides Sendtner	SA	1977		R		–	SA, NA, E.Eu, S
Sorghum virgatum (Hackel) Stapf	P	1922	H, Ir, Co	C, Ex	M	E	Af
Tagetes minuta L.	N	1970	F, W, O	R	m	E	E.Af, SAf, E.Eu, SA
Tetragonia tetragonioides (Pall.) O. Kuntze	A	1932	C, G	R	–	–	T, ST
Trianthema portulacastrum L.	SA	1967	F	R	–	E, Ar	SA, NA, In
Verbesina encelioides (Cav.) Benth. & Hooker f. ex A. Gray		1971	W	NC	–	E, Ar	Af, NA, C. & N.Eu

Table 1. (Continued).

Species	Geograph-ical origin	First record from Israel	Habitat and crops	Distribution and expansion	Status as a weed	Presence in the Middle East	World distribution
Wedelia glauca (Ort.) Hoffm.	SA	1968		Ac			SA
Xanthium italicum Moretti	N	1945	H	N	–	T, Ar	S.E.Eu
Xanthium spinosum L.	SA	1912	W, H, R	NC	–	All	Cos, ST
Xanthium strumarium L.	N, P	1921	W, F, CO	C, Ex	M	All	Cos, ST

Geographical origins and distributions		Habitats and crops		Distributions and expansion		Presence in the Middle East	
Af	Africa	C	Coastal sands	C	Common	A	Saudi Arabia
Au	Australia	Co	Cotton	Ca	Casual	All	All over
Cos	Cosmopolitan	F	Fields	Ex	Expanding, colonizing	C	Cyprus
Eu	Europe	FP	Fish ponds		species	E	Egypt
In	India (non-tropical area)	G	Gardens	N	Not common	Iq	Iraq
M	Mediterranean	Gr	Grains	NC	Present in natural	Ir	Iran
N	Neotropical (Central and South	H	Wet or water habitats		communities	J	Jordan
	America)	I	Irrigated fields	R	Rare	L	Lebanon
NA	North America (non-tropical area)	L	Lawns			S	Syria
P	Paleotropical	N	Nurseries			T	Turkey
SA	South America	O	Orchards				
SAf	South Africa	S	Interior sands	Status of weediness		W	= West
ST	All the subtropics	V	Vegetables			E	= East
TA	Temperate Asia	W	Waste grounds	M	Major weed	S	= South
WM	West Mediterranean	Wa	Walls	m	Minor weed	N	= North
						C	= Central

144

of agricultural products. Examples are *Solanum elaeagnifolium, Verbesina encelioides, Euphorbia geniculata, Abutilon theophrastii* and most of the *Amaranthus* species.

When the first record of an adventive plant in the region is within Israel, it can be concluded that the plant was a direct import from its country of origin, e.g. *Aster subulatus* (Danin 1976) *Cenchrus incertus, C. echinatus, Euphorbia prostrata, Panicum antidotale* and *Solanum cornutum*.

Some species were introduced as forage plants throughout the Middle East and have escaped cultivation. There is no accurate evidence for their first appearance in Israel, e.g. *Eleusine indica, Panicum miliaceum* and *Chloris gayana*. Other dated species from Australia are *Atriplex suberecta, A. semibaccata, A. holocarpa, Maireana brevifolia* and *Bassia scoparia. Heterotheca subaxilliaris* was introduced into Israel from Texas in 1975 as a dune stabilizer and has since escaped aggressively to sandy roadsides, dunes and fields.

In some cases there is not enough available information to decide on the route of a species to Israel. On the whole, there are more neophytes (ca. 60%) that were introduced from neighbouring countries than these which have arrived directly.

Chronology

There is a constant increase in the number of adventive species in Israel. From ca. 10 species at the beginning of the XXth century it raises to 80 species in the 1970's and will exceed 140 species in the 1990's. There is an overall correlation between the periods of large agricultural development and the invasion of adventive species, e.g. after the independence of Israel (1948) and the subsequent intensification of agricultural and land development activities.

Adventive plants as colonizing species

Population Dynamics:
(a) Accidental species
(b) Species with limited distribution
(c) Colonizing species in man-influenced habitats
(d) Colonizing species of natural habitats
Colonization abilities:
(a) Introduction
(b) Seed production and dispersal
(c) Germination and establishment
(d) Number of generations per year
(e) Perennial colonizers

For the purpose of our discussion we can recognize the following four patterns of population dynamics:

A. *Accidental species*

Fourteen species were recorded only once or twice and disappeared. These species are: *Alternathera pungens, Amaranthus muricatus, Argemone mexicana, Diplotaxis tenuifolia, Eragrostis sarmentosa, Galenia secunda, Gisekia pharanacioides, Mantisalca salmantica, Oenothera laciniata, Solanum capsicoides, Solanum laciniatum, Solanum sarrachoides, Wedelia glauca,* and *Euphorbia nutans.* It is most likely that these species were unable to establish a second generation and to enlarge their distribution although some (e.g. *Argemone mexicana, Amaranthus muricatus, Alternathera pungens*) are known as aggressive weeds elsewhere (Holm *et al.* 1977). *Oenothera rosea* was first (1980, 1981) recorded as an accidental species (only one plant in each locality) and spread only during 1987–88. This species indicates that this category could be temporal since adventive species in general have the tendency to occur locally during one or a few years and then to disappear (Ridley 1930). Their ephemeral existence usually seems to be due to temporarily favourable situations or an unusual combination of climatic conditions. This phenomenon can be exemplified by several species that have been almost 'dormant' for 10–30 years and then began to spread rapidly, such as: *Panicum capillare, Abutilon theophrastii, Solanum elaeagnifolium, Oenothera rosea, and Euphorbia lasiocarpa.* All these species have been treated firstly as 'accidental', illustrating that this status could be temporal while plants which failed to establish initial populations will remain to be recorded as 'accidental'.

B. *Species with limited distribution*

This 'labile' status includes species in the 'lag' period before massive invasion or those which are known for many years only from several localities with minor changes. It is not an ecologically homogeneous group but more an artificial classification based on actual changes in the field or on the status of our present knowledge. In some cases worldwide known invasive species (e.g. *Fallopia convolvulus, Physalis angulata, Phytolacca americana, Solanum cornutum*) are known in Israel, sometimes for many years, only from very few localities.

C. *Colonizing species in man-influenced habitats*

Abutilon theophrastii – expanding over the past decade mainly in cotton fields in the northern part of the country. As a malvaceous species, it escapes the systemic herbicides used in cotton (Trifluralin and Diuron). This weed has

146

a long period of seed production (up to three months), high seed yield (up to 6,000 seeds per plant), considerable ability of seedling emergence from depth (up to 40cm), maintenance of two to three generations per year and a similarity to the crop's phenodynamics.

Because of all this, we expect a massive enlargement of its population and its becoming a major weed. The dispersal of this species is caused also by the cotton combine, since the awned mericarps become attached to the machine and are distributed by it. It is a notorious weed of cotton in Africa, Australia, and North America (Holm *et al.* 1977).

Achyranthes aspera – In Israel, as in Egypt (Täckholm, 1974) and in East Africa (Ivens 1975) it is a minor weed of nurseries, orchards, gardens and is quite rare in irrigated fields. It maintains a fairly stable population in orchards and gardens in the northern regions of the country.

Amaranthus albus – A minor weed of fields and roadsides although it is known as a troublesome weed in many countries (Reed 1970).

A. retroflexus – The most common species of this genus in Israel. Especially troublesome in irrigated fields and gardens.

Amaranthus rudis – Appeared in the early 1970's in the Northern Jordan Valley near fish ponds. It is now common from the Hula plain south to the Beit-Shean Valley, sometimes invading cotton fields (Joel & Liston 1986).

Anoda cristata – Since it was first recorded in Israel in 1981 (Mienis 1982) it has spread considerably in cotton fields throughout various parts of the country and also near fish ponds in the Dan Valley (Joel & Liston 1986), which suggests at the possible origins of the seeds. This species spread rapidly as a weed in cotton fields in all the southern states of the U.S.A. and also in Central America and Australia (Holm *et al.* 1979).

It spread due to heavy use of selective herbicides (Buchanan 1974, Chandler & Oliver 1979). Being a number of the Malvaceae as is cotton, it imposes a special threat while using a systemic herbicide as has already occurred for *Abutilon theophrastii*.

Araujia sericifera – Known as an ornamental plant in Israel (Zohary & Fahn 1981). It has invaded orchards in the Coastal Plain (1976) and Esdraelon Valley (1984) (Joel & Liston 1986). It spreads rapidly and threatens to be a major weed as has already occurred in South Africa (Holm *et al.* 1979).

Atriplex holocarpa – Appears, to a limited extent, along roadsides in the Northern Negev not far from the localities of its intentional introduction as a forage plant. Several plants have been recorded in the Arava Valley, 100 km south of its seminal population.

Aster subulatus – an aggressive colonizer which expands rapidly, mainly along roadsides and waste areas (Danin 1976). It also appears in irrigated fields, nurseries, and greenhouses, but until now only as a minor weed. Its perennial habit and lack of vegetative multiplication seems to limit its distribution under intensively cultivated situations, although under favourable conditions it can behave as an annual. This species deserves close observation in order to realize the potential weediness of the annual plants. In 20 years

(1960–1980), this species has spread all over the country especially along roadsides.

Bassia indica – Spread massively in the Northern Negev during the last 40 years in waste places and along roadsides. It has lately invaded the Central Coastal Plain.

Bassia scoparia – Known in various parts of the country, especially along roadsides. It has expanded aggressively around fish ponds in the Beit Shean Valley (Northen Jordan Valley) during the last decade, especially in dried ponds, and has penetrated into fields.

Bidens pilosa – This worldwide important weed (Holm *et al.* 1977) is still in rapid expansion all over Israel, mainly in nurseries and gardens. To judge from its behaviour in other parts of the world there is a chance that it could massively invade irrigated crops. Its rapid growth, high seed production, efficient seed disperal, accomplishment of up to four generations per year, and long seed viability (Ivens 1975, Holm *et al.* 1977) are favourable adaptations for colonizing new areas. We expect its expansion to continue so that it might become a major weed in the near future, especially if it continues to be spread with stock from nurseries and greenhouses.

Bromus catharticus – A minor weed of grain fields with local distribution. Introduced into the United States as a forage plant and escaped to become a weed of ruderal habitats and roadsides (Gould 1975).

Cassia obtusifolia – Found in dried fish ponds in the Northern Jordan Valley in 1985 (Joel & Liston 1986). Since it is known as a major weed in South and Central America, USA, Malaysia, and India (Holm *et al.* 1979) it is of prime importance to control it in its first steps of invasion before a large population is established.

Cenchrus echinatus – An aggressive worldwide weed (Holm *et al.* 1979). It has been in the process of rapid colonization since 1970, and is a major weed especially in citrus orchards. Its presence in other parts of the country should be inspected and controlled to prevent a build-up of massive populations.

C. incertus – A minor weed of light soils near the seashore. A troublesome weed in the southern United States in cultivated fields, lawns, and gardens (Reed 1970).

Chenopodium ambrosioides – A minor weed all over the country especially in waste places and roadside, sometimes invades gardens to a limited extent. A known cosmopolitan weed (Holm *et al.* 1977).

Chloris gayana – Probably introduced in the late 1930's for forage, is today largely distributed in the hot parts of the country. It has become a minor weed of lawns, roadsides, and gardens. The same situation also occurred in Texas after its introduction there for the same purpose (Gould 1975).

Chloris virgata – Locally distributed today in irrigated fields. Based on its existence as a weed in Egypt (Täckholm 1974) and Iraq (Bor 1968) it is expected to expand in the future especially along roadsides, as is the case in the United States (Gould 1975).

148

Conyza albida – In Israel since 1957, widespread in the Mediterranean in ruderal communities (Danin 1977). Today it is a common weed of roadsides and wet fields and is still in the process of rapid spreading (Danin 1976). as an aggressive colonizer in SouthEastern United States (Munz & Keck 1970).

C. bonariensis – A common weed all over the Middle East in various habitats such as fields, gardens, roadsides, ruderal areas and river banks (Danin 1976). An aggressive species which spreads steadily also into natural habitats. Known in gardens in Europe since the 18th century (Michael 1977).

Conyza canadensis – Collected first in the Upper Jordan Valley (1939), but now spread throughout the country; one of the first to appear on open roadsides, unattended fields, nurseries and waste ground. A known cosmopolitan weed (Michael 1977).

Cuscuta campestris – According to Fienbrun-Dothan (1978) it was introduced into Israel after the First World War but its first records in the herbarium are from the early 1950's (Heller & Dafni 1983). Today it is widespread all over the country. It seems that its invasion was made possible by the parallel increase of its hosts *Amaranthus retroflexus* and *A. blitoides*, as was suggested by Le Floch *et al.* (this volume).

Datura innoxia – Has been known for many years as a common ruderal plant. In the past decade it has slowly penetrated into cotton fields in the north of the country. Because it has a life cycle similar to cotton in germination and growth dynamics, it is likely that this species will multiply its presence within this crop and gradually become a major weed. Some plants complete their life cycle as a winter plant (generally it flowers during the summer), including seed production (Heller & Dafni 1983).

Datura ferox – Known from several localities in vegetables and cotton (Heller & Dafni 1983). It is a well-known weed in South Africa, Australia, and India (Holm *et al.* 1979). First collected in 1952 as an accidental plant and has spread since the early 1980's especially in cotton fields (Mienis 1982). Known as a weed in South Africa, South America, Australia and India (Holm *et al.* 1977).

Dinebra retroflexa – A minor weed of wet habitats in the northern part of Israel. Became a troublesome weed in cotton during the last decade (Joel 1986). Known from the United States (Reed 1972) from similar habitats.

Eleusine indica – A worldwide major weed (Holm *et al.* 1977), which appears here as a minor weed in gardens, lawns and irrigated fields. It is distributed almost all over the country, mainly in small populations, and thus does not cause serious disturbances.

Eragrostis echinochloidea – Has expanded during the last decade (since 1980) into irrigated gardens in the Arava valley. Feinbrun-Dothan (1986) comments: 'It represents another case of an adventitious plant of irrigated habitats in a desert area along the Rift Valley, and was presumably introduced by migratory birds'.

Eragrostis japonica – Has expanded slowly since the early 1930's in northern Israel. A tropical weed in Asia, Africa, S.E. Asia, listed as a potential weed in the United States (Reed 1977).

Eragrostis palmeri – First collected in 1970 in the coastal plain and has expanded steadily especially along roadsides.

Eragrostis prolifera – Has expanded moderately along the coastal plain, especially during the past decade.

Euphorbia hirta – In Israel since 1929, expansion during the last 20 years mainly though nurseries. A similar lag in distribution of *E. geniculata* for about 20 years (1940–1960), but now a dominant weed in some cotton fields. A rapid spread during the last decade also in *E. prostrata* mainly in gardens and cotton fields. A troublesome worldwide weed in various crops (Holm *et al.* 1977).

Euphorbia geniculata – Since it was first recorded in 1941, this species has spread slowly, but since the early 1960's it has been rapidly expanding in cotton fields. Today it is a major weed and is still in massive expansion. This species produces up to four generations per year under local conditions. An individual plant can produce up to 500 seeds, 60% of the seeds are products of self-pollination and seedlings can renew themselves after top removal even in early stages. All these characteristics (Dafni & Karnieli 1979) explain its rapid colonization. Chemical treatments have recently been introduced to control it (Kleifeld *et al.* 1979).

E. lasiocarpa – has been spreading extremely rapidly in gardens in the last 15 years. Recorded by Dafni & Heller (1980) as *E. hypericifolia*.

E. serpens – The first record of this species is from 1947. It is widespread in gardens, along trickle pipes, penetrating into cotton fields, pavement crevices and greenhouses in almost every part of Israel. Recorded by Dafni & Heller (1980) as *E. maculata*.

E. prostrata – very common all over the country usually with the *E. maculata*, despite the fact that the first collection of it was carried out by us in 1978. The delay in the identification of these two species excludes the possibility of reconstructing their spread into Israel. The last two species are prostrate plants which create dense mats and compete strongly, mainly with young seedlings. Both are possible candidates for the status of major weeds.

Fallopia convolvulus – Based on its occurrence in Cyprus (Americanos 1972) and in Iran (Maddah & Mirkamaly 1973), Heller and Dafni (1983) predicted its appearance here. It was found by Joel in 1985 (Joel 1986) in a potato field and by us in a nursery. This species is well known as a worldwide weed in various crops such as: potatoes, vegetables, cotton, etc. Its occurrence here is a real threat since it is regarded as one of the 'world's worst weeds' by Holm *et al.* (1977).

Galinsoga parviflora – Cosmopolitan major weed (Holm *et al.* 1977) which has been under rapid expansion in Israel during the past two decades. It spreads especially along winter vegetables grown under plastic tunnels, greenhouses, nurseries, gardens and lawns. It seems that it has been spread with runners from strawberry nurseries, mainly in the sea-coast belt.

Helianthus annuus – This potential noxious weed (Heiser 1965) spreads today

150

along roadsides and near fish ponds. It is reasonable to postulate that imported fish food from the U.S.A. is the seed source and thus explains the typical distribution of the species here. In the last 10 years, it has also appeared along roadsides which can serve as a 'standby' habitat before invasion into fields. Further information on the population dynamics is needed to forecast the behaviour of this species and its real threat from an agricultural viewpoint.

Ipomoea hederacea – Known today from several localities in the Dan Valley and Beit Shean Valley near fish ponds and cotton fields (Joel & Liston 1986). It is known as an invasive weed which was widely spread in tropical and subtropical regions. Today it is present only in its pioneering state and it is advisable to control it in this stage.

Ipomoea triloba – First recorded in 1984, and is already known from cotton fields in several independent localities (Joel & Liston 1986). Known as a major weed in Central and North America, Philippines and Australia (Holm *et al.* 1979).

Lantana camara – Common ornamental and fence plant. Although it is an alarming weed in many tropical and subtropical countries, and has the reputation of being one of the 'world's worst weeds' (Holm *et al.* 1977), it rarely invades agricultural fields and is generally restricted to gardens and fences. The present situation does not confer any safety against unexpected invasion as has occurred, for example, in the Cape Province in the Java (Taylor 1975).

Nicotiana glauca – Common all over the country (even in the deserts) especially in waste places and along roadsides. Expands steadily with the construction of new roads. Observed as a colonizing species in abandoned quarrier in northern Israel.

Nicandra physalodes – This species was known as a very rare plant found only in a single botanical garden since the 1950's. During 1986-7 it has appeared in several independent localities. One plant was recorded near Haifa (northern Israel) in a margarine factory which imports oil plants from East Africa, where the plant is already an established weed (Ivens 1967). Its appearance 20km south of Jerusalem as a roadside weed could be alarming. In Australia this species has spread from gardens over the past 50 years in a wide range of soil types, habitats, and crops and is even recognized as an important pest in maize crops in Queensland (Horton 1979, Hawton 1976). The same process of escape from gardens and invasion into fields has occurred also in the United States (Steyemark 1963) and India (Maheshwari 1968).

Oenothera rosea – Until 1987 it has been known only through accidental records from two localities. During 1987–1988 it has spread rapidly (thousands of plants) in one nursery and also invaded the surrounding natural vegetation and, to a lesser extent, other nurseries in the area. It is known as a garden escapee which has been naturalized in India (Maheshwari 1962).

Opuntia ficus-indica – Common all over the country especially in waste places and as a fence plant. Spontaneous propagation was hardly seen under natural conditions; most of its distribution is, thus, manmade.

Oxalis corniculata – Spread since the beginning of this century, today it is commong in almost every lawn or nursery.

O. pes-caprae – First recorded in 1960 and spread in orchards, gardens and roadsides throughout the country. Only one of the heterostylic types exists here; the expansion therefore must be only by vegetative means.

O. europaea – It has spread since 1984 (Joel 1985) into nurseries in northern Israel. This North American species (despite its name) is a common garden weed in Britain (Young 1958).

Panicum antidotale – A minor weed of irrigated fields mainly in the Arava Valley (Southern Israel). Known as a forage plant in the SouthWestern United States and as a weed in Central America.

P. capillare – Has spread rapidly during the last decade especially in lawns, gardens, and irrigated fields all over the country.

Parthenium hysterophorus – It was first recorded in Israel in 1980 by Dafni & Heller (1982) who predicted its potential invading ability. Today it' is spreading rapidly into various crops in the Esdraelon Valley (Joel & Liston 1986) and also near roadsides in the Carmel coast and Western Galilee. This species is notorious for its aggressiveness and alarming spreading abilities, especially in North America, Central America, the West Indies, and South America. Introduced into Africa, Australia, and southwest Asia (Picman & Picman 1984).

Paspalum dilatatum – Probably introduced for forage and in less than a decade has become a major weed (Oren & Israeli 1979), mainly throughout the seashore lowland. Its wide distribution in many regions exemplifies a rapid spread which might be eplained by its large ecological amplitude and vigorous reproduction (Holm *et al.* 1977). It is expected that this species will spread more in the near future. An active chemical control programme is taking place (Oren & Israeli 1979) and that for a species that was completely unknown in Israel only 20 years ago!

P. paspaloides – A minor weed of irrigated fields, lawns, gardens and wet habitats, which has been introduced and become a weed in many countries (Bor 1968).

Ricinus communis – Common all over the country in waste places and along roadsides. Expands with the construction of new roads especially in the coastal plain.

Rorippa prostrata – has spread rapidly in several nurseries in various parts of the country since 1976 (Joel, pers. comm.).

Setaria glauca, S. viridis and *S. verticillata* – All these species are well established as weeds, especially in orchards and gardens. *S. verticillata* is a major weed in corn fields. All these species are serious weeds on a worldwide scale (Holm *et al.* 1979).

Sesbania sesban – Has spread slowly especially as a garden escapee along the Jordan Valley. A widespread weed of the Old World Tropics introduced into Tropical America, E. Asia, E. Africa, S. of W. Africa. Listed as a potential weed of the United States (Reed 1977).

152

Sorghum virgatum – This species has become a major weed in some restricted areas in various parts of the country. It is a spreading weed in cotton fields as well as in vegetable gardens. The first record from Israel is from 1922. It has not been recorded for about 60 years and then it began to spread massively.

Solanum cornutum – Locally distributed in some irrigated vegetable fields and in gardens in various places. This species is known as a casual weed in California (Munz & Keck 1970).

Solanum elaeagnifolium – Aggressive species, known as a colonizing invader from various parts of the world such as South Africa (Siebert 1975), Australia (Smith 1975), California (Munz & Keck 1970) and Greece (Yannitsaros & Economidou 1974). It spreads mainly along roadsides, sexually as well as vegetatively, creating dense populations. This species is currently penetrating gardens, but it is likely that it will invade fields and orchards. Its ability for renewal after cutting (Smith & Cooley 1972) and propagating from rhizome segments enables it to become a major weed which would be very complicated to control.

Tagetes minuta – Was known as an ornamental plant for many years. Two decades ago it escaped from cultivation and spread rapidly in orchards and along roadsides. It is now locally distributed in the north of the country and it is expected that it will spread, due to its weedy character (Morton & Zallinger 1976). This species is known as a spreading weed also in Yugoslavia (Trinajstic 1974), Greece (Yannitsaros & Economidou 1974) and in East Africa (Ivens 1975). It is recommended to control it chemically now before a possible massive invasion of other parts of the country.

Trianthema portulacastrum – Has expanded during the last five years along the coastal plain and in the Esdraelon Valley to become a troublesome weed. This species was firstly recorded in Egypt in 1978. Based on its ecology and distribution in tropical areas as a weed (Holm *et al.* 1979:479) it is forecast to become a major weed, as it is in Central America (Guatemala and Nicaragua) and in Thailand (Holm *et al.*, l.c.), especially in cotton fields.

Verbesina encelioides – Firstly recorded in 1970 and has spread slowly along roadsides in the coastal plain of Israel. A known weed of California (Munz & Keck 1970) and a colonizing species of exposed habitats in Zimbabwe (Wild 1968).

Xanthium strumarium – A typical plant of stream banks (in Israel) which penetrates into irrigated fields. In some cotton fields it becomes a major weed because its life cycle is similar to cotton and because of its seed dispersal by the cotton combine.

D. *Colonizing species of natural habitats*

Colonizing species occupying natural habitats are limited to four examples.

Oenothera drummondii – Very widespread and established in sandy coastal communities in spite of the fact that it arrived only at the end of the last century. Also recorded from Morocco, Egypt (1928) and Spain (1968) (W. Dietrich, pers. comm.), likely from the Israeli stock.

Aster subulatus – An aggressive invader of some wet habitats, competing succesfully with the local natural vegetation (Danin 1976) as well as with ruderals; in less than 20 years it has spread all over the country and is still expanding.

Conyza bonariensis – Aggressive ruderal species successfully invading natural habitats (Danin 1976).

Heterotheca subaxillaris – In 1975 it was introduced to stabilize mobile dunes. Five grams of seed were sown in the coast of Galilee (D. Tsuriel, Pers. Comm.), and the same amount in the Sharon plain (60km N of Tel-Aviv). It is now spreading from year to year along roadsides and gardens aggressively and into natural habitats on sands – up to 30km from its pioneer population. Since this species is known to originate in North Carolina as a colonizing weed (Awang & Monaco 1978) it is a pity that such a species was intentionally introduced.

It can be seen from these examples that only a few adventive species has become naturalized. The common suggestion (Polunin 1960, p. 118) is that adventive species are inferior in their competing ability to indigenous species, unless there is human interference, such as in disturbed habitats. In Britain, for example, only four species out of 130 introduced by the import of wool have become naturalized (Salisbury 1961).

Colonization abilities

A. *Introduction*

Adventive plants are usually regarded as weeds (Mulligan 1965, Baker 1974). The typical habitats favouring their establishment are temporal and disturbed. Here, density independent r-selection is promoted (Gadgil & Solbrig 1972, Pianka 1972, Mc Naughton 1975).

The following may be regarded as preadaptations for colonizing ability: extension of seed germination over a long period, wide amplitude of modificational plasticity and adaptability, fast reproduction even under adverse conditions, large quantity of seeds, wide seed dispersal, and a genetic system balanced between flexibility and stability (Ehrendorfer 1965). Allard (1965) stresses that the great majority of the most successful colonizing species worldwide tend towards predominant self-fertilization. While agricultural techniques favour annuals on cultivated fields, perennials are tolerated on roadsides and ruderal habitats particularly if they are strong in vegetative reproduction.

154

B. *Seed production and dispersal*

The rapid establishment of a large population originating from one or a few propagules is an important advantage for colonization if the environment does not restrict the expansion (Stebbins 1965). Many adventive plants are capable of rapid spread because they combine high reproductive efforts with the production of many lighweight seeds, e.g., *Galinsoga parviflora*: up to 400,000 seeds (Usami 1976), *Eleusine indica*: up to 135,000 (Holm *et al.* 1977), *Conyza bonariensis*: 7,500–23,000 (Danin 1976), *Euphorbia hirta*: about 3,000 (Holm *et al.* 1977), and *Parthenium hysterophorus*: 15,000 (Haseler 1976).

Teleochory is an efficient adaptation for reaching new habitats even without the aid of human activity. 60% of our adventive species have mechanisms for such long range dispersal which suggests that man is responsible for the dispersal of the remaining 40% of the species. It has been noted previously (Polunin 1960, p. 120) that many weeds have antiteleochoric seed dispersal, such as *Capsella bursa-pastoris, Stellaria media*, and *Poa annua* among many others, and have become cosmopolitan with the aid of man.

C. *Germination and establishment*

Weeds are characterized by having rather unspecific requirements for germination, considerable rates of dormancy, long viability and rapid growth of seedlings (Baker 1974). The same features are probably crucial for colonizing ability (Ehrendorfer 1965). *Abutilon theophrastii, Euphorbia geniculata, Cenchrus echinatus, Conyza bonariensis* and *Datura stramonium* are repesentative examples among Israeli neophytes for these germination and establishment strategies.

D. *Number of generations per year*

The ability to produce several generations per year characterizes annual colonizing weeds (Baker 1974). More specifically, the same adaptations are regarded as typical for r-selected species and are of obvious advantage for the occupation of new and open habitats while the population density is still low (Gadgil & Solbrig 1972). In such species most of the energy is allocated for reproduction since competition with other plants is minimal (Harper 1977, p. 700). Such adventives are *Amaranthus palmeri, A. retroflexus, A. graecizans, Euphorbia geniculata, E. prostrata, Abutilon theophrastii, Conyza bonariensis, C. canadensis, C. albida, Aster subulatus* and *Tagetes minuta*. All of these produce between two to four generations per year under favourable conditions in Israel (Dafni, unpubl.).

E. *Perennial colonizers*

Such colonizers are: *Solanum elaeagnifolium, Chloris gayana, Paspalum dilatatum, P. paspaloides, Oxalis pes-caprae,* and *Achyranthes aspera.* Density seems not to be a crucial selection factor for their successful establishment since they rely strongly on vegetative reproduction. This can be regarded as an adaptation to repeated damage, as it occurs in cultivated field and greatly favours vegetative over sexual reproduction (McNaughton 1975).

It is generally accepted that expanding species are characterized by a lag period of some generations before massive distribution begins. Lag periods of 10–30 years are recorded in several adventives in Israel. Baker's (1965) explanation is that some recombination and introgression is needed in order to reach a level of variability that allows expansion beyond the limited distribution of the founder population. Salisbury (1953) stressed the need for masses of available propagules as a precondition for a rapid spread. In contrast, rapid colonization was recognized almost immediately after the initial discovery in Israel of adventives like *Paspalum dilatatum, Cenchrus echinatus* and *Conyza albida.* The difference in time needed for rapid distribution can be explained by the importance of having new open habitats in addition to a population's capacity for reproduction.

Some neophytes are present in quite steady populations, mainly along roadsides and in ruderal habitats, without showing increasing spread. It is likely that an opening of new niches could result in further expansion. Roadsides are known as 'standby' habitat for many weeds before their intrusion into fields (Good 1964 p. 379, Danin 1976). Species like *Atriplex suberecta, A. semibaccata, Chloris gayana, Chenopodium ambrosioides, Panicum miliaceum, Phytolacca americana, Bassia indica* and *B. scoparia* are potential colonizing species, some of which are classified as weeds.

Agricultural aspects

Cultivated fields are temporally disturbed habitats maintained by human activities. Agrotechnical treatments create open spaces for colonization which are quite free of competition and serve as ideal places for the establishment of adventive species. International agricultural cooperation and the improvement of transportation means have caused waves of weed immigration throughout the world, and some of these weeds have become cosmopolitan.

The following adventive species are well established in Israel as widespread weeds which have economic importance: *Amaranthus retroflexus, A. palmeri, A. graecizans, Cenchrus echinatus, Conyza bonariensis, Eleusine indica, Euphorbia geniculata, E. prostrata, Oxalis corniculata, Paspalum dilatatum, P. paspaloides, Xanthium strumarium, X. spinosum, Abutilon theophrastii, Bidens pilosa, Datura stramonium* and *Chloris gayana.* Other aggressive weeds still occur in restricted populations only, but could pose a threat in the future and therefore

156

should be controlled, e.g., *Tagetes minuta, Solanum cornutum, Fallopia convolvulus, Helianthus annuus* and *Physalis angulata*. The rapid development of house plant nurseries during the last decade was accompanied by the large importation of peat moss from Europe which was presumably contaminated by seeds of adventive species. Some of the most impressive invasions (e.g., *Oenothera rosea, Rorippa prostrata, Fallopia convolvulus*) have originated in nurseries. Since their products are diffused all over the country the chances of the propagation of these species are immense. Another source of adventive weed is in low grade grains (e.g., *Sorghum bicolor*) which are imported from the United States as food used in fish ponds. It is easy to point to the first establishment of many aggressive weeds (e.g., *Parthenium hysterophorus, Helianthus annuus, Sida acuta, Amaranthus rudis* etc.) directly near the feeding area on the banks of fish ponds.

In other Mediterranean countries the problem of adventive weeds is not so serious. Aymonin (1965) surveying the adventive flora of France, found only a few neophytes that reached the status of weeds. In Portugal, only 17 out of the 246 adventive species have become weeds (Pinto da Silva 1971).

Floristic analysis of the alien flora of the Middle East reveals the presence of noxious weeds which are still not present in Israel e.g., *Striga asiatica, S. densiflora, Sonchus arvensis, Cirsium arvense,* and *Euphorbia hyssopifolia*. Taking into consideration the close ecological similarities between fields throughout the region and the intensive exchange of agricultural products since the last two decades, it is quite reasonable to predict the future appearance of these species in Israel.

There are some species which appeared in the beginning of this century but which have spread aggressively only in recent decades. Neophytes which came after the 1940's generally showed shorter lag periods than the earlier invaders. All this is apparently due in Israel to the rapid agricultural development, obvious from the cultivation of new areas, intensification and improvement of irrigation systems, heavy use of herbicides and immense enlargement of greenhouse areas. These massive changes offer new open habitats and are responsible for the fact that not less than two thirds of the adventive species are weeds. This is consistent with Baker's (1974) view that changes in agricultural practices enable the invasion of alien weeds to take place especially when there is a sharp transition from extensive to intensive cultivation.

In the countries around Israel agricultural activities have been intensive for many centuries, in comparison to the relative neglect of agriculture in Israel until the last 40 years. It is hardly surprising therefore that most (60%) of the adventive weeds were first observed in neighbouring regions. But at present there are more adventive weeds in Israel than in the rest of the region. It can be seen that most of the adventive weeds of Iraq, Lebanon and Syria have reached Israel, but not five adventive weeds from the Nile Valley in Egypt. Their habitat type is drastically reduced in Israel (Dafni & Agami 1976), making their introduction unlikely. But the analysis of the Middle East

157

flora reveals the presence of other aggressive weeds which could reach Israel easily and rapidly. It is of great importance to recognize them before they build up massive populations.

References

Ahronson, E. 1985. *Dodonaea viscosa* - a new garden escape in the Negev. *Rotem* 16: 33–40 (In Hebrew).

Allard, R.W. 1965. Genetic systems associated with colonizing ability in predominantly self-pollinated species. In H.G. Baker & G.L. Stebbins (eds.), *The genetics of colonizing species*. Academic Press, New York, pp. 50–76.

Americanos, P.G. 1972. The Weed Complexes of *Citrus* Groves in Cyprus. *Agric. Res. Inst. Min. Agric. Nat. Resour. Tech. Bull.* (Nicosia) 9: 32 pp.

Anonymous 1972. Weeds and Weed Control at the Haft Tappeh Cane Sugar Project, 1950–71. Iranian Ministry of Water and Power. (Mimeo.)

Awang, M.B. & Monaco, J.J. 1978. Germination, growth, development and control of Camphorweed (*Heterotheca subaxillaris*). *Weed Sci.* 26: 51–57.

Aymonin, G. 1965. The phenomenon of 'adventicity'. 2e Colloque sur la Biologie des Mauvaises Herbes, p. 18, Grignon (In French).

Baker, H.G. 1965. Characteristics and modes of origin of weeds. In H.G. Baker & G.L. Stebbins (eds.), *The genetics of colonizing species*. Academic Press, New York, pp. 147–169.

Baker, H.G. 1972. Migration of weeds. In D.H. Valentine (ed.), *Taxonomy, Phytogeography and Evolution*. Academic Press, New York, pp. 327–347.

Baker, H.G. 1974. The evolution of weeds. *Ann. Rev. Ecolo. Syst.* 5: 1–24.

Boissier, E. 1867–1888. *Flora Orientalis*. 5 Vols. & Suppl. Basel-Geneva-Leiden.

Bor, N.L. 1968. *Gramineae In Flora of Iraq* Vol. 9. Ministry of Agriculture Baghdad.

Bornmüller, J. 1898. Ein Beitrag zur Kenntnis der Flora von Syrien und Palästina. *Verh. Zool.-Bot. Ges. Wien* 48: 544–653.

Buchanan, G.A. 1974. Weed plague cotton growers from the Carolinas to California. *Weeds Today* 5: 6–70, 20.

Chandler, J.M. & Oliver, L.R. 1979. Spurred *Anoda*: A potential weed in southern crops. U.S. Sci. Ed. Admin., *Agric. Rev. & Man*, Southern Ser., 2, 19 pp.

Chaudhary, S.A., Parker, C. & Kasasian, L. 1981. Weeds of central, southern and eastern Arabian Peninsula, *Trop. Pest. Manag.* 27: 181–190.

Dafni, A. & Agami, M. 1976. Extinct plants of Israel. *Biol. Conser.* 10: 49–52.

Dafni, A. & Heller, D. 1980. The threat posed by alien weeds in Israel. *Weed Res.* 20: 277–283.

Dafni, A. & Heller, D. 1982. Adventive flora of Israel: phytogeographical, ecological and agricultural aspects. *Plant Syst. Evol.* 140: 1–18.

Dafni, A. & Karnieli, Erga 1979. Observations on the life-cycle of *Euphorbia geniculata*. *Israel Weed. Sci. Conf.* 7: 11–12.

Danin, A. 1976. Notes on four adventive *Compositae* in Israel. *Notes Royal Bot. Gard. Edinb.* 34: 403–410.

Danin, A. 1977. On three adventive species of *Conyza (Compositae)* in Greece. *Candollea* 31: 107–109.

Davis, P.H. 1965–1985. *Flora of Turkey and the East Aegean Islands*. 9 Vols., University Press, Edinburgh.

Dinsmore, J.E. 1912. *The Jerusalem catalogue of Palestina plants*. Ed. 3. Jerusalem.

Edgecombe, W.S. 1970. *Weeds of Lebanon*. 3rd ed. American Univ. of Beirut, 457 pp. Beirut.

Ehrendorfer, F. 1965. Dispersal mechanisms, genetic systems and colonizing abilities in some flowering plant families. In H.G. Baker & G.L. Stebbins (eds.), *The genetics of colonizing*

158

species. Academic Press, New York, pp. 331-352.

Eig, A., Zohary, M. & Feinbrun, N. 1948. *Analytical flora of Palestina*, ed. 1 (In Hebrew). Jerusalem: Palestine Journal of Botany.

Feinbrun-Bothan, N. 1978, 1986. *Flora Palaestina Vols. 3 & 4*. Text + Plates. Jerusalem: Israel Acad. Sci. and Human.

Gadgil, M.D. & Solbrig, O.T. 1972. The Concept of r- and k-selection: evidence from wild flowers and some theoretical considerations. *Amer. Nat.* 106: 14-31.

Good, R. 1964. *The geography of flowering plants*. Longmans, 3rd ed. London.

Gould, F.W. 1975. *The grasses of Texas*. Texas University Press.

Gruenberg-Fertig, I. 1966. List of Palestine plants with data on their geographical distribution. Ph.D. thesis, The Hebrew University, Jerusalem.

Hariri, G. 1972. Weeds of Syria. Pamphlet No. 3. Dept. of Plant Protection, Fac. of Agric., Univ. of Aleppo, 15 p. (Mimeo).

Harper, J.L. 1965. Establishment, aggression and cohabitation in weedy species. In H.G. Baker & G.L. Stebbins (eds.), *The genetics of colonizing species*. Academic Press, New York, pp. 245-266.

Harper, J.L. 1977. *Population biology of plants*. Academic Press, New York.

Haseler, W.H. 1976. *Parthenium hysterophorus* in Australia. *PANS* 22: 515-517.

Hawton, D. 1976. Control of apple of Peru (*Nicandra physaloides*) in maize on the Atherton Tablelend, Queensland. *Austral. J. Exp. Agri. Anim. Husb.* 16: 765-770.

Heiser, C.B. 1965. Sunflower, weeds and cultivated plants. In H.G. Baker & G.L. Stebbins (eds.), *The genetics of colonizing species*. Academic Press, New York, pp. 391-399.

Heller, D. & Dafni, A. 1983. *Adventive plants of Israel*, Israel Society for Protection of Nature. Tel Aviv. 148 pp. (in Hebrew).

Holm, G.L., Plucknett, D.L., Pancho, J.V. & Herlberger, J.P. 1977. *The world's worst weeds – distribution and biology*. East-West Center, Honolulu.

Holm, L., Pancho, J.V., Herlberger, J.P. & Plucknett, D.L. 1979. *A Geographical atlas of world weeds*. John Wiley, New York, Chicester, Brasbane, Toronto.

Horton, P. 1979. Taxonomic account of *Nicandra (Solanaceae)* in Australia. *J. Adelaide Bot. Gard.* 1: 351-356.

Hussain, S.M. & Kasim, M.K. 1976. Weeds and their control in Iraq. *PANS* 22: 399-404.

Ivens, G.W. 1975. *East African weeds and their control*. Oxford University Press, Nairobi.

Joel, D.M. 1985. New weeds in nurseries in Israel. *Hassadeh* 75: 1437-1438. (In Hebrew).

Joel, D.M. 1986. *New graminae* in our fields. *Hassadeh* 76: 1032-1033. (In Hebrew).

Joel, D.M. & Liston, A. 1986. New adventive weeds in Israel. *Isr. J. Bot.* 35: 215-223.

Joel, D.M. & Sando, Z. 1986. *Polygonum convolvus* – a new alien weed in Israel. *Hassadeh* 76: 1056-1057 (In Hebrew).

King, L.J., 1966. *Weeds of the world*. Interscience Publishers Inc., New York.

Kleifeld, Y., Blumenfeld, T., Barguti, A. & Sachs, Y. 1979. Control of *Euphorbia geniculata* in cotton fields. *Israel Weed Sci. Conf.* 7: 80-81.

Kosinova, J. 1974. Studies on the weed flora of cultivated land in Egypt. 4. Mediterranean and tropical elements. *Candollea* 29: 281-295.

Kosinova, J. 1975. Weed communities of winter crop in Egypt. *Preslia* 47: 58-74.

Maddah, M.B. & Mirkamaly, H. 1973. Weed of wheat fields in Arak. *Iranian J. Pl. Pathol.* 9: 12-14.

Maheshwari, J.K. 1962. *Studies in the naturalized flora of India*. In P. Maheshwari, M. Johri & I.K. Vasil (eds.), *Proceedings of the summer school of botany*, Darjeeling, pp. 156-170.

Maheshwari, J.K. 1968. Studies on the alien flora of Rajasthan. In Symposium on Natural Resources of Rajasthan. *Jodhpur* 23-26, October 1968. Mimeo.

McNaughton, S.J. 1975. r- and k-selection in *Typha*. *Amer. Natur.* 109: 251-261.

Meikle, R.D. 1977, 1982. *Flora of Cyprus*. 1 & 2. Kew: Royal Bot. Gardens, Bentham Moxon Trust.

Michael, P.W. 1977. Some weedy species of *Amaranthus* and *Conyza/Erigeron* naturalized in the Asian-Pacific region. *Proc. 6th. Asian-Pacific Weed Science Conference* 1: 87-95.

159

Mienis, H.K. 1982. *Anoda cristata*: a new weed in Israel. *Phytoparasitica* 10: 264.

Morton, J.F. & Zallinger, J.D. 1976. *Herbes and Spices*. Golden Press, New York.

Mouterde, P. 1966, 1970. *Nouvelle flore du Liban et de la Syrie*. 1 & 2, Text + Atlas. Beirut: Impr. Cathol.

Mulligan, G.A. 1965. Recent colonization by herbaceous plants in Canada. In H.G. Baker & G.L. Stebbins (eds.), *The genetics of colonizing species*. Academic Press, New York, pp. 127–144.

Munz, A. & Keck, D.D. 1970. *A California flora*. University of California Press, Berkeley.

Oliver, D. & Lambert, B. 1974. Today's weed: spurred *Anoda*. *Weeds Today* 5: 22.

Oren, Y. & Israeli, A. 1979. Control of *Paspalum dilatatum* by glyphosate with addition of ammonium sulphate. *Israel Weed Sci. Conf.* 7: 24.

Parker, C. 1973. Weeds in Arabia. *PANS* 19: 345–352.

Parker, C. 1977. Prediction of new weed problems, especially in the developing world. In G.R. Sagar & J.M. Cherret (eds.), *Origins of pests, parasites diseases and weed problem, 18th Symp. Brit. Ecol. Soc.*, Blackwell Scientific Publications, Oxford, pp. 249–264.

Pianka, E. 1972. r- and k-selection or b- and d-selection. *Amer. Nat.* 106: 581–588.

Picman, J. & Picman, A.K. 1984. Autotoxicity in *Parthenium hysterophorus* and its possible role in control of germination. *Biochem. Syst. Ecol.* 12: 287–292.

Pinto da Silva, A.R. 1971. Les plantes synanthropiques au Portugal continental et aux Acore *Boissiera* 13: 297–303.

Polunin, N. 1960. *Introduction to plant geography*. Logmans, London.

Post, G.E. (2. ed. by Dinsmore, J.R.), 1932–1933. *Flora of Syria, Palestine and Sinai*. 1–2. American University, Beirut.

Rechinger, K.H. (Ed.) 1963. *Flora Iranica (appearing in parts by families)* Akademische Druk- und Verlaganstalt, Graz.

Reed, C.F. 1970. *Selected Weeds of the United States*. U.S.D.A., Agriculture Handbook, No. 366, Washington.

Ridley, H.N. 1930. *The dispersal of plants throughout the world*. Reeve, Ashford, Kent.

Rijn, Van P.J. 1962. Weed problems in the Kimberleys. *Jour. Agric. west Austral.* 9: 211–214.

Salisbury, E.J. 1953. A changing flora as shown in the study of weeds of arable land and waste places. In J.E. Lousley (ed.), *The changing Flora of Britain*, Botanical Society of the British Isles, Oxford, pp. 130–133.

Salisbury, E.J. 1961. *Weed and aliens*. Collins, London.

Siebert, M.W. 1975. Candidates for the biological control of *Solanum elaeagnifolium* Cav. (*Solanaceae*) in South Africa. 1. Laboratory studies on the biology of *Gratiana lutescens* Boh. and *G. pallidula* Boh. (Coleoptera, Cassididae). *J. Entomol. Soc. S. Africa* 38: 297–304.

Smith, D.T. & Cooley, A.W. 1972. Seed aspect of two perennials – woody leaf bursage and silverleaf nightshade. *Proc. 25th Ann. Meet. Southern Weed Sci. Soc.*, 443.

Smith, K.R. 1975. A new system of weed surveying and its use on silver-leaf nightshade. *J. Agric. S. Australia* 78: 35–39.

Stebbins, G.L. 1965. Colonizing species in Californian species. In H.G. Baker & G.L. Stebbins (eds.), *The genetics of colonizing species*. Academic Press, New York, pp. 173–192.

Steyemark, J.A. 1963. *Flora of Missouri*. Iowa State University Press, Ames, Iowa.

Täckholm, V. (in collab. with Boulos, L., El-Hadidin, N., El-Gohary, M., Amin, A.), 1974. *Student's Flora of Egypt* 2nd. ed., Cairo University, Beirut.

Taylor, H.C. 1975. Weeds in the south western Cape vegetation. *South. Afr. Forestry Journal* 93: 32–36.

Trinajstic, I. 1974. A report on the distribution of *Tagetes minuta* L. in Yugoslavia. *Acta Botanica*

160

Croatica 33: 231–235 (In Serbo-Croat) *Weed Abstr.* 25: 1635.

Tristram, H.B. 1884. *The survey of Western Palestine; the fauna and flora of Palestina.* London.

Usami, Y. 1976. Ecological studies on weeds in mulberry fields, 2. Auto-ecology of *Galinsoga parviflora* Cav. *Weed Res. Japan* 32: 76–80 (In Japanese).

Whigham, D.F. 1984. The effect of competition and nutrient availability on the growth and reproduction of *Ipomoea hederacea* in an abandoned field. *J. Ecol.* 72: 721–730.

Wild, H. 1968. *Weeds and aliens in Africa: The American immigrant.* Publ. Univ. Coll. Rhodesia.

Yannitsaros, A. & Economidou, E. 1974. Studies on the adventive flora of Greece. I. General remarks on some recently introduced taxa. *Candollea* 29: 111–119.

Young, D.P. 1958. *Oxalis* in the British Isles. *Watsonia* 4: 51–59.

Zohary, M. 1941. *Weeds of Palestine and their control.* Hassade, Tel Aviv, (In Hebrew).

Zohary, M. 1962. *Plant life in Palestine, Israel and Jordan.* Academic Press, New York.

Zohary, M. 1966, 1972. *Flora Palaestina*, Vols. 1 & 2 Text & Plates. Israel Acad. Sci. and Hum., Jerusalem.

Zohary, M. & Fahn, A. 1981. *The cultivated plants of Israel.* 2nd. ed. Tel Aviv (In Hebrew).

Postscriptum:

Since this chapter was sent for typesetting some new adventive species were revealed or identified (although collected previously). This list includes the following species:

Chloris andropogonoides Fourn.
Coronopus didymus (L.) Sm.
Euphorbia supina Rafin.
Lobularia maritima (L.) Desv.
Polygonum aviculare L.
Sporobolus indicus (L.) R. Br.

The following species are known from Israel for many years as ornamental or experimental plants and invaded lately uncultivated areas (some are also recorded from Israel by: Danin, A. 1989. *Willdenowia* 19, 1:27–48):

Acacia saligna (Labill.) Wendl. f. (= A. cyanophylla Lindl.)
Ailanthus altissima (Mill.) Swingle
Einadia nutans (R. Br.) A.J. Scott
Enchylaena tomentosa R. Br.
Melia azedarach L.
Rosmarinus officinalis L.
Schinus terebinthifolia Raddi
Tetraclinis articulata (Vahl) Masters.

Including these recent additions the adventive flora of Israel comprise one hundred and thirty-seven species.

9. Man and vegetation in the Mediterranean area during the last 20,000 years

JEAN-LOUIS VERNET

Abstract

Chronological ecology based on studies of plant macro remains, especially prehistoric charcoal, provides evidence concerning the changes in western Mediterranean vegetation during the past millennia. Comparisons are based on present vegetation zones as defined by Ozenda (1975). From the last glacial period to the present time, climatic differences between warm and cool vegetation zones were of about 8°C in the south of France but less extreme in more southern regions. Pines and junipers were characteristic of the late Pleistocene and early Holocene (late Paleolithic and part of Mesolithic). The late Mesolithic and the early Neolithic are periods of forest optimum. During the Neolithic, deciduous and evergreen oaks had a role of varying importance among all the present Mediterranean zones. Man's influence on the vegetation became significant in the middle Neolithic (south of France) or earlier (south of Spain) and may be characterized by the presence of plants such as *Buxus sempervirens*, *Quercus ilex*, *Pinus halepensis* and *Erica*. The Chalcolithic, the Roman period, and the Middle Ages are also important periods of man's influence.

Introduction

Just like palynology, the study of plant macro remains allow us an eco-chronological approach to ancient Mediterranean floras and vegetations. This paper is especially concerned with prehistoric charcoal analysis, which has allowed during the last 15 years many answers concerning man's impact on vegetation (see e.g. Vernet & Thiébault 1987). Prehistoric charcoal analysis includes the study of wood directly collected and used by man. Combustion generally facilitates the conservation of wood. In the Mediterranean area woody vegetation components are abundant and grow either in forests or in various non-sylvatic formations. Many taxa (up to 30) are found in each stratified charcoal deposit; this observation strongly supports the hypothesis that the

F. di Castri, A. J. Hansen and M. Debussche (eds.), Biological Invasions in Europe and the Mediterranean Basin. 161–168. © 1990, *Kluwer Academic Publishers, Dordrecht.*

162

wood was not selected by man. Charcoal may be considered more as random samples than as selective collections. However, samples collected in particular situations such as fire pits, show a low number of taxa because of their short time of use (Heinz, unpubl.). Thus, charcoal analysis may be considered as a complement of palynology taking into account ecological and ethnological components of man's environment. The charcoal deposits studies to date span the last 20 millennia, the period of the development of modern man and the transition from hunting-gathering to agriculture. This time period included particularly the last Glacial maximum and the post-Glacial warming.

Climatic influences on vegetation

Charcoal analysis provides evidence that between 20,000 and 11,000 years B.P. the periglacial vegetation under the latitudes of Southern France was very closely related to the present mountain vegetation (Vernet 1973, 1980; Bazile-Robert 1981). The typical taxa were *Pinus sylvestris, Betula pendula, Hippophae rhamnoides* and *Sambucus racemosa*. During periods characterized by a warmer temperature (dated 20,000, 16,000 and 13,000 B.P. according to Bazile *et al.* 1986) many thermophilous taxa came back from their refuges (Jalut *et al.* 1975). In these warm periods with a 6–8°C range of temperature variation, vegetation was quite similar to the present vegetation observed in the meso-mediterranean or supra-mediterranean zones (see CLIMAP 1976; Sabatier & Van Campo 1984; Vernet, 1986). Mediterranean vegetation zones of lower altitude were replaced by forest-steppe or pre-forest communities mainly composed of *Pinus sylvestris*.

Another important climatic aspect may be deduced from birch ecology. The existence of this tree (*Betula* cf. *pendula*) in lower zones during the last Glacial up to 10,000 of 11,000 B.P. is indicative of moist summers. However, it is difficult to know if summer was the most rainy season. In this case, the climate can not be assumed to have been Mediterranean. However, from 15,000 to 10,000 B.P. (late Glacial) thermophilous periods were more frequent. This first transition period is correlated with increasingly dry summers. A second transition phase took place between 10,000 and 8,000 B.P. with the occurence of pines, mainly *Pinus sylvestris* and junipers, probably *Juniperus phoenicea, J. oxycedrus* and *J. communis*. This phase preceded the establishment of deciduous oak forest (late Mesolithic and early Neolithic) and may be related to mountain or supra-mediterranean zones.

This information raises the following points:
– the decrease and disappearance of birch during the late Glacial from lower Mediterranean elevations may be correlated with an extension of the species in the Atlantic area since 13,000 B.P. (Van Campo 1984) as a result of the increase in temperature and moisture (change from a continental climate to an oceanic one in non-Mediterranean area). In other words, 13,000 B.P. may be regarded as a time of divergence between Mediterranean and Atlantic

bioclimatic history;

- the mountain mediterranean *Juniperus* vegetation has been recognized and exploited by man between 12,000 and 8,000 B.P. while the palynological optimum was generally dated from 13,000 to 14,000 B.P. in the south of France (Triat-Laval 1979; Jalut *et al.* 1982). Likewise, the beginning of deciduous oak vegetation (8,000 B.P.) based on charcoal study is correlated with the palynological dominance of this vegetation;
- finally, transition between forest-steppes to deciduous oak forests, the early Holocene post-Glacial climax, is the last main phytoclimatic division at the 42–44°N lattitude. Later, the upper Holocene vegetations record only human influences.

Human impacts on vegetation

In the 43–44°N latitude human action on the vegetation is first recorded during middle Neolithic, with the Chasséens civilization (6,000 B.P.) (Vernet & Thiébault 1987). Vegetation opening is characterized by typical indicators, such as *Buxus sempervirens*, and an increase of *Quercus ilex*. Archaeological data from the eastern Pyrénées show that, during this period, hunting was decreasing and losing importance in the prehistoric economy and was being replaced by sheep and goats breeding. Later on, especially between 4,500 B.P. and Roman times, these activities, breeding and human action on vegetation, show different degrees of development. Based on these observations we may consider that present 'garrigues' did not occur before the late Neolithic (ca. 4,500 B.P.). By Roman times and the Middle Ages an important period of change took place. During Roman times olive trees and vineyards have taken a decisive role while *Pinus halepensis* importance was increasing. Both beech (*Fagus sylvatica*) and also fir (*Abies alba*) charcoal have been found in early Middle Ages excavations in alluvial plains. Now accurate palynological evidence (Planchais 1982) has recorded beech disappearance after 1,300 B.P. in the Languedoc plain of southern France. Therefore, mesophytic river forests would have been destroyed only in the Middle Ages between the 10th and the end of 12th century. This hypothesis is supported by the Middle Age history when new soils were constituted along the rivers. The valorization of these new soils was helped by 'moulin à paissière' a technique of systematic irrigation for cultivated areas and natural meadows. We must also take into account the production of new tools used to cultivate heavy soils which were, until then, unexploited. In Languedoc, the forest was definitely fixed in its major plant components between the 10th and 12th century. At this time, last glacial relicts disappeared from plains and a new soil patterning was initiated (Durand & Vernet 1987).

At the north boundary of the Mediterranean area the main thermophilous components disappeared (e.g. *Quercus ilex*). At Choranche-Coufin (French Alpine foreland) an oak phase is recorded during Mesolithic and early Neolithic.

164

During late Neolithic and posterior times human activity is mainly revealed by ash exploitation. The charcoal ash maximum may be correlated with the occurence of Bovidae bones (Thiébault 1988). Foliage exploitation for domestic animals known nowadays in this area goes back to the Mesolithic. Another feature is the importance of *Taxus baccata* between 5,000 and 2,000 B.P. This species has also been important between 6,500 and 5,000 B.P., as shown by pollen diagrams (Clerc 1985). On the other hand, the data from the Sarrasins cave (French Alpine foreland) (Thiébault 1988) reveal the disappearance of *Taxus* and forest exploitation during about 1,000 years (Figure 1) with the succession of the following stages:

(a) oak stage with *Fraxinus excelsior, Taxus baccata, Abies, Acer campestre* (late Neolithic);
(b) ash stage with *Taxus* decreasing, *Abies* disappearance, *Fagus sylvatica* increasing with *Laburnum anagyroides* (Chalcolithic, early Bronze age);
(c) *Corylus avellana* stage with oak increasing; yew disappearance (middle and late Bronze age);
(d) new oak stage with *Corylus* and *Fraxinus* (late Bronze age and Iron age).

These charcoal stages may reflect a major exploitation from late Neolithic to early Bronze age followed by a reforestation illustrated by hazel progression during middle and late Bronze age until the forest regeneration.

The north and middle of Catalonia in Spain represents a transitional area between the south of France charcoal model and more southern arid areas. In the Cova del Frare cave (Ros-Mora 1985) a postglacial succession quite similar with Languedoc has been identified. At this locality a deciduous oak stage (early and late Neolithic) is followed by a *Quercus ilex* stage (Chalcolithic and Bronze age). Pines (*Pinus halepensis*) were not recorded before early Bronze age, while *Taxus baccata* has a relative importance during Neolithic. This may be related to the particular bioclimatic features of this area in the Iberic Penisula.

The typical thermo-mediterranean conditions (nowadays) of the region of Valencia (southeastern Spain) present new sequences. In Cova de l'Or we have recorded an early Neolithic phase with *Quercus ilex* and scarce *Quercus faginea* (Vernet *et al.* 1983). These results are similar to the displacements observed in the last Glacial. Climacic vegetation is mainly composed of *Quercus ilex* formations and not deciduous oaks as in northern areas. Therefore, this scheme is more meso-mediterranean than supra-mediterranean. Our conclusions are corroborated by the pollen analysis from Padul, at the foot of Sierra Nevada (southern Spain) (Pons & Reille 1986). For these authors, early Holocene climatic vegetation is a thermophilous formations with *Quercus ilex* and *Pistacia*. Moreover wild olive tree (*Olea europaea* var. *sylvestris*) found in charcoal samples played an important role, particularly from ca. 7,500–8,000 B.P. to present time. In Padul, the beginning of the continuous pollinic graph of *Olea* has been dated to 7, 840 B.P. (Pons & Reille 1986). Charcoal analysis in Cova de l'Or recorded an olive tree increasing and the beginning of a *Pinus halepensis* presence at ca. 6,800 B.P. In Padul, a *Pinus* pollens

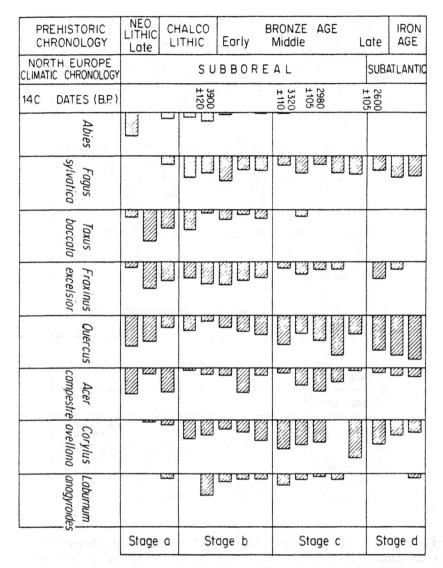

Figure 1. Charcoal analysis diagram of Les Sarrasins cave (French Alpine foreland) (after Thiébault 1983).

development may be related to this scheme. This history can be explained by the increase of human influence.

Pollen analysis and geomorphological studies in cave deposits are in accordance with our results (Fumanal-Garcia & Dupré-Ollivier 1986), particularly with data concerning the degradation of vegetation. Our results also demonstrate the great importance of *Pinus halepensis* and *Olea europaea* var. *sylvestris* in the middle Neolithic. This evidence was corroborated by another

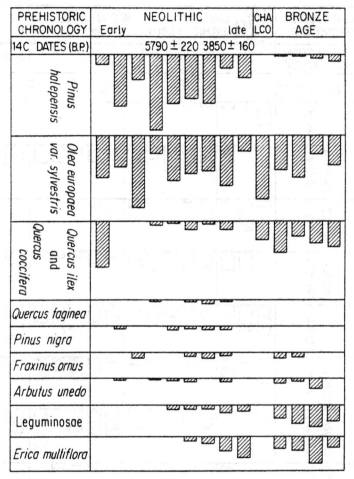

Figure 2. Charcoal analysis diagram of Recambra cave (southeastern Spain) (after Grau-Almero 1984).

charcoal diagram at Recambra, a site nearby Cova de l'Or (southeastern Spain) (Grau-Almero 1984) (Figure 2). Here, there is evidence of an early middle Neolithic *Pinus halepensis* charcoal stage with *Olea europaea* var. *sylvestris*, scarce *Quercus faginea* and *Pinus nigra*. However, with Recambra we know post-Neolithic evolution. Therefore, during the Chalcolithic and the Bronze age the *Pinus halepensis* charcoal remains become more scarce and *Quercus faginea* and *Pinus nigra* disappear. These species were replaced by plants from *Rosmarino-Ericion* (e.g. *Erica* and Leguminosae) and *Quercus coccifera*. This history, which culminated in the present vegetation, represents a regressive evolution in contrast to that of the northern boundary where a cyclic evolution has been recorded.

In thermophilous area in southern France, particularly in Provence, a

167

progression of *Pinus halepensis* during the early Neolithic has been identified. This evolution is similar to the thermo-mediterranean model but may differ in the scarcity or absence of *Quercus ilex* and *Olea europaea* var. *sylvestris*. In this area, *Quercus pubescens* may be present and abundant in coastal regions and inland.

Conclusion

Charcoal analysis provides evidence of preponderant climatic influences in the vegetation transformation from Paleolithic time to the early Neolithic. From this time on, man's influence prevails in Mediterranean vegetation changes. There was a decrease of glacial influences with latitude (Vernet 1986) such as in North Africa the difference cold/warm must have been almost nonexistent as expressed in vegetation zones.

From the anthropic point of view, middle Neolithic populations were the first mediators of vegetation degradation. These were followed by Chalcolithic, Bronze and Iron age, Roman and Middle Age people. Finally, forest potentialities of the Mediterranean area are certainly important in meso-mediterranean zones of subhumid climate and in the surrounding mountains. This is quite different in thermo-mediterranean zones of semi-arid and arid climates, where degradation is more pronounced.

References

Bazile, F., Bazile-Robert, E., Debard, E. & Guillerault, P. 1986. Le Pléistocène terminal et l'Holocène en Languedoc rhodanien; domaine continental, littoral et marin. *Revue de Géologie dynamique et de Géographie physique* 27(2): 95–103.

Bazile-Robert, E. 1981. Flore et végétation des gorges du Gardon à la moyenne vallée de l'Hérault, de 40,000 à 9,500 B.P. d'après l'anthracoanalyse. *Paléobiologie continentale* 12(1): 79–90.

Clerc, J. 1985. Première contribution à l'etude de la végétation tardiglaciaire et holocène du piémont dauphinois. *Documents Cartographie écologique* 28: 65–83.

CLIMAP project members 1976. The surface of the ice-age earth. *Science* 191: 1131–1137.

Durand, A. & Vernet, J.L. 1987. Anthracologie et paysages forestiers médiévaux, à propos de quatre sites languedociens. *Annales du Midi*, 99(180): 397–405.

Fumanal-Garcia, M.P. & Dupré-Ollivier, M. 1986. Aportaciones de la sedimentologia y de la palinologia al conocimiento del paleoambiente valenciano durante el Holoceno. *Proceedings of the symposium 'Quaternary climate' in western mediterranean*, Madrid, 325–343.

Grau-Almero, E. 1984. *El hombre y la vegetacion del Neolitico a la edad del Bronce valenciano en la Safor, provincia de Valencia, segun el analisis antracologico de la cova de la Recambra.* Tesis licenciatura, Facultad de Geografia e Historia, Valencia, 130 pp.

Jalut, G., Delibrias, G., Dagnac, J., Mardones, M. & Bouhours, M. 1982. A palaeoecological approach to the last 21,000 years in the Pyrenees, the peat bog of Freychinède (alt. 1350m, Ariège, south France). *Palaeogeography, Palaeoclimatology, Palaeoecology* 40: 321–359.

Jalut, G., Sacchi, D., & Vernet, J.L. 1975. Mise en évidence d'un refuge tardiglaciaire à moyenne altitude sur le versant nord-oriental des Pyrénées. *C.R. Académie des Sciences, Paris*, 280, D: 1781-1784.

Ozenda, P. 1975. Sur les étages de végétation dans les montagne du bassin méditerranéen. *Documents*

168

cartographie écologique, XVI: 1-32.

Planchais, N. 1982. Palynologie lagunaire de l'étang de Mauguio, paléoenvironnement végétal et évolution anthropique. *Pollen et Spores* 24(1): 93-118.

Pons, A. & Reille, M. 1986. Nouvelles recherches pollenanalytiques à Padul (Granada), la fin du dernier glaciaire et l'Holocène. *Proceedings of the symposium 'Quaternary climate in western Mediterranean'* Madrid, 405-420.

Ros-Mora, M.T. 1985. *Contribucio antracoanalitica a l'estudi de l'entorn vegetal de l'home, del Paleolitic superior a l'edat del Ferro a Catalunya.* Tesi licenciatura, Univers. autonoma de Barcelona, 198 pp.

Sabatier, M. & Van Campo, M. 1984. L'analyse en composantes principales de variables instrumentales appliquée à l'estimation des paléoclimats de la Grèce il y a 18000 ans. *Bull. Société botanique de France* 131: 85-96.

Thiébault, S. 1983. *L'homme et le milieu végétal à la fin du Tardiglaciaire et au Postglaciaire, analyse anthracologique de six gisements des Préalpes sud-occidentales.* Thèse 3e cycle, Université Paris I, 215 pp.

Thiébault, S. 1988. L'homme et le milieu végétal. Analyse anthracologique de six gisements des Préalpes au Tardi-et au Postglaciaire *D.A.F.*, 15, 112 pp.

Triat-Laval, H. 1979. *Contribution pollenanalytique à l'histoire tardi et postglaciaire de la végétation de la vallée du Rhone.* Thèse Aix-Marseille III, 343 pp.

Van Campo, M. 1984. Relations entre la végétation de l'Europe et les températures de surface océaniques après le dernier maximum glaciaire. *Pollen et Spores* 26(3-4): 497-518.

Vernet, J.L. 1973. Etude sur l'histoire de la végétation du sud-est de la France au Quaternaire d'après les charbons de bois principalement. *Paléobiologie continentale* 4(1): 1-90.

Vernet, J.L. 1980. La végétation du bassin de l'Aude au Tardiglaciaire et au Postglaciaire d'après l'analyse anthracologique. *Review of Palaeobotany and Palynology* 30: 33-55.

Vernet, J.L. 1986. Changements de végétation, climats et action de l'homme au Quaternaire en Méditerranée occidentale. *Proceedings of the symposium 'Quaternary climate in western Mediterranean'*, Madrid, 535-547.

Vernet, J.L., Badal-Garcia, E. & Grau-Almero, E. 1983. La végétation néolithique du sud-est de l'Espagne, Valencia, Alicante, d'après l'analyse anthracologique. *C.R. Académie des Sciences*, Paris, 296(3): 669-672.

Vernet J.L. & Thiébault, S. 1987. An approach to northwestern Mediterranean recent prehistoric vegetation and ecologic implications. *Journal of Biogeography* 14: 117-127.

10. Plant invasions in Southern Europe from the Paleoecological point of view

A. PONS, M. COUTEAUX, J.L. DE BEAULIEU and M. REILLE

Abstract

Paleoecology distinguishes between natural invasions and anthropogenic invasions. The first are coordinated processes which show the permanent ecological role of plants. The second result from environmental changes (transgressions, resurgences or propagation) or from an introduction by man.

Introduction

Invasions are by nature historical phenomena. Therefore, their description and comprehension are among the main objectives of paleoecology. Paleoecology distinguishes between 'natural' invasions which mainly correspond to climatic changes and which occurred before the time when man had a direct impact on ecosystems, and 'anthropogenic' invasions, which are related to man's action and are therefore more recent.

Natural invasions

There are abundant paleoecological and especially pollen analytical data available concerning invasions. An enumeration would be tedious. It seems more convenient to present here two constant characteristics of natural invasions which are particularly clear in invasions related to the two climatic improvements that occurred at the outset of the last glaciation.

A marked climatic improvement took place some 13,000 years ago. It led throughout Europe to the replacement of a tundra, steppe and ice cover by forest vegetation. Everywhere, except at high elevation, there appeared the same succession of invasions: first *Juniperus*, then *Betula*, lastly *Pinus*. But the most detailed analyses (Usinger 1982, Beaulieu & Reille 1983, Gaillard 1984) show that each successive genus was represented by different species according to regions.

F. di Castri, A. J. Hansen and M. Debussche (eds.), Biological Invasions in Europe and the Mediterranean Basin.
169–177. © 1990, *Kluwer Academic Publishers, Dordrecht.*

170

These natural invasions are coordinated at two levels because:
- the order of the invasions mainly depends on the biological characteristics of the genera (reproduction, dispersal and establishment ability)
- consequently, the ecological peculiarities of the species (e.g. tolerance levels) is but of secondary importance in the selection of the invaders.

Some 10,000 years ago, after a transitory phase of climatic degradation which has led to the reestablishment of a treeless vegetation, a second climatic improvement took place, which brought about a complex series of trees invasions. These invasions represent the vegetal history of the first part of our postglacial times.

The postglacial history of some arboreal taxa in southeastern France, for example, demonstrates on the basis of 80 sites and with the support of nearly 300 14C dates (Beaulieu *et al.* 1982) the permanent role of species through their invading process.

Some invaders expanded from well known glacial 'refuges', following a simple scenario and at a regular speed. This is the case for *Betula, Pinus* and *Corylus.* During their first expansion, owing to their precocious reproductive maturity and/or efficient and long distance dispersal, and to low edaphic requirements, these trees could invade the whole surface available because they did not encounter any well-established precursor. They are 'pioneer' plants whose main permanent role is that of accompanying taxa.

On the contrary, other invaders, whose past refuges are not well localized, followed obscure paths before finding conditions suitable for expansion, which they sometimes reached long after leaving their refuges. For example, *Quercus* cf. *pubescens*, which has been present in the lower Rhône valley since the Bolling times (ca. 12,000 B.P.), became well established there only 4 millennia later, i.e. at a relatively late Boreal time. *Fagus*, which arrived in Auvergne, France, at ca. 6,000 B.P., began to expand only towards 4,700 B.P. *Abies*, which has been present in Velay, France, since 5,600 B.P., became abundant only at ca. 3,200 B.P.

It is obvious that poor dispersal ability added to difficulties in coping with competing taxa, as well as particular edaphic requirements, led these taxa to adopt an invading behavior which was not always timid, but always complex. Some of them managed to develop vast and new ecosystems. These taxa were destined to constitute stable vegetation types: we can call them 'climacic' plants.

Anthropogenic invasions

Invasions attributable to man characterize the last 5–6 millennia. None of them are known to have occurred before the Neolithic when man became sedentary, stock-breeder and farmer (Pons 1984). These invasions are of two kinds: the first are natural consequences of man-induced changes in environment, the second are the results of voluntary or involuntary introduction of plants whose expansion thereafter man may or may not have controlled.

Invasions associated with human changes in environment

Among the numerous and diverse invasions that naturally result from anthropogenic modifications in environment and ecosystems, as evidenced by paleoecology, there appears three types of processes: *transgressions, resurgences* and *propagations.*

Transgression occurs when one or several plants from a nearby natural habitat invade areas in which they were formerly absent or played but a secondary role.

Two examples illustrate this type of invasion.

– *Erica arborea*, occupied in Corsica the steep region of the Cap and the foothills of the mountain range in the center of the island until the 5th millennium B.P., while deciduous forests and mixed formations with yew (*Taxus baccata*), as in the mountain zone, prevailed in the eastern deep soiled plain. The anthropogenic clearing of these formations led as early as the 5th millennium, but especially during the historical times and the last 7 centuries, to the transgression of *Erica arborea* on the coastal alluvial plains where the species now forms vast maquis (Reille 1975, 1984).

– *Quercus ilex* occupied very early in the Mediterranean region the present stage of the semi-arid climate. Oak forests with *Quercus ilex* prevailed at Padul (near Granada, Spain) as early as the Lateglacial interstadial and again after the cold and arid episode of the younger Dryas, at the beginning of the Postglacial (Pons & Reille 1986). In a same manner, in the present stage of the subhumid mediterranean climate, *Quercus ilex* became established at the beginning of the Postglacial (about 9,000 B.P.), as is evidenced in the pollen diagram of the narrow site – which therefore reflects the vegetation of a limited area – of Tourves (near Toulon, southern France) in the calcareous hills region. But it occupied, within the regional vegetation, areas with hardest edaphic conditions, where the deciduous *Quercus* could not compete (Pichard 1987). From these places, the transgression of *Quercus ilex*, and of formations in which it was the dominant element, was possible later on the greatest part of the subhumid stage, because of the reduction and even the disappearance of deciduous *Quercus* forests, as a result of human action. Finally, this general transgressive process led to the establishment of most sclerophilous forests either of *Querces ilex* and *Q. coccifera* in Provence, France (Triat-Laval 1978) and in Istria, Yugoslavia (Beug 1977), or *Q. ilex* and *Buxus sempervirens* in Languedoc, France (Vernet 1973), *Q. ilex* (including *Q. rotundifolia*) and *Q. suber* in the Rif, Morocco (Reille 1977) and in Kroumiria, Tunisia (Ben Tiba & Reille 1982).

It is important to note that some ecosystems resulting from transgressions, especially ancient ones, can easily be misinterpreted by the ecologist who analyzes the present state of things. The natural maquis of Cap, Corsica, and the anthropogenic maquis of the eastern plain in Corsica may be confused. Moreover, in the greatest part of the mediterranean subhumid bioclimatic stage, forests mostly made up of *Quercus ilex* can easily be interpreted in

172

terms of climax. These mistakes may have grave consequences, even at the practical level of management or protection.

There is a *resurgence* when man created conditions favourable to the development of plants which were once wide spread but became localized in the more or less remote past.

The range contraction and expansion of *Pinus sylvestris* on the south-eastern border of the Massif Central, France, serves as an example. There, at the end of the temperate Lateglacial stage (about 11,000 B.P.) and again at the beginning of the Postglacial (toward 10,000 B.P.), *Pinus sylvestris* formed small isolated populations in a sparse woodland (Beaulieu *et al.* 1984). But as soon as forest cover was established (at about 9,000 B.P.), this pine receded and was confined to small areas scattered in the most arid stations (Reille & Pons 1982). More recently, it invaded areas abandoned during the agricultural recession that followed the fall of the Roman Empire after the gallo-roman culmination in the 3rd and 4th centuries. As an heliophilous and edaphically accommodating species, it had a preponderant role in secondary formations before it was supplanted by natural forest taxa such as *Fagus sylvatica* and deciduous *Quercus* species. A second resurgence took place in the middle of the 19th century when, immediately after the demographic maximum, there was a recession of traditional agricultural activities. The use of pine in the reafforestation of uncultivable slopes and its invasion of abandoned fields let to an omnipresence of *Pinus sylvestris* which, though temporary, could induce the false notion of a *Pinus* climax (Carles 1957).

There is a *propagation* when man's action in a region favoured plants that arrived via an intrinsically natural migration. This is the case of *Picea abies* in a large part of the southern French Alps. This eastern species came there recently following two converging migration routes (an ancient one through the Italian versant and the Maritime Alps and a more recent one through the Jura and the northern Alps. It was able to expand following its arrival (from 3,000 to 1,000 B.P.) owing to clearings made by man in the former forest cover.

Invaders introduced by man

There is little paleoecological information available concerning invasions related to an introduction by man. There are two reasons for this. The first comes from a difficulty in identifying the pollen of herbaceous plants to the species level. Generic determination does not always bring valuable information regarding vegetal transports. The second reason lies in the difficulty in interpreting the most recent pollen records in sedimentary deposits. Sedimentation rates are often not sufficient to document events as rapid as vegetal invasions resulting from human transport. Also, human activities have often rearranged sediment layers.

Numerous humid areas bordering the Mediterranean, where sedimentation

La Chaumette - Brameloup, Lozère.

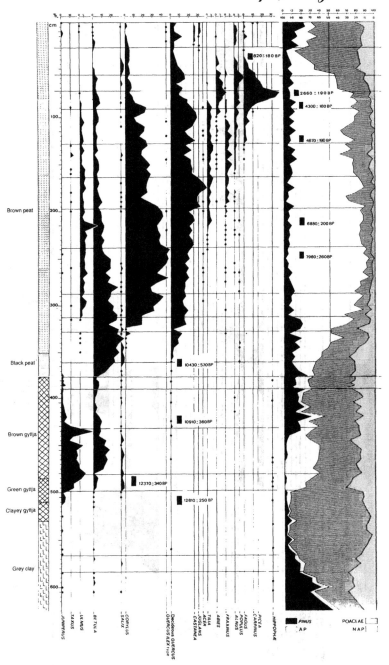

Figure 1. This diagram concerning the Mont d'Aubrac (from Beaulieu and Reille, in Beaulieu *et al.* 1985) clearly shows that herbs related to cultures (*Rumex, Polygonum, Euphorbiaceae, Chenopodiaceae*) as well as *Calluna*, which forms vast heaths in the region – all these vegetals producing very little and poorly dispersed pollen – appear as invaders concurrently with the first continuous anthropogenic activity some 3,000 years ago (Subboreal). This is well reflected in the decrease of the Arboreal Pollen and the first record of cereals pollen. But some of these plants appeared in the region during the first warm phase at the end of the last glaciation (the

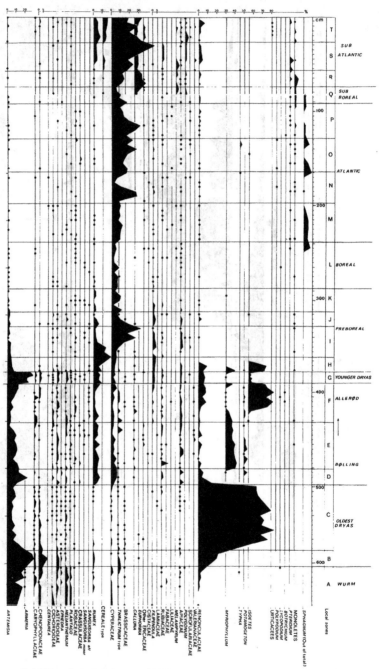

so-called Würm glaciation), between 15,000 and 13,000 B.P. (Oldest Dryas). Thereafter, these plants began to expand during the first marked temperate episode, between 13,000 and 10,000 B.P. (Bölling-Alleröd) and until the beginning of the Postglacial (Preboreal), at ca. 9,000 B.P. The domination of forest ecosystems overshadowed these heliophytes, of which only a few pollen grains are occasionally recorded. However, after an interval of 6 millennia, new vast areas were again open to them as a result of anthropogenic forest clearance and cultivation.

provides valuable paleoecological information, have been drained for several millennia. Peat bogs, which represent the best paleoecological records, have been diversely exploited throughout Europe. Thirdly, deforestation can lead to re-deposition on lacustrine deposits or even the complete turnover of sediment layers (Coûteaux 1985).

Even when botanical identification of fossil pollen is possible (e.g. for *Castanea sativa* and *Pinus sylvestris*), the sediment record often only indicates the time when the species was established. It does not permit the tracing of the different steps of its transport or invasion.

Paleoecology, however, provides explanations concerning the invading ability of species imported for reafforestation. It is the case of *Cedrus* which disappeared from southern Europe during one of the middle Pleistocene cycles (a cycle which is comparable to the present one). Therefore, this tree lived there under conditions similar to the present and this explains why its use in reafforestation has been such a success in this century. Similarly, one can also easily understand the success of *Picea abies* in the reafforestation of the Massif Central, France, if one keeps in mind that it has existed in this region until the last Pleniglacial, hardly 30 millennia ago (Beaulieu & Reille 1985).

Invasions by ruderal and segetal plant species

Paleoecology highlights a phenomenon that is partly related to a resurgence and partly to introduction by man over the last several millennia: the gradual expansion of herbaceous or shrub species. These plants are either segetal species that very seldom escape from cultures, or species occupying ruderal habitats such as pastures, paths, fallows, pastured woods and shrublands. These assemblages are generally found in places rich in nitrates and their expansion is mainly related to deforestations. Moreover, they all belong to taxa which formed a steppe vegetation during the Pleniglacial and the Lateglacial (e.g., *Artemisia, Caryophyllaceae p.p., Chenopodiaceae, Poaceae p.p., Plantago, Polygonum, Rumex*). In fact, heliophily is the main common ecological characteristic of these invaders (Coûteaux & Pons 1987). Limitations in the botanical determination of pollen do not make it possible to distinguish between:
- species from former steppe formations for which the clearing of forests by man brought about a resurgence after a survival period of 3 to 8 millenia in 'refuges' (the 'steppo-ruderal species' described by Coûteaux (1969)); numerous diagrams illustrate this process (Figure 1);
- species more or less involuntarily introduced by man from the cradle of agriculture, i.e. the Near East, where discontinuous forest or shrub formations prevailed during the Postglacial until the first agricultural expansion (Bottema & Van Zeist 1981).

It is obvious, however, that the resurgence potentiality of steppo-ruderal

176

species was proportionate, in each region, to the number, distribution, and richness of their extra-forest refuges during the first part of the Postglacial. Because of the complexity of its physical environment, the whole Mediterranean area could have provided a wide range of refuges, and a rich flora could have developed there long before the last climatic cycle (Quezel 1985). Thus, some taxa that had been long restricted to narrow and edaphically well characterized refuges, could generate much more dynamic biotypes (especially caryotypes) when vast areas were opened to them.

All the conditions were combined in Mediterranean regions for numerous resurgent steppo-ruderal species to join heliophytes whose seeds were mixed with those of cultures.

Conclusion

Finally, paleoecology shows that natural invasions are coordinated processes during which plants reveal their ecological potentiality. Invasions resulting from man-induced changes in the environment and the ecosystems can work as transgressions, resurgences or propagation; they bring about deep modifications that cannot be evidenced without the help of historical data. Invasions resulting from voluntary or involuntary transports, hence relatively recent, are not easily apprehended through paleoecology, chiefly because ever since man began to clear land for cultivation, he has, so doing, also removed his own traces. In the process of heliophytes expansion that accompanied the settlement of man, the resurgence of steppo-ruderals seems to have played, at least in Mediterranean regions, a preponderant role in comparison with anthropogenic propagations.

References

Beaulieu, J.L. de, Coûteaux, M., Pons, A. & Triat-Laval, H. 1982. Première approche d'une histoire postwürmienne de quelques taxons arboréens dans le sud-est de la France. *Rev. Paléobiologie. Genève* vol. spéc.: 11–24.

Beaulieu, J.L. de & Reille, M. 1983. Paléoenvironnement tardiglaciaire et holocène des lacs de Pelléautier et Siguret (Hautes-Alpes). 1) Histoire de la végétation d'après les analyses polliniques. *Ecologia Mediterranea* 9: 19–36.

Beaulieu, J.L. de, Pons, A. & Reille, M. 1984. Recherches pollenanalytiques sur l'histoire de la végétation des Monts du Velay, Massif Central, France. *Dissertationes Botanicae* 72: 45–70.

Beaulieu, J.L. de, Pons, A. et Reille, M. 1985. Recherches pollenanalytiques sur l'histoire tardiglaciaire et holocène de la végétation des Monts d'Aubrac (Massif Central, France). Elsevier Sc. Pub, B.V., Amsterdam. *Review of Paleobotany and Palynology* 44: 37–80.

Beaulieu, J.L. de & Reille, M. 1985. L'intérêt paléoécologique du remplissage sédimentaire des maars du Velay occidental. *C.R. Acad. Sc. (Paris)* 301 sér. II: 443–448.

Ben Tiba, B. & Reille, M. 1982. Recherches pollenanalytiques dans les montagnes de Kroumirie (Tunisie septentrionale): premiers résultats. *Ecologia mediterranea* 8: 75–86.

Beug, H.J. 1977. Vegetationsgeschichtliche Untersuchungen im Küsternbereich von Istrien. *Flora*

166: 357–381.

Bottema, S. & Van Zeist, W. 1981. Palynological evidence for the climatic history of the Near East, 50,000–60,000 B.P. *Coll. Intern. C.N.R.S.*, 598 Préhistoire du Levant. Maison de l'Orient, Lyon, Edit. C.N.R.S., Paris, pp. 111–132.

Carles, J. 1957. *Carte de la végétation de la France*, C.N.R.S. Toulouse, notice détaillée de la feuille 59, 40 pp.

Coûteaux, M. 1969. Recherches palynologiques en Gaume, au pays d'Arlon, en Ardenne méridionale (Luxembourg belge) et au Gutland (Grand Duché de Luxembourg). *Acta geographica Lovaniensia* 8: 1–193.

Coûteaux, M. 1985. Recherches pollenanalytiques au Lac d'Issarlès (Ardèche, France) : évolution de la végétation et fluctuations lacustres. *Bull. Soc. Roy. Bot. Belgique* 177: 197–217.

Coûteaux, M. & Pons, A. 1987. La signification écologique du pollen d'*Artemisia* dans les sédiments quaternaires. I. Le problème. II. L'éventualité d'un rôle d'*Artemisia* palustre. *Bull. Soc. Bot. France* 134, Lettres bot.: 283–292.

Gaillard, M.J. 1984. Etude palynologique de l'évolution tardi- et postglaciaire de la végétation du Moyen-Pays Romand (Suisse). *Dissertationes Botanicae* 77: 1–322.

Pichard, S. 1987. Analyse pollinique d'une séquence tardi- et postglaciaire de Tourves (Var) en Provence orientale. *Ecologica Mediterranea* 23: 29–42.

Pons, A. 1984. Les changements de la végétation de la région Méditerranéenne durant le Pliocène et le Quaternaire en relation avec l'histoire du climat et de l'action de l'homme. *Webbia* 38: 427–439.

Pons, A. & Reille, M. 1986. Nouvelles recherches pollenanalytiques à Padul (Granada): la fin du dernier glaciaire et l'Holocène. In: F. Lopez (ed.), *Quaternary climate in Western Mediterranean*, Univ. Autonoma Madrid, pp. 407–420.

Quezel, P. 1985. Definition of the Mediterranean region and the origin of its flora. In: *Plant conservation in the Mediterranean area*, Geobotany 7, Dr. W. Junk pub. pp. 9–25.

Reille, M. 1975. *Contribution pollenanalytique à l'histoire postglaciaire de la vegétation de la montagne Corse*. Thèse Université Aix-Marseille III. 206 pp.

Reille, M. 1977. Contribution pollenanalytique à l'histoire holocène de la végétation des montagnes du Rif (Maroc septentrional). In: Recherches françaises sur le Quaternaire, INQUA 1977, suppl. *Bull. A.F.E.Q.* 50: 53–76.

Reille, M. 1984. Origine de la végétation actuelle de la Corse sud-orientale; analyse pollinique de cinq marais côtiers. *Pollen et Spores*, 26: 43–60.

Reille, M. & Pons, A. 1982. L'histoire récente de *Pinus sylvestris* L. en Margeride (massif Central, France) et la signification de cette essence dans la végétation actuelle. *C.R. Ac. Sc. (Paris)* 294, sér. III: 471–474.

Triat-Laval, H. 1978. *Contribution pollenanalytique à l'histoire tardi- et postglaciaire de la végétation de la basse vallée du Rhône*. Thèse doctorat Aix-Marseille III. 343 pp.

Usinger, H. 1982. Pollenanalytische Untersuchungen an spätglazialen und präborealen Sedimenten aus dem Meerfelder Maar (Eifel). *Flora* 172: 73–409.

Vernet, J.L. 1973. Etude sur l'histoire de la végétation du sud-est de la France au Quaternaire, d'après les charbons de bois principalement. *Paléobiologie continentale* 4: 1–93.

11. Mediterranean weeds: exchanges of invasive plants between the five Mediterranean regions of the world

MARILYN D. FOX

Abstract

The five regions of the world sharing a climate like that of the Mediterranean Basin have exchanged, and continue to exchange, weedy plant species. This exchange is seen to be of two forms: an earlier **primary** invasion of the other four regions by aggressive annual weeds from the Mediterranean Basin and a later **secondary** invasion by woody species, often between the four regions and from them back to the Mediterranean Basin. Of the woody invaders two interesting groups are the conifers and the succulents. There is also a **tertiary** invasion *within* each of the mediterranean regions of native species that have become more invasive as a result of human disturbance.

The five regions are seen to comprise three groups. The pivotal 'crossroads' (di Castri 1981) of the Mediterranean Basin itself, the other more recent (Pleistocene) group of Chile and California, and the older (Gondwanan) group of South Africa and southern Australia. As well as sharing important evolutionary and biogeographic traits (Naveh & Whittaker 1980, di Castri 1981), the two subsidiary groups are seen as sharing important patterns of settlement and subsequent trade. California and Chile, as well as having other strong links with the Mediterranean Basin, were discovered and first settled by people from the Mediterranean Basin and this early contact must have dictated the rate and extent of invasions. The two older, southern regions, Australia and South Africa, on the other hand, were settled by people from northern Europe and only latterly had direct trade links with Mediterranean countries.

The prognosis for the future of invasions in the mediterranean regions is for a reduction in agrestal weeds but an increase in community weeds, particularly woody secondary invasions.

Introduction

The five regions of the world sharing a climate like that of the Mediterranean

F. di Castri, A. J. Hansen and M. Debussche (eds.), Biological Invasions in Europe and the Mediterranean Basin.
179–200. © 1990, Kluwer Academic Publishers, Dordrecht.

180

Basin have exchanged, and continue to exchange, weedy plant species. This exchange is seen to be of two forms: an earlier **primary** invasion of the other four regions by aggressive annual weeds from the Mediterranean Basin and a later **secondary** invasion by woody species, often exchanged between the four regions and from them back to the Mediterranean Basin. There is also a **tertiary** invasion within each of the mediterranean regions, of native species that have become more invasive as a result of human disturbance.

The prognosis for the future of invasions in the mediterranean regions is for a reduction in agrestal weeds but an increase in 'community' weeds, particularly woody secondary invasions. In regions where the maintenance of high native plant richness is desired, there will be considerable cost involved in weed management.

Convergence in mediterranean biotas

There are at least two major aspects of the regions of the world with mediterranean-type climates (see Figure 1) that have considerable phytogeographic significance. One is that these five regions of the world are widely spaced on separate continents but share essentially the same climate (Aschmann 1973a). Their geographic separation also dictates a biotic separation: to a large extent the biotas of the five regions have evolved from different evolutionary precursors (Raven 1971). This fact and the observation that there is, however, considerable convergence in form and function of the regions' biotas has been a productive avenue of research (di Castri & Mooney 1973, Mooney 1977, Cody & Mooney 1978, di Castri *et al.* 1981, di Castri 1981, Margaris & Mooney 1981, Kruger *et al.* 1983, Dell *et al.* 1986). The second major aspect is that of these regions, the Mediterranean Basin has been the source of many invasive organisms that have now become naturalised in the other regions, and subsequently there has been limited but continuing exchange of organisms between all five regions. In the Preface to *Mediterranean Type Ecosystems – Origin and Structure* (di Castri & Mooney 1973), considering the convergence of the regions, the editors note that 'one other general feature is the massive interchange of cultivated and weed species of plants that has occurred between the five areas of the world that have a mediterranean-type climate, with the Mediterranean basin region itself as a major source'.

Some authors have dealt with the Mediterranean basin as a major source of adventive plants (Ames 1939, Vavilov 1951, Baker 1965a, 1965b, Sakamoto 1982), while others have studied one or a subset of the five regions, often compared with the Mediterranean Basin. A recent paper (Macdonald *et al.* 1988) has reviewed introduced species in nature reserves in the four mediterranean areas (excluding the Mediterranean Basin). Cowling & Campbell (1980) made comparisons of California, Chile and South Africa in terms of the structure of the vegetation. Naveh (1967) and Barbour *et al.* (1981) made the comparison between Israel and California while Raven (1971) has compared

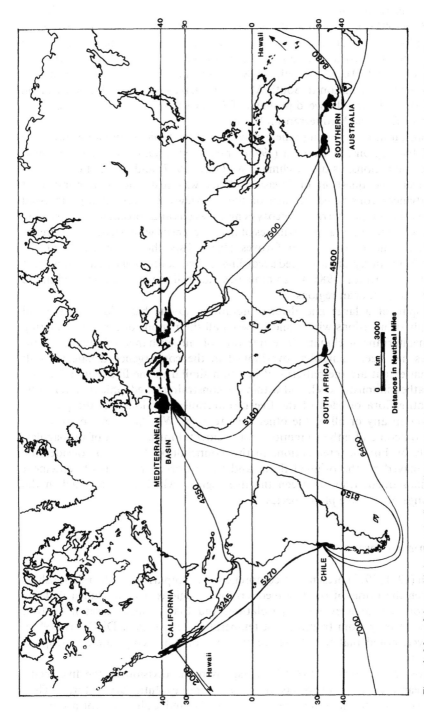

Figure 1. Map of the world showing the five mediterranean regions (after di Castri 1981). Distances between the regions is shown in nautical miles (1 nautical mile = 1,853 km) (from several sources).

182

California with Chile, as did Cody & Mooney (1978). Axelrod (1973) and Rundel (1981) have described the history of the mediterranean ecosystems in California and Chile respectively. A comprehensive comparison of paired sites in South Africa and southwestern Australia was reported by Milewski & Cowling (1984). Macdonald (1984) made an ecological analysis of the Australian contribution to South Africa's invasive alien flora. Wells & Stirton (1982a) considered the weed problem of South African pastures and the influx of plants from the Mediterranean Basin.

Some authors have comprehensively treated the topic of human disturbance to all five regions. Aschmann (1973b) presented a comprehensive treatment of the five regions, as did Aschmann & Bahre (1977) and Fox & Fox (1986a). However, the question of invasive plants was just one of a number of disturbances considered in each of these reviews. Groves (1986) addressed the invasion of mediterranean ecosystems by weeds and considered the invasion process and the biological attributes of invasive weeds with particular reference to management. Table 1 in Groves (1986) lists the major weeds of the mediterranean regions arranged according to their country of origin and present distribution. Kloot (1985a) also tabulates Australian plants naturalised in the other mediterranean regions.

In spite of a large and diverse literature, no-one has addressed the full panoply of invasions within and between all the regions of the world sharing the mediterranean climate. The question of the exchange of invasive plant species has been somewhat overlooked in the convergence debate as well; to what extent are the patterns of invasion similar in the five regions? Taken at a rather simplistic level, and within the constraints of edaphic requirements, the entire flora of each of the five mediterranean regions has the potential to grow in any or all of the other four regions. Whether they do so or not is the outcome of either intentional or accidental introduction of propagules, usually by human intervention. In this contribution I wish to consider the roles played by the original floras and the historical patterns of settlement and subsequent trade between the five regions which have resulted in the exchange of invasive plant species.

The original floras

Raven (1971, 1973) presented a comprehensive comparison of the relationships between the floras of the five mediterranean regions. He concluded that the flora of each area evolved in isolation, and that in most cases it evolved almost entirely from tropical, not temperate, predecessors. Di Castri (1981), however, noted that the floras are mixtures of both temperate and tropical elements.

Naveh & Whittaker (1980) have espoused the division of the five mediterranean regions into two sets, based on shared evolutionary histories. They proposed that the Mediterranean Basin, Chile and California form one set.

These regions share a relatively recent evolutionary history (Raven 1973, Solbrig *et al.* 1977), have low species diversities (except when increased by human disturbance), high richness of annuals, and rejuvenation of soils following orogenic movements and glacial climatic changes. The other older set comprises the Cape Region of South Africa and southern Australia. Both regions have had a long evolutionary association with oligotrophic soils and now exhibit high species diversities, but with a conspicuous absence of annuals.

Di Castri (1981) in his review of the five regions also suggests the same two clusters but points out that the highest absolute similatities are between California and Chile as one set, and South Africa and southern Australia as the other, with the Mediterranean Basin, itself a very heterogeneous region, in a group by itself. The main points of similarity or contrast, with respect to invading species, are summarized below from di Castri (1981).

Mediterranean Basin complex

Whereas its principal phylogenetic and biogeographic affinities are with California, in terms of the history of human degradation, the Mediterranean Basin is most like Chile. Di Castri (1981) emphasized the internal heterogeneity of the Mediterranean Basin, and the north-south and east-west divisions that can be identified.

Compared with the other mediterranean regions, the Mediterranean Basin has been home to people with their husbanded animals and cultivated plants for a much longer period (le Houerou 1981). The Mediterranean region has been the centre of origin of many cultivated species as well as the centre of dispersion of annual weeds.

California – Chile complex

Both regions have relatively young mountain systems and have active volcanic and earthquake zones. The remarkable similarity of terrain has often been noted (e.g. Bradbury 1981). There is a winter concentration of rainfall augmented in both regions by coastal fog. Since the Pleistocene there have been some interchanges of taxa and the floras are mixtures of elements of tropical and temperate origins. However, the two floras are fundamentally different (Raven 1973) and have evolved in isolation.

Chile, because of former Gondwanan links, has greater phylogenetic affinities with the other mediterranean regions of the Southern Hemisphere than does California. Endemism is higher in Chile and there is greater diversity of growth forms. Possibly because of the longer history of human exploitation, the Chilean vegetation is more open, more diverse and more stratified. There are more spiny plants and succulents in the Chilean flora, possibly related to the long history of grazing. The grasses and weedy annuals have greater cover and

184

are distributed more widely in Chile than in California. This feature is no doubt also linked with the higher nutrient status of the Chilean soils.

Both areas were colonized by the Spaniards, but the 200 year delay in settling California (Aschmann & Bahre 1977) has meant that human impact is greater in Chile. Mooney *et al.* (1972) maintain that the lack of convergence between Chile and California is due to their different treatments by people.

Australia – South Africa complex

The vegetation of both of these mediterranean regions has developed on ancient, stable, basement complexes which exhibit more subdued landforms. The soils are oligotropic, very deficient in phosphorus, nitrogen and micro-nutrients. In both regions an exceptionally poor, acidic (sandy) soil supports heathlands (fynbos in South Africa) while relatively more fertile soils support mediterranean shrublands. However, because of the relatively richer soils and flat topography, it is this latter formation that has been extensively cleared for cropping in both regions.

Interestingly, both regions' floras still exhibit the summer growth, relic of their tropical origin (Specht 1973, Goldblatt 1978), a feature not found in the other mediterranean regions with their spring growth pulse. The introduced species thereby have the competitive advantage, maintaining their normal spring growth (Edmonds & Specht 1980), and growing after the winter rain. In central Europe, however, the Mediterranean annual weeds have shifted to a summer growing season (Kornas 1983).

Contrasting these two regions, Australia is flatter and the mediterranean region is in two major disjunct regions. The shrubland (mallee) of mediterranean southern Australia is structurally diverse with an overstory of tall (2–5 m) woody shrubs of *Eucalyptus*. The present-day mediterranean shrubland (renosterveld) of South Africa is a structurally simple disclimax community (Boucher & Moll 1981). The original community may have been structurally more comparable with that in Australia.

In considering the exchange of invasive species these groupings of mediterranean regions gain even greater validity. However, before addressing this exchange it is necessary to review two other important features of the mediterranean regions viz: high endemism and the occurence of annuals.

High endemism

High levels of endemism is a general pattern in all the mediterranean regions. In the Mediterranean Basin 10% of the genera and 40% of species are endemic (Raven 1971), while for California the figures are 6% and 48% respectively (Raven & Axelrod 1978). The degree of endemism in the Chilean flora is reputedly higher than that for California. The Cape region is renowned for its exceptionally high incidence of endemism with 21% endemic genera and 73% endemic species (Goldblatt 1978). While in the mediterranean region of Western Australia at least 68% (and possibly as high as 80%) of the 3600 species are endemic (Beard 1969).

Annuals

Raven (1971) contends that annual plants reach the zenith of their diversity in areas with a mediterranean climate. Orshan (1953) gave a value of 30% annuals for the Mediterranean Basin, but in areas subject to intense grazing they often constitute half the total species (Zohary 1962). By comparison, the world average is *ca.* 13% (Raven & Axelrod 1978). The representation of annuals in the Californian Floristic Province is 27% (Raven & Axelrod 1978), similar to values for the Mediterranean Basin. Of the weed flora of California, almost 60% are annual (Robbins *et al.* 1941). The invasive Mediterranean annual grasses in California now form stable annual grasslands (Jackson 1985) whereas in their native habitats they are ruderal species. However, whereas a rich annual flora is characteristic of California and the Mediterranean, there are relatively few annuals in the Cape flora with only *ca.* 10% (Adamson & Salter 1950). In a survey of the weed flora of the Cape of Good Hope Nature Reserve, Macdonald *et al.* (1987) gave the relative composition of the indigenous and alien floras in six growth-forms. Very few (2.7%) of the indigenous flora are trees or shrubs; however, 36.9% of the alien flora are; and whereas only 7.5% of the indigenous species are annual herbs, 39.7% of the aliens are. There also seems to be a paucity of annuals in the

Table 1. The percentage composition of the mediterranean weed floras of three target areas: California (Robbins *et al.* 1941), South Africa (Stirton 1978, Grabandt 1985) and southern Australia (Lamp & Collet 1979), with the weed flora of South Australia in square brackets (Kloot 1987b). The annual/perennial ratios for four of the contribution regions are given (*Barbour *et al.* 1981, **Macdonald *et al.* 1987, ***Kloot 1987a). The figures in parentheses in the body of the table are the proportions of annuals in each component.

	Contributing mediterranean region				
Target area	Mediterra-nean Basin (1.06)	California (0.43)*	Chile (- -)	South Africa (0.08)**	Southern Australia (0.05)***
California	47%	–	22%	13%	18%
(45)	(0.75)	–	(0.60)	(0.17)	(0.75)
Southern Australia	42%	22%	13%	22%	–
(99)	(0.67)	(0.41)	(0.39)	(0.14)	–
[353]	[59%]	[7%]	[9%]	[24%]	–
South Africa	44%	13%	13%	17%	13%
(39)	(0.47)	(0.40)	(0.60)	–	(0.08)

186

flora of the mediterranean regions of southern Australia (Hopper 1979). Kloot (1987a) presents the annual: perennial ratios for four vegetation formations in southern Australia, comparing the native and introduced floras. Of the two mediterranean formations (the other two are semi-arid), there are markedly more annuals in the introduced flora. Of the two, there are more annual weeds in the mallee than in the scleromorphic heathlands.

Table 1 shows the percentage composition of the weed flora (as treated in published sources) of three of the mediterranean regions for the respective other four source regions. The incidence of annuals is given in parentheses. Over 40% of mediterranean weeds in each region are from the Mediterranean Basin, and of these at least half are annual. Although the Californian flora has contributed relatively fewer weeds to both southern Australia and South Africa, approximately 40% of these are annuals. The similar low contributions from Chile are predominantly annual. However, the weeds that have originated in South Africa are principally perennial species with only 15% annuals. Of interest are those Australian plants that have become weeds in California and South Africa. Whereas the majority of those naturalised in California are annuals, the overwhelming pattern is for Australian perennial species to establish in South Africa. This somewhat mimics the original floras of each region, with many annuals in California and relatively few in South Africa.

Recent mixing of the floras

The Mediterranean Basin

The role of the Mediterranean Basin as the origin of many adventive plant species has been expounded by many authorities. A predisposition to weediness would have been imparted by the rise and fall of advancing ice sheets, and then the creation of open habitats by the procedures of early agriculture and grazing would have confirmed that trend. Anderson (1967) first suggested that the high intensity of human disturbance, particularly the husbanding of grazing and browsing animals, was a positive force in the evolution of weedy annuals (see also Aschmann 1973b, Baker 1974, le Houerou 1981). Stebbins (1965) summarized these evolutionary forces by stating 'It seems to me that in the Mediterranean area nothing could become a weed unless it could resist cattle, sheep, goats, donkeys, cows and women that cut wood!'.

There is some difficulty in addressing the old Mediterranean weeds, the true archaeophytes (Thellung 1918–19). These plants could and did move northwards with the expansion of the Roman Empire, eastwards to a more limited extent with the Ottoman Empire, and around the Mediterranean with any number of trade or military contacts. Added to the considerable detective work required to unequivocally state an origin to what are now cosmopolitan weeds, are the equally considerable taxonomic problems involved (McNeill 1976, 1982). Many of the plants found to be useful to humans still have wild or weedy populations (McNeill 1976); the two morphs, crop and weed share the same genotype but have quite distinct ecologies. Nearly all field and garden crops and some tree crops have weed races that belong to the same biological species as the crop, examples include wheat, rice, maize and sorghum (Harlan 1982).

Another taxonomic problem arises from the movement of plants by humans to new localities. Some plants, introduced to the Mediterranean Basin, were first described there by European systematists from adventive populations, ignorant of the actual origin of the plant (Sykora 1988). This has happened elsewhere as well, *Lampranthus tegens* (*L. caespitosus*) (previously *Mesembryanthemum*) was described in Australia long before being described in its native country, South Africa (Brown *in* Ewart & White 1908).

Some of the problems associated with the taxonomic status and original distribution of weeds can be illustrated by an example used by Groves (1986). Groves presented a table of mediterranean weeds, and included the aquatic weed *Myriophyllum brasiliense* as an example of a Chilean species now established in the Mediterranean Basin, California, South Africa and Australia. Apart from the question of the appropriateness of an aquatic weed in such a context, that older species name has been subsumed by the earlier name *M. aquaticum* (Verdcourt 1973). Verdcourt gives the origin as 'South America' but Orchard (1981) is more specific and considers its original distribution to have encompassed Brazil, Uruguay, Argentina and Chile. Its present distribution is essentially throughout the tropical and warm temperate regions of the world, including Central and North America, Europe, Africa, Asia, Australia and the Pacific (Orchard 1981).

Michael (1972) in a paper on the history of weed research in Australia, presented a more comprehensive account of the taxonomic problems of both distinguishing alien from indigenous plants and subsequently untangling the nomenclatural confusion in some weed groups. On a wider scale, Baker (1972) also addressed these and other sources of confusion in the weed literature.

Whereas there appears to have been considerable invasion of Mediterranean plants into contiguous, more continental zones of Europe to the north (Kornas 1983, 1988a, Sykora 1988), there has been less movement into the arid zones to the south and east. Rather there has been a tendency for arid-adapted plants to find niches in the anthropogenic landscape, particularly open areas (Raven 1971, di Castri 1973, Naveh & Dan 1973) of the Mediterranean.

The social history of discovery and settlement of the other four parts of the world with similar climate has played a very great role in the subsequent biotic invasions. Useful sources for distributions of world weeds are Holm *et al.* (1977) and Holm *et al.* (1979), however both should be used in conjunction with regional weed floras. The principal sources used for the invasive species in each mediterranean region are:

Mediterranean Basin
– Casasayas (1990), Guillerm *et al.* (1990), Quezel *et al.* (1990),

188

Chile
- Philippi (1881), Matthei (1963), Vallejo (1968);
California
- Smiley (1922), Robbins *et al.* (1941), Stebbins (1965), Munz (1974);
South Africa
- Goldblatt (1978), Stirton (1978), Neser & Cairns (1980), Macdonald & Jarman (1984), Grabandt (1985), Taylor & Macdonald (1985), Taylor *et al.* (1985);
Australia
- Orchard & O'Neill (1959), Meadly (1965), Michael (1972, 1981), Parsons (1973), Lamp & Collet (1979) and Williams (1985).

I will consider first the invasion of each region by plants from the Mediterranean Basin.

Chile

In the 16th Century the Spaniards began their conquest of South America and first began settling Chile in the late 1530's, establishing Santiago in 1541. The tendency then to introduce crops from one holoclimate to the other was a natural corollary of settlement. Both the crop plants, and the weed seed mixed with them, would have found environmental conditions and the absence of their usual predators, ideal for vigorous growth. The fact that there was early massive invasion by Mediterranean annuals (Raven 1971, Aschmann 1973b) is not surprising. These annuals had already been sieved by over 10,000 years of association with human activities in the Mediterranean Basin.

South Africa

The second of the mediterranean regions to be explored and settled by Europeans was the Cape Region of South Africa in 1650. However, it was settled by Dutch and other northern Europeans who attempted to implement their practices of agriculture and farming. They did begin growing grapes, and recognised the climatic analogy with the Mediterranean, but the subsequent rate and degree of invasion would have been very different had the Cape been settled by the Spaniards.

California

Whereas California had been discovered in 1542, the Spaniards only began settling the mediterranean region of North America at about the time eastern Australia was discovered by James Cook in 1770. The first of a series of

189

missions was established in San Diego in 1769 and settlement was confined to the mission areas until 1824. The Gold Rush of 1848 saw a rapid increase in population and a concommitant increased demand for food crops (Mooney *et al.* 1972). Not only was the pattern of invasion by annuals from the Mediterranean Basin repeated there (Aschmann & Bahre 1977), but because of trade and other contact between the two Americas, there was secondarily the exchange of some Chilean and Californian species (Raven 1971). Raven reports that there are 130 important herbaceous species in common, and the most abundant ones are the 'European' weeds.

An analysis of the Californian weed flora (Raven & Axelrod 1978) indicates that of 674 introductions, 559 are form the Old World. Of the 115 New World species, most come from temperate North America and Tropical America, only 38 species are from South America and some sub-set of those would be Chilean.

Southern Australia

The last European contact and settlement was to southern Australia in 1828 but, as with South Africa, these settlers came principally from northern Europe. However, not only did Australia receive Mediterranean weeds indirectly from British sources, but the fairly rapid expansion of vines in South Australia and the expansion of wheat and sheep industries promoted contact with Mediterranean countries. Very soon, by accident (such as with *Chondrilla juncea*) or design (such as the improvement of pastures with *Trifolium* species) (Morley & Katznelson 1965), many plants of Mediterranean origin were naturalised in southern Australia. Kloot (1987a) has analysed the origins of the 903 alien plant species naturalised in South Australia (the settled part of which is mediterranean). Of these 292 (32%) originated in the Mediterranean Basin. Specht (1972) had earlier analysed a smaller weed flora (654 species) of South Australia and found that 32% came from the Mediterranean Basin, 13% from South Africa, 5% from Chile and 4% from California.

Contact between regions

Figure 1 shows the dispositions of the five mediterranean regions of the world and the sea distances between them in nautical miles.

Contact between Australia and the Americas was probably promoted by the Gold Rushes of California (1848) and Australia (1867), but subsequently maintained by trade. Traditionally, as is still the case, this trade is primarily with North America and may account for the dearth of biotic exchanges with South America. There are 18 Australian species in the Californian weed flora (Robbins *et al.* 1941). This is still increasing and Baker (1972) commented on the (then) recent introduction and spread of *Senecio minimus* in coastal California. Some Chilean weeds had however established very early after

190

settlement, *Physalis peruviana* was reported from Sydney in 1802 (Kloot 1985b). *Xanthium spinosum* became prevalent in eastern Australia in the 1840's after being introduced on horses from Valparaiso (Maiden 1917) and/or with garden seed from Europe (Michael 1972) which it had invaded from Chile.

Before the opening of the Suez Canal in 1869 trade to and from England from Australia was around the Cape of Good Hope and this promoted contact with South Africa. Quite early on there was exchange of plant material, early settlers were encouraged to bring seed and cuttings to their new home, and the last land-fall at the Cape provided many plant propagules (Kloot 1985b). The number of geophytes introduced to southern Australia from South Africa such as *Homeria* species and *Watsonia bulbillifera* is interesting given the richness of the endemic (Ausralian) geophytic flora (Pate & Dixon 1981, 1982). Paradoxically some of the most invasive elements from each country were intentionally introduced, often as sand binders.

This history of contact is reflected in the representation of each region's flora in the weeds of South Australia (Kloot 1984). Plants from South Africa constitute 14% of the weed flora, 6% from North America (particularly California), 8% from Chile and less than 1% from the mediterranean region of Western Australia. In addition, 309 species (34%) had natural ranges that include one of the mediterranean regions. Over 80% of the weed flora of South Australia has mediterranean origins.

Although on the same seabord of connected continents (Figure 1), there has been only one relatively short but intense period of contact between Chile and California (Mooney *et al.* 1972). This was at the time of the Californian Gold Rush when demands for wheat and other produce promoted trade between the two, with ships sailing around the Horn to Valparaiso and on to California. Farm produce and seed was transported from Chile to California and resulted in some weed spread (Mooney *et al.* 1972). However, with the establishment of the trans-continental railway in North America and then the opening of the Panama Canal in 1914 further contact was minimal.

There appears to have been only limited contact also between the mediterranean regions of South America and South Africa. Because of their separate European connections, and direct sea lanes from each to these (see Figure 1) via Cape Horn and Cape of Good Hope respectively, there was no incidental contact. Again, from the evidence of naturalised aliens in each region, the impact has been less than with other regions.

Similarly the Californian-South African contact has been restricted by separate patterns of trade and transportation. Contact between California and South Africa would have been sporadic. This is reflected in the fact that the weed flora of California (Raven & Axelrod 1978) contains only 30 species (5%) from central and southern Africa. Similarly few examples of Californian species established in South Africa were found.

What this leads to then is shown schematically in Figure 2 with the pivotal role of the Mediterranean Basin shown and the four other mediterranean regions arranged around it. California shares the strongest link with the

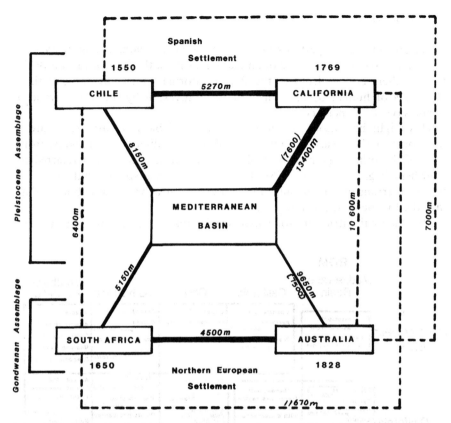

Figure 2. Similarities among the five mediterranean regions, the thickness of the connecting lines is propotional to the degree of similarity. The dates of European settlement of the four 'colonial' mediterranean regions are given, as well as the distance between regions in nautical miles (1 nautical mile = 1,853 km), the shorter distance for California-Mediterranean Basin is via the Panama Canal, the shorter distance for Australia-Mediterranean Basin is via the Suez Canal. The two groupings of regions, 'Pleistocene' and 'Gondwanan' are from Naveh and Whittaker (1980).

Mediterranean, followed by Chile, South Africa and Australia.

The two pair-wise comparisons, the two Americas (Chile and California) with their common Spanish heritage, and the two austral regions, southern Australia and the Cape Region of South Africa, are most similar. Although these two groups can be interpreted by the evolutionary and edaphic constraints mentioned by Naveh & Whittaker (1980) and by the additional criteria summarized by di Castri (1981), I think that in terms of invasions the common patterns of settlement and trade have greater significance.

Primary invasions

There are interesting patterns of invasion in each region, but in every case the initial wave of invasions is of annual weeds from the Mediterranean Basin to the other regions. Even in the Baja California which has had less than 100 years of human occupation there is an invasion by the Mediterranean annuals (G. Long pers. comm.).

This might be considered the *primary* level of biotic invasions. Figure 3 lists some of the invasive plants established in mediterranean regions, from each of the other regions. The table attempts to use species not referred to elsewhere (e.g. in Groves 1986) and attempts to use species that are from the mediterranean region only, rather than from a broad distribution that includes the mediterranean region.

The diagonal entries are for native species that have become invasive and

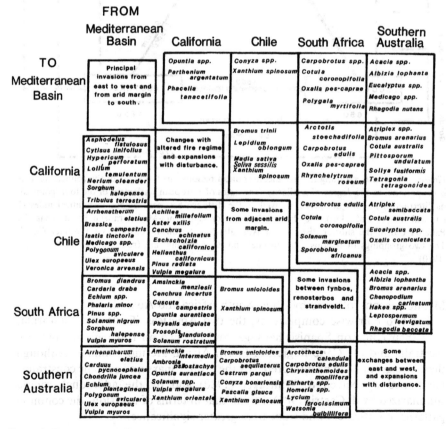

Figure 3. Examples of invasive plants species from each of the five mediterranean regions that have invaded other mediterranean regions. Diagonal entries are for invasive native species within each region. See text for sources used.

state some of the causes of this. For example, the California Poppy (*Eschscholzia californica*), as well as proving weedy in other mediterranean regions, has expanded its range in California due to the overgrazing, followed by under-grazing, by introduced sheep and cattle (Baker 1972). Stebbins (1965) com-prehensively covered the ecology and genetics of native Californian species that have become weedy there. Of the many species listed by Stebbins, most have proved to be invasive of the other mediterranean regions as well.

The success of the Mediterranean annuals in the two sets of regions has different bases. In the more recent set, of California and Chile, there was already a rich annual flora but is was out-competed by the successful invaders pre-adapted to the rigours of grazing and cropping. In the older (Gondwanan) set, of South Africa and southern Australia, not only was there no rich endemic annual flora, but the invaders had the additional edge of an earlier growth pulse.

Secondary invasions

There is a second level of invasion, that of perennial (woody) weeds and this is apparently stronger between the four other mediterranean regions and from them back to the Mediterranean Basin. For example, in the Cape region of South Africa there is a serious problem caused by introduced woody plants from Australia. The principal weeds are the legumes, *Acacia longifolia*, *A. saligna* and *A. cyclops* (Milton 1980) and the *Hakea* species (Proteaceae), *H. suaveolens* and *H. tenuifolia*, as well as *Leptospermum laevigatum* and *Eucalyptus lehmannii* (Myrtaceae). Roux (1964) has in fact suggested that Australian acacias are behaving like annual weeds. Specht (pers. comm. *in* di Castri 1981) has suggested that shrubs and trees from Australia have established in the South African heathlands (fynbos) not in the mediterranean shrublands (renosterbos), and that the reverse flow of shrubs, geophytes and annuals from South Africa have established in the Australian shrublands (mallee), not in the heathlands. The South African fynbos and Western Australian shrublands are far richer in woody species than any other mediterranean region (Naveh & Whittaker 1980).

Two recent papers (Taylor & Macdonald 1985, Taylor *et al.* 1985) report on the progress of invasive alien woody plants in the Cape of Good Hope Nature Reserve based on two surveys in 1966 and 1976. Of the fifteen species listed in the second survey, nine are native to southern Australia; one, *Cupressus macrocarpa*, from California; and three species of *Pinus* from the Mediterranean region. Only two species are not of mediterranean origin, one from tropical South America and one from India. Some woody weeds have originated in the Mediterranean Basin and invaded the other regions. *Calicotome spinosa* and *Teline monspessulana* are naturalised in California (Munz 1974). *Nerium oleander* is also in California. *Eucalyptus globulus* is tending to behave as a weed in California (Baker 1972) and several eucalypts are invasive in all

194

mediterranean regions (Aschmann 1973b). The Californian species *Prosopis glandulosa* var. *torreyana*, Mesquite, is a naturalised weed in Western Australia (Meadly 1965) and South Africa (Stirton 1978).

Another feature of many of the woody invaders of southern Australia are the numbers that have fleshy fruits. Debussche & Isenmann (1990) has considered the significance of introduced fleshy fruits in the Mediterranean flora and their effect on bird populations. The introduced shrubs in Australia are also spread by birds, but around towns usually by introduced birds. Examples from the mediterranean region of Australia include the olive *Olea africana* (Mediterranean), *Chrysanthemoides monilifera* and *Lycium ferocissimum* (South Africa), *Nicotiana glauca* (South America) and *Cestrum parqui* (Chile).

The Mediterranean Basin is relatively rich in conifers compared to the other mediterranean regions. Conifers are virtually absent from the floras of the three southern regions. Interestingly it is to these three that *Pinus radiata* from California has become a problem, and *P. pinaster* and *P. halepensis* from the Mediterranean Basin is a woody invader in the Cape Region of South Africa (e.g. Richardson & Brown 1986). *Cupressus macrocarpa* from California is also naturalised in the Cape Region (Taylor & Macdonald 1985).

Tertiary invasions

The third level of biological invasions are those occurring within a mediterranean region e.g. from east to west within the Mediterranean Basin (Guillerm & Maillet 1982, Guillerm *et al.* 1990), and to a lesser extent, from north to south (Le Floc'h *et al.* 1990), also from northern Mexico into California, plus native species within California (Stebbins 1965), some examples of east-west invasions within mediterranean Australia (Kloot 1984), and local movements in South Africa (Wells & Stirton 1982b). Quezel *et al.* (1990) have reviewed recent introductions to the Mediterranean Basin from the other four regions.

Future invasions

What of the future? With increasing technology for seed sorting and increasingly sophisticated herbicides the agrestal weeds' days are numbered. A recent paper by Kornas (1988) documents the decline in those weeds with diaspores adapted to dispersal with crop seeds. What we are seeing instead is a rise in the incidence of 'community' weeds, plants finding unexploited resources in disturbed communities (Fox & Fox 1986b). These are more insidious, and are often 'woody' weeds: the hakeas in the South African fynbos and the *Chrysanthemoides* in coastal Australian communities; in time this may well include more species of Australian *Acacia* around the Mediterranean Basin, more

Australian *Eucalyptus* in California, more widespread invasions by conifers, etc. These cannot be solved with the use of herbicides, they require modification of management practices such as the use of fire – especially in areas where the maintenance of a natural composition is highly desirable. Beyond that we may have to accept continued mixing of the regions' biotas.

Frenkel (1970) shows the exponential increase in number of naturalised species in the Californian flora and Groves (1986) has compared this with the shorter record of introductions in Australia. The trend in both regions is for an influx of from four to six new species to be naturalised each year, many of which will come from other mediterranean regions.

We are seeing the blurring of distinctions established in evolutionary time in parts of the world that now share similar climate. All the fascinating similarities of community structure, based on floras with quite distinct phylogenies, may in time be lost in an homogenised soup of mediterranean weeds.

Acknowledgements

The first draft of this contribution was written while the author was a Visiting Scholar in the Department of Ecology and Evolutionary Biology at the University of Arizona.

References

Adamson, R.S. & Salter, T. (eds.) 1950. *Flora of the Cape Peninsula.* Juta, Capetown. 889 pp.

Ames, O. 1939. *Economic annuals and human cultures.* Botanical Museum of Harvard University. Cambridge, Massachusetts.

Anderson, E. 1967. *Plants, man an life.* University of California Press, Berkeley, California. 25 pp.

Aschmann, H. 1973a. Distribution and peculiarity of mediterranean ecosystems. In: F. di Castri & H.A. Mooney, (eds.), *Mediterranean type ecosystems – Origins and structure.* Springer-Verlag, Berlin, pp. 11–19.

Aschmann, H. 1973b. Man's impact on the several regions with mediterranean climates. In: F. di Castri & H.A. Mooney (eds.), *Mediterranean type ecosystems – Origins and structure.* Springer-Verlag, Berlin, pp. 363–390.

Aschmann, H. & Bahre, C. 1977. Man's impact on the wild landscape. In: H.A. Mooney (ed.), *Convergent evolution in Chile and California.* Dowden, Hutchinson & Ross, Inc., Stroudsburg, Pennsylvania, pp. 73–84.

Axelrod, D.I. 1973. History of the Mediterranean ecosystem in California. In: F. di Castri & H.A. Mooney (eds.), *Mediterranean type ecosystems – Origins and structure.* Springer-Verlag, Berlin, pp. 225–277.

Baker, H.G. 1965a. *Plants and civilization.* Wadsworth Publishing Co. Inc., California. 183 pp.

Baker, H.G. 1965b. Characteristics and modes of origin of weeds. In: H.G. Baker & G.L. Stebbins (eds.), *The genetics of colonizing species.* Academic Press, New York and London, pp. 147–172.

Baker, H.G. 1972. Migrations of weeds. In: D.H. Valentine (ed.), *Taxonomy phytogeography and evolution.* Academic Press, London, pp. 327–347.

196

Baker, H.G. 1974. The evolution of weeds. *Annual Review of Ecology and Systematics 5:* 1–24.

Barbour, M.G. Shmida, A., Johnson, A.F. & Holton, B. 1981. Comparison of coastal dune scrub in Israel and California: physiognomy, association patterns, species richness, phytogeography. *Israel Journal of Botany* 30: 181–198.

Beard, J.S. 1969. Endemism in the Western Australian flora at the species level. *Journal of the Royal Society of Western Australia* 52: 18–20.

Boucher, C. & Moll, E.J. 1981. South African mediterranean shrublands. In: F. di Castri, D.W. Goodall & R.L. Specht (eds.), *Mediterranean-type shrublands.* Elsevier Scientific Publishing Company, Amsterdam, pp. 233–248.

Bradbury, D.E. 1981. The physical geography of the Mediterranean lands. In: F. di Castri, D.W. Goodall & R.L. Specht (eds.), *Mediterranean-type shrublands.* Elsevier Scientific Publishing Company, Amsterdam, pp. 53–62.

Casasayas Fornell, T. 1990. Widespread adventive plants in Catalonia. In: F. di Castri, A.J. Hansen & M. Debussche (eds.), *Biological invasions in Europe and the Mediterranean Basin.* Kluwer Academic Publishers, Dordrecht, pp. 85–104.

Cody, M.L. & Mooney, H.A. 1978. Convergence versus nonconvergence in the Mediterranean climate ecosystems. *Annual Review of Ecology and Sytematics* 9: 265–321.

Cowling, R.M. & Campbell, B.M. 1980. Convergence in vegetation structure in the mediterranean communities of California, Chile and South Africa. *Vegetatio* 43: 191–198.

Debussche, M. & Isenmann, P. 1990. Introduced and cultivated fleshy fruited plants: consequences on a mutualistic Mediterranean plant-bird system. In: F. di Castri, A.J. Hansen & M. Debussche (eds.), *Biological invasions in Europe and the Mediterranean Basin.* Kluwer Academic Publishers, Dordrecht, pp. 319–416.

Dell, B., Hopkins, A.J.M., & Lamont, B.B. eds. 1986. *Resilence in mediterranean-type ecosystems.* Dr W. Junk, Dordrecht. 168 pp.

di Castri, F. 1973. Climatographical comparisons between Chile and the western coast of North America. In: F. di Castri, & H.A. Mooney (eds.), *Mediterranean type ecosystems – Origins and structure.* Springer-Verlag, Berlin, pp. 21–36.

di Castri, F. 1981. Mediterranean – type shrublands of the world. In: F. di Castri, D.W. Goodall & R.L. Specht (eds.), *Mediterranean type shrublands.* Elsevier Scientific Publishing Company, Amsterdam, pp. 1–52.

di Castri, F., Goodall, D.W. & Specht, R.L. (eds.), 1981. *Mediterranean – type shrublands.* Elsevier Scientific Publishing Company, Amsterdam, 643 pp.

di Castri, F. & Mooney, H.A., eds. 1973. *Mediterranean – type ecosystems – Origins and structure.* Springer-Verlag, Berlin. 402 pp.

Edmonds, S.J. & Specht, M.M. 1980. Dark Island heathland, South Australia: faunal rhythms. In: R.L. Specht (ed.), *Heathlands and related shrublands. B. Analytical studies.* Ecosystems of the World, 9B. Elsevier, Amsterdam, pp. 155–27.

Ewart, A.J. & White, J. 1908. Article XXV – Contibutions to the Flora of Australia, No, 10. *Proc. Roy. Soc. Vic.* 21 (N.S.): 540–549.

Fox, M.D. & Fox B.J. 1986a. Resilience of animal and plant communities to human disturbance. In: B. Dell, A.J.M. Hopkins & B.B. Lamont (eds.), *Resilience in mediterranean-type ecosystems.* Dr W. Junk Publishers, Dordrecht, pp. 39–64.

Fox, M.D. & Fox B.J. 1986b. The susceptibility of natural communities to invasion. In: R.H. Groves & J.J. Burdon (eds.), *Ecology of biological invasions: An Australian perspective.* Australian Academy of Science, Canberra, pp. 57–66.

Frenkel, R.E. 1970. Ruderal vegetation along some California roadsides. *University of California Publication in Geography.* 20: 1–163.

Goldblatt, P. 1978. An analysis of the flora of Southern Africa: It's characteristics, relationships and origins. *Annals of the Missouri Botanic Garden.* 65: 369–436.

Grabandt, K. 1985. *Weeds of crops and gardens in Southern Africa.* Seal Publishing, Johannesburg. 135 pp.

Groves, R.H. 1986. Invasion of mediterranean ecosystems by weeds. In: B. Dell, A.J.M. Hopkins

& B.B. Lamont (eds.), *Resilience in mediterranean - type ecosystems*. Dr W. Junk Publishers, Dordrecht, pp. 129–145.

Guillerm, J.L., le Floc'h, E., Maillet, J., & Boulet, C. 1990. The invading weeds within the western Mediterranean Basin. In: F. di Castri, A.J. Hansen & M. Debussche (eds.), *Biological invasions in Europe and the Mediterranean Basin*. Kluwer Academic Publishers, Dordrecht, pp. 61–84.

Guillerm, J.L. & Maillet, J. 1982. Western Mediterranean countries of Europe. In: W. Holzner & M. Numata (eds.), *Biology and ecology of weeds*. Dr W. Junk Publishers, The Hague, pp. 227–243.

Harlan, J.R. 1982. Relationships between weeds and crops. In: W. Holzner & M. Numata (eds.), *Biology and ecology of weeds*. Dr W. Junk Publishers, Dordrecht, pp. 91–96.

Holm, L., Plucknett, D.L., Pancho, J.V., & Herberger, J.P. 1977. *The world's worst weeds – Distribution and biology*. The University Press of Hawaii, Honolulu.

Holm, L., Pancho, J.V. & Herberger, J.P., & Plucknett, D.L. 1979. *A geographical atlas of world weeds*. John Wiley and Sons, New York. 391 pp.

Hopper, S.D. 1979. Biogeographical aspect of speciation in the southwest Australian flora. *Annual Review of Ecology and Systematics* 10: 399–422.

Jackson, L.E. 1985. Ecological origins of California's Mediterranean grasses. *J. Biogeography* 12: 349–361.

Kloot, P.M. 1984. The introduced elements of the flora of southern Australia. *Journal of Biogeography* 11(1): 63–78.

Kloot, P.M. 1985a. The spread of native Australian plants as weeds in South Australia and in ohter Mediterranean regions. *J. Adelaide Bot. Gard.* 7(2): 145–157.

Kloot, P.M. 1985b. Plant introductions to South Australia prior to 1840. *J. Adelaide Bot. Gard.* 7(3): 217–231.

Kloot, P.M. 1987a. The naturalised flora of South Australia 3. Its origin, introduction, distribution, growth forms and significance. *J. Adelaide Bot. Gard.* 10(1): 99–111.

Kloot, P.M. 1987b. The naturalised flora of South Australia 4. Its manner of introduction. *J. Adelaide Bot. Gard.* 10(2): 223–240.

Kornaś, J. 1983. Man's impact upon the floras and vegetation in Central Europe. In: W. Holzner, M.J.A. Werger & I. Ikusima (eds.), *Man's impact on vegetation*. Dr W. Junk Publishers, Dordrecht, pp. 277–286.

Kornaś, J. 1988. Speirochoric weeds in arable fields: from ecological specialization to extinction. *Flora* 180: 83–91.

Kornaś, J. 1990. Plant invasions in Central Europe: historical and ecological aspects. In: F. di Castri, A.J. Hansen & M. Debussche (eds.), *Biological invasions in Europe and the Mediterranean Basin*. Kluwer Academic Publishers, Dordrecht, pp. 19–36.

Kruger, F.J., Mitchell, D.T., & Jarvis, J.U.M. (eds.) 1983. Mediterranean-type ecosystems: The role of nutrients. Springer-Verlag, Berlin 552 pp.

Lamp, C. & Collet, F. 1979. *A field guide to weeds in Australia*. Inkata Press, Melbourne. 376 pp.

Le Floc'h, E., Le Houerou, H.N., & Mathez, J. 1990. History and patterns of plant invasion in Northern Africa. In: F. di Castri, A.J. Hansen & M. Debussche (eds.), *Biological invasions in Europe and the Mediterranean Basin*. Kluwer Academic Publishers, Dordrecht, pp. 105–133.

Le Houerou, H.N. 1981. Impact of man and his animals on Mediterranean vegetation. In: F. di Castri, D.W. Goodall & R.L. Specht (eds.), *Mediterranean-type shrublands*. Elsevier Scientific Publishing Company, Amsterdam. pp. 479–521.

Macdonald, I.A.W. 1984. The Australian contribution to Southern Africa's invasive alien flora: an ecological analysis. *Proceedings of the Ecological Society of Australia* 14: 225–236.

Macdonald, I.A.W. & Jarman, M.L., eds. 1984. *Invasive alien organisms in the terrestrial ecosystems of the fynbos biome, South Africa*. South African National Scientific Programmes Report No. 85.

Macdonald, I.A.W., Clark, D.L. & Taylor, H.C. 1987. The alien flora of the Cape of Good

198

Hope Nature Reserve. *South African Journal of Botany* 53(5): 398–404.

Macdonald, I.A.W., Graber, D.M., DeBenedetti, S., Groves, R.H. & Fuentes, E.R. 1988. Introduced species in Nature Reserves in Mediterranean-type climatic regions of the world. *Biological Conservation* 44: 37–66.

Maiden, J.H. 1917. Weeds of New South Wales, *Agriculture Gazette of N.S.W.* Miscellaneous Publication No. 1951, 489–497.

Margaris, N.S. & Mooney, H.A. eds. 1981. *Components of productivity of mediterranean regions – basic and applied aspects.* Dr W. Junk, Dordrecht.

Matthei, O.R. 1963. *Manual ilustrado de las malezas de la Provincia de Nuble.* Universidad de Concepcion, Chile. 116 pp.

McNeill, J. 1976. The taxonomy and evolution of weeds. *Weed Research* 16: 399–413.

McNeill, J. 1982. Problems of weed taxonomy. In: W. Holzner & M. Numata (eds.), *Biology and ecology of weeds.* Dr W. Junk Publishers, Dordrecht, pp. 35–45.

Meadly, G.R.W. 1965. *Weeds of Western Australia.* Department of Agriculture, Western Australia. 173 pp.

Michael, P.W. 1972. The weeds themselves – early history and identification. *Proceedings of the Weed Society of N.S.W.* 3: 18.

Michael, P.W. 1981. Alien plants. In: R.H. Groves (ed.), *Australian vegetation.* Cambridge University Press, Cambridge. pp. 44–64.

Milewski, A.V. & Cowling, R.M. 1984. Anomalies in plant and animal communities in similar environments at the Barrens, Western Australia, and the Caledon Coast, South Africa. *Proceedings of the Ecological Society of Australia* 14: 199–212.

Milton, S.J. 1980. Australian acacias in the S.W. Cape: pre – adaptation, redation and success. In: S. Neser & A.L.P. Cairns (eds.), *Proceedings of the Third National Weed Conference of South Africa.* A.A. Balkema, Rotterdam, pp. 69–78.

Mooney, H.A., ed. 1977. *Convergent evolution in Chile and California. Mediterranean climate ecosystems.* Dowden, Hutchinson and Ross, Inc., Stroudsburg Pennsylvania. 244 pp.

Mooney, H.A., Dunn, E.L., Shropshire, F. & Song, Jr, L. 1972. Landuse history of California and Chile as related to the structure of the sclerophyll scrub vegetations. *Madrono* 21: 305–319.

Morley, F.H.W. & Katznelson, J. 1965. Colonization in Australia by *Trifolium subterraneum.* In: H.G. Baker & G.L. Stebbins (eds.), *The genetics of colonizing species.* Academic Press, New York, pp. 269–282.

Munz, P.A. 1974. *A Flora of Southern California.* University of California Press, Berkeley. 1086 pp.

Naveh, Z. 1967. Mediterranean ecosystems and vegetation types in California and Israel. *Ecology* 48: 445–459.

Naveh, Z. & Dan, J. 1973. The human degradation of mediterranean landscapes in Israel. In: F. di Castri & H.A. Mooney (eds.), *Mediterranean-type ecosystems, origing and structure.* Springer-Verlag, Berlin, pp. 373–390.

Naveh, Z. & Whittaker, R. 1980. Structural and floristic diversity of shrublands and woodlands in N. Israel and other mediterranean areas. *Vegetatio* 41: 171–80.

Neser, S. & Cairns, A.L.P. eds. 1980. *Proceedings of the Third National Weeds Conference of South Africa.* A.A. Balkema Rotterdam.

Orchard, A.E. 1981. A revision of South American *Myriophyllum* (Haloragceae), and its repercussions on some Australian and North American Species. *Brunonia* 4: 27–65.

Orchard, H.E. & O'Neill, J.M. 1959. *Declared weeds of South Australia and their control.* Bulletin No. 453, Department of Agriculture, South Australia. 115 pp.

Orshan, G. 1953. Notes on the application of Raunkiaer's life forms in arid regions. *Palestine J. Bot., Jerusalem Series.* 6: 120–122.

Parsons, W.T. 1973 *Noxious weeds of Victoria.* Inkata Press, Melbourne. 300 pp.

Pate, J.S. & Dixon K.W. 1981. Plants with fleshy underground storage organs – a Western Australian survey. In: J.S. Pate and A.J. McComb (eds.), *The biology of Australian plants.* University of Western Australia Press, Nedlands, pp. 181–215.

199

Pate, J.S. & Dixon K.W. 1982. *Tuberous, cormous and bulbous plants.* University of Western Australia Press. 268 pp.

Philippi, A.F. 1881. *Plantarum Vascularium Chilensium.* Imprenta Nacional. Calle de la Bandera, Num. 29. Santiago, Chile. 377 pp.

Quezel, P., Barbero, M., Bonin, G. & Loisel, R. 1990. Recent invasions around the Mediterranean Basin. In: F. di Castri, A.J. Hansen & M. Debussche (eds.), *Biological invasions in Europe and the Mediterranean Basin.* Kluwer Academic Publishers, Dordrecht, pp. 51–60.

Raven, P.H. 1971. The relationships between 'Mediterranean' floras. In: D.H. Davis, P.C. Harper & I.C. Hedge (eds.), *Plant life of south-west Asia.* The Botanical Society of Edinburgh, pp. 119–134.

Raven, P.H. 1973. The evolution of Mediterranean floras. In: F. di Castri, & H.A. Mooney (eds.), *Mediterranean-type ecosystems, origin and structure.* Springer-Verlag, Berlin, pp. 213–224.

Raven, P.H. & Axelrod, D.I. 1978. *Origin and relationships of the California flora.* University of California, Publications in Botany 72.

Richardson, D.M. & Brown, P.J. 1986. Invasion of mesic mountain fynbos by *Pinus radiata. South African Journal of Botany* 52(6): 529–536.

Robbins, W.W., Bellue, M.K. & Ball, W.S. 1941. *Weeds of California.* California State Department of Agriculture, California. 491 pp.

Roux, E.R. 1964. The Australian acacias in South Africa. In: D.H. Davis (ed.), *Ecological studies in southern Africa.* Dr W. Junk, Dordrecht, pp. 137–142.

Rundel, P.W. 1981. The matorral zone of central Chile. In: F. di Castri, D.W. Goodall & R.L. Specht (eds.), *Mediterranean-type shrublands of the world.* Elsevier Scientific Publishing Company, Amsterdam, pp. 175–201.

Sakamoto, S. 1982. The Middle East as a cradle for crops and weeds. In: W. Holzner & M. Numata (eds.), *Biology and ecology of weeds* Dr W. Junk Publishers, Dordrecht, pp. 97–109.

Smiley, F.J. 1922. *Weeds of California and methods of control.* Department of Agriculture, California.

Solbrig, O.T., Cody, M.L., Fuentes, E.R., Glanz, W., Hung, J.H. & Moldenke, A.R. 1977. The origin of the biota. In: H.A. Mooney, (ed.), *Convergent evolution in Chile and California, Mediterranean climate ecosystem.* Dowden, Hutchinson & Ross, Stroudsburg, Penn., pp. 13–26.

Specht, R.L. 1972. *Vegetation of South Australia.* Second Edition, South Australian Government Printer, Adelaide, 328 pp.

Specht, R.L. 1973. Structure and functional response of ecosystems in the mediterranean climate of Australia. In: F. di Castri & H.A. Mooney (eds.), *Mediterranean-type ecosystems, origin and structure.* Springer-Verlang, Berlin, pp. 113–120.

Stebbins, G.L. 1965. Colonizing species of the native California flora. In: H.G. Baker & G.L. Stebbins (eds.), *The genetics of colonizing species.* Academic Press Inc., New York, pp. 173–195.

Stirton, C.H. 1978. *Plant invaders: Beautiful, but dangerous.* The Department of Nature and Environmental Conservation, Capetown. 175 pp.

Sykora, K.V. 1990. History of the impact of man on the distribution of plant species. In: F. di Castri, A.J. Honsen & M. Debussche (eds.), *Biological invasions in Europe and the Mediterranean Basin.* Kluwer Academic Publishers, Dordrecht, pp. 37–50.

Taylor, H.C. & Macdonald, S.A. 1985. Invasive alien woody plants in the Cape of Good Hope Nature Reserve. I. Results of a first survey in 1966. *S. Afr. J.. Bot.* 51(1): 14–20.

Taylor, H.C., Macdonald, S.A. & Macdonald, I.A.W. 1985. Invasive alien woody plants in the Cape of Good Hope Nature Reserve. II. Results of a second survey from 1976 to 1980. *S. Afr. J. Bot.* 51(1): 21–29.

Thellung, A. 1918–19. Zur Terminologie de Adventiv- und Ruderalflora. *Allg. Bot. Z. Syst. Karlsruhe.* 24: 36–42.

Vallejo, A. 1968. *Malezas de Chile.* Boletin Tecnico 34, Servicio Agricola y Granadero, Santiago.

200

Vavilov, N.I. 1951. The origin, variation, immunity and breeding of cultivated plants. Translated from the Russian by K. Starr Chester. *Chronica Botanica* 13(1/6): 1–366.

Verdcourt, B. 1973. A new combination in *Myriophyllum* (Haloragaceae). *Kew Bulletin* 28(1): 36.

Wells, M.J. & Stirton, C.H. 1982a. Weed problems of South Africa pastures. In: W. Holzner & M. Numata (eds.), *Biology and ecology of weeds*. Dr W. Junk Publishers, Dordrecht, pp. 429–448.

Wells, M.J. & Stirton, C.H. 1982b. South Africa. In: W. Holzner & M. Numata (eds.), *Biology and ecology of weeds*. Dr W. Junk Publishers, Dordrecht, pp. 339–343.

Williams, O.B. 1985. Population dynamics of Australian plant communities with special reference to the invasion of neophytes. In: J. White (ed.), *The population structure of vegetation*. Dr. W. Junk Publishers, Dordrecht, pp. 623–635.

Zohary, M. 1962. *Plant life of Palestine, Israel and Jordan*. Ronald Press, New York.

PART THREE

Animal invasions

12. The invasion of Northern Europe during the Pleistocene by Mediterranean species of Coleoptera

G.R. COOPE

Abstract

Investigations of subfossil Coleoptera show that during the last few hundreds of thousands of years, species have remained evolutionarily constant. They have, however, altered their geographical distributions extensively in response to the climatic changes of the glacial/interglacial cycles. It can be demonstrated that many species which have today a Mediterranean range, extended their northern limits during the warmer interglacial and interstadial interludes, at least as far as the British Isles. These species comprised groups of varied ecological preferences and some have such restricted present day distributions that they have been viewed as 'endemic' Mediterranean species. Patterns of present day geographical distributions are thus the result of dynamic interaction between the species and the numerous and intense climatic changes that caused continuous biogeographic adjustments. This complex history can only be understood adequately by reference to the subfossils which provide the only objective evidence of where species lived in the geologically recent (and ecologically relevant) time.

Introduction

Invasions by Mediterranean species of beetle have a long history dating back in time to long before the period of human intervention. The evidence for this comes from the rich and varied subfossil remains that occur in deposits of freshwater and terrestrial sediments laid down during the Pleistocene period, particularly in deposits that date from the last few hundred thousand years. The systematic investigation of these remains has been carried out for a long time but has increased greatly during the last few decades (Coope 1979).

The fossils are often so well preserved that they may be identified to the species level by precise matching of their exoskeletons with those of modern specimens. The abundance of these fossils provides evidence on the structure and composition of past insect communities. One of the most interesting

F. di Castri, A. J. Hansen and M. Debussche (eds.), Biological Invasions in Europe and the Mediterranean Basin. 203–215. © 1990, *Kluwer Academic Publishers, Dordrecht.*

discoveries of this investigation is that species of insect do not show any evidence of evolutionary change during the Pleistocene period and because fossil assemblages often resemble those of the present day, it seems likely that there has been little or no alteration of their environmental requirements either. Our fossil evidence suggests that this specific constancy has lasted for the last million years at least and for probably longer that this.

Throughout this time, however, there have been changes in the geographical distributions of species in response to changes in location of acceptable environments. Some species have altered their ranges by thousands of kilometres during the last few tens of thousands of years.

Biotal response to the glacial/interglacial cycles

In northern Europe the Pleistocene period has been characterised by the frequent and intense climatic oscillations of the glacial/interglacial cycles. During the predominantly glacial events there were also warmer interludes termed interstadials that were sometimes at least as warm as the present day but were of shorter duration than the true interglacials. All these climatic

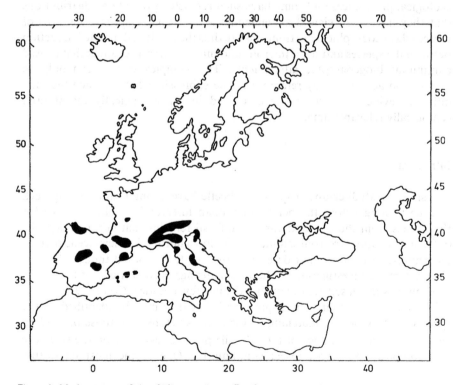

Figure 1. Modern range of *Asaphidion cyanicorne* Pand.

changes had a profound influence on the flora and fauna of the times. Both plants and insects responded to the stress of these climatic changes not by evolving out of trouble but by shifting their geographical ranges and thus tracking acceptable conditions as these moved across the continents.

It can now be shown that, during the prolonged periods when the climate was in a glacial mode, many arctic species extended their ranges to lower latitudes and lower altitudes, living in the extensive tundra areas south of the continental ice sheets. The response of these cold adapted species is widely known and well documented. What is not so generally recognised is that during the interglacial and interstadial periods (always much shorter than the glacial episodes) southern species extended their ranges far to the north of where they occur today. In fact the fossil record shows that the present day distributional patterns are but the latest stage in a complex history of successive waves of invasions and local extinctions, in which most species remain the same but in which their geographic settings are constantly changing in response to the dictates of the glacial/interglacial fluctuations.

During some of the warmer periods, of both interglacial and interstadial status, species which today have Mediterranean distributions, reached as far north as the British Isles. Such species include plants, vertebrates, molluscs

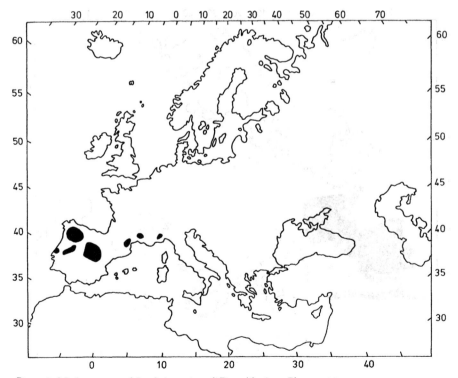

Figure 2. Modern range of *Bembrion grisvardi* Dew./*ibericum* Pioc.

and insects but I am going to concentrate here on the beetles (the Coleoptera) because they are robust enough to have left an abundant subfossil record and also because they illustrate so clearly the large scale biogeographic response forced upon species by the drastic events of the Pleistocene climatic changes.

Case histories will be discussed here of a number of species which have broadly Mediterranean geographical ranges at the present day but for which there is good fossil evidence for their presence in Britain during the Upper Pleistocene. It must be emphasised, however, that this is not an exhaustive list but merely intended to provide representative groups of species, each group with very different habitat requirements. Thus the presence in Britain during geologically recent past of species with currently Mediterranean distributions is not confined to any ecologically restricted group of species.

Mediterranean Carabidae in the British Pleistocene

The Carabidae comprise a family of ground beetles which include specialist carnivores, general scavengers and a few graminivorous species. They inhabit a wide diversity of terrestrial habitats, from seemingly bare soils to swamps

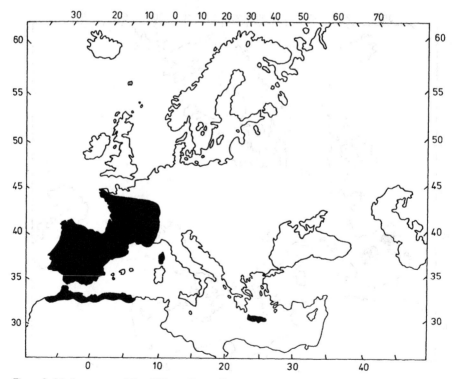

Figure 3. Modern range of *Bembidion callosum* Kust.

and deep forests.

Asaphidion cyanicorne Pand. (modern range; Figure 1); *Bembidion grisvardi* Dew./*ibericum* Pioc. (modern range; Figure 2.) and *Bembidion callosum* Kust. (modern ranges; Figure 3.) are three species of ground beetle that live in rather open habitats with sparse vegetation. They occurred in Britain in the period directly after the sudden climatic warming at the end of the Last (Würm, Weichselian) Glaciation, sometime shortly after 13,000 years ago. At this time the temperatures in Britain were at least as warm as those of the present day (Atkinson *et al.* 1987) and the landscape was only thinly vegetated, largely because of the poor nutritional status of the soil (van Geel *et al.* 1984; Pennington 1986). Suitable habitats for these species must have been more widespread than they are at the present day, notwithstanding the cultural steppes and deserts imposed by modern agricultural practices. Seemingly these species require more than an expansion of open habitats to enable them to invade more northerly parts of Europe at the present time. Associated with the three relatively southern species mentioned above, there was a species of *Synomus* (= *Metabletus*) that does not resemble any northern or central European species but which matches the eastern Mediterranean *S. parallellus* Ballion. This is another bare ground species and it has been found as a fossil

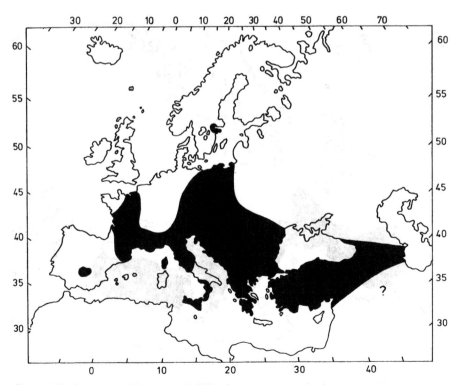

Figure 4. Modern range of *Oodes gracilis* Villa.

from Lateglacial deposits in the Isle of Man (Joachim 1978) and I have a specimen from deposits of the same age from central Ireland.

Two important points arise from the occurrence of these Mediterranean species of beetle so far north of their present day geographical limits. Firstly, the climate immediately after the termination of the Last Glaciation must have warmed up very rapidly indeed. Mean annual temperatures would appear to have risen in Britain by about 16°C (Atkinson *et al.* 1987). Secondly, the pioneer vegetation, which was thin and patchy, provided habitats that were readily heated by the sun and thus may have provided good mimics of the present day Mediterranean environments. Later the vegetation succession gradually led to the closing of these open habitats and the elimination from northern Europe of this suite of beetle species.

During the Last Interglacial period (= Eemian) about 120,000 years ago, several species of southern ground beetle lived in abundance as far north as the British Isles. These species do not depend on the transitory pioneer habitats discussed above. Such a species is *Oodes gracilis* Villa (modern range; Figure 4) which is well adapted to aquatic environments where it lives in reed swamps and is able to crawl about under water on submerged plant stems. This species only occurs in northern Europe in isolated places where

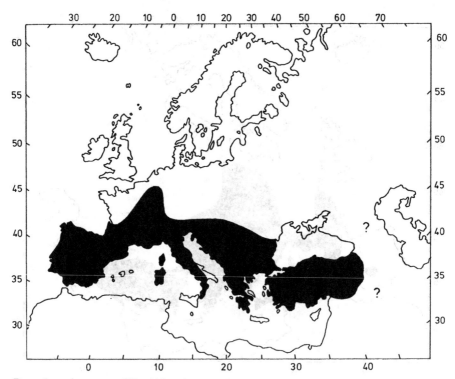

Figure 5. modern range of *Bembidion elongatum* Dej.

aspect ensures abnormally high summer temperatures (Lindroth 1943). Its presence in Britain during the Last Interglacial implies that the climate was warmer than that of the present day. It is interesting to note that this species also occurred in Britain in postglacial time (the present interglacial) living in southern England during the bronze age when the climate was warmer than now and swamps were much more widespread.

Another common species of the British insect fauna of the Last Interglacial is *Bembidion elongatum* Dej. (modern range; Figure 5). This is also a species of marshy habitats and both this and the preceding species may be in part restricted in their modern ranges by drainage of wetlands.

Mediterranean dung beetles in the British Pleistocene

The beetles of the family Scarabaeidae are an enormous group of specialist dung feeders that is very rich in species in warmer parts of the world. The present day fauna of Britain is thus relatively meagre in comparison, for instance to that of the Mediterranean. During the warmer climatic interludes, the dung beetle fauna of Britain was greatly enriched by an influx of species that are

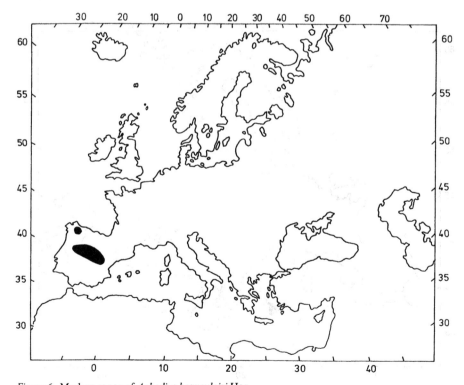

Figure 6. Modern range of *Aphodius bonvouloiri* Har.

today restricted in Europe to the more southerly regions.

One of the most abundant dung beetles in Britain during the warm interstadial in the middle of the Last Glaciation was the currently exclusively Spanish species *Aphodius bonvouloiri* Har. (modern range; Figure 6) (Girling 1974). At this time large herds of bison inhabited the open plains of central and southern England together with mammoth, wooly-rhinoceros and horse. There was thus ample habitat for dung beetles. This species of *Aphodius* was associated with a diverse assemblage of other warm temperate insect species (Coope &, Anugus 1975). It is interesting to note that when the climate deteriorated shortly afterwards, *Aphodius bonvouloiri* disappeared from the British fauna and the most common dung beetle then was *Aphodius holdereri* Reitt.; a species that is today restricted to high altitude Tibet and the neighbouring parts of north-western China (Coope 1973).

During the Last Interglacial the insect fauna of Britain was also characterised by a wealth of southern dung beetles such as *Onthophagus furcatus*, *Caccobius schreberi* Linn. and *Oniticellus fulvus* Goez. Though these all are dominantly southern European, the most exotic and most abundant dung beetle at that time was *Onthophagus massai* Barr. (modern range; Figure 7). Were it not for the fossil record of this species from various sites in Britain, the present

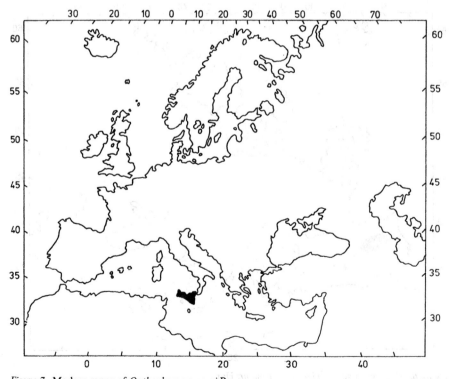

Figure 7. Modern range of *Onthophagus massai* Barr.

day geographical distribution of this species might mislead us into believing that, since it is confined to the Island of Sicily, it must have originated there. Such cases of endemism could just as well be the location of the last stand of a species as its place of birth. It is thus interesting to note that the Spanish and Sicilian species *Aphodius carpetinus* Gra. has been found as a fossil in deposits dating from the same interglacial period in eastern England (L. Holdridge pers. comm.).

Mediterranean tree feeding beetles from the British Pleistocene

A large number of beetle species are specialist feeders on the highly nutritious wood just beneath the bark of trees. Remains of such species are often common in deposits that accumulated in forest surroundings. They can provide information on the types of trees living at the time.

Brachytemnus submuricatus Schonh. (modern range; Figure 8), feeds under the bark of *Populus nigra* and *Salix alba*. It was a very common species in Britain during the last Interglacial (Girling 1974) and it was also abundant during the Hoxnian (= Holsteinian) Interglacial (Shotton and Osborne 1965).

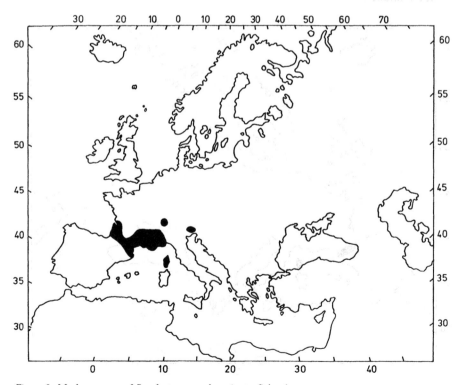

Figure 8. Modern range of *Brachytemnus submuricatus* Schonh.

The restriction of this species to southern Europe cannot be due to the absence of its food plants since suitable trees are widespread throughout. It seems likely that some climatic factor today prevents the beetle from occupying the whole of the range of its potential hosts. A similar case can be made out for *Platypus oxyurus* Duft. (modern range; Figure 9) which has also been found in the Hoxnian Interglacial in Britain (Shotton and Osborne 1965). This species feeds under the bark of *Abies*. Here again the restricted modern range must be attributed to some climatic factor since *Abies* occurs over a large part of Europe today. A third bark beetle is worthy of attention in this context. The rare weevil *Scolytus koenigi* Sch. (modern range; Figure 10) attacks species of *Acer* and related trees. Fossils of this species were recovered from Trafalgar Square, London, (Coope, unpubl.) where they were obtained from deposits of the Last Interglacial.

Associated with fossils of these tree dependent beetles were frequently found the remains of their present day host plants either as pollen or macroscopic structures such as seeds and leaves. They show that the beetles have had similar food preferences for many hundreds of thousands of years. Few of these phytophagous beetles are absolutely loyal to a particular host species and can transfer their allegiance to alternatives when their preferred option is not available.

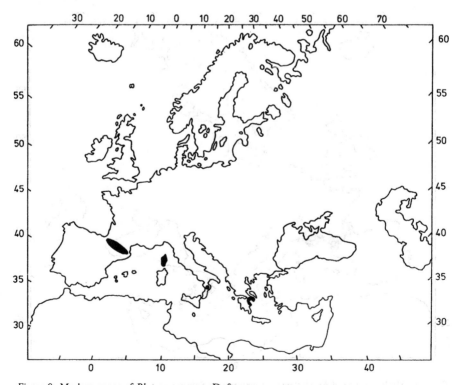

Figure 9. Modern range of *Platypus oxyurus* Duft.

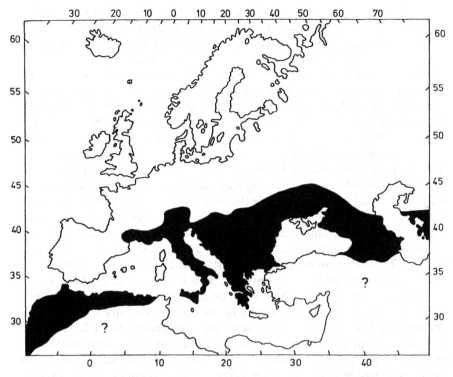

Figure 10. Modern range of *Scolytus koenigi* Sch.

Conclusion

These 'case histories' show that whenever the climates of the past became warm enough, species whose ranges are currently Mediterranean thrived at least as far north as the British Isles and may indeed have extended further north than this, but fossil evidence is not yet available.

Furthermore, the present day ranges of these currently Mediterranean beetle species cannot be used by themselves to interpret past biogeographical histories. Nor should the restricted present day distributions of species, often referred to as 'endemics', be taken as adequate evidence of the place of the evolutionary origin of such species unless there are compelling reasons for believing that the species could not have migrated from elsewhere to the place where it is living today. During the glacial periods, beetles occurred in Britain that have remarkably limited ranges in Asia (Coope 1973). The fossil record shows that it is unsafe to infer that they originated in Asia.

The study of subfossil palaeoentomology thus shows that biogeography is a dynamic response to the large scale climatic fluctuations of the geologically recent past. The patterns of distributions of animal and plant species is the result of a complex history that cannot be understood completely without the evidence of this fossil record. To ignore the fossil data would be like trying to reconstruct the plot of a film from a study of its last frame. After all, the fossils provide us with the only objective evidence of where a species lived in the period before biological records were devised.

From an evolutionary point of view, the subfossil record shows that almost all insect species so far identified, and there are several thousand of them, have remained evolutionarily stable for hundreds of thousands of generations; though at the same time many vertebrates evolved rapidly or became extinct altogether. It would seem that, for insect species at least, adaptation should be seen as fitting the members of a species not only to the conditions of the present day but also to those of the recent past with which the species has had to contend. For example, a species must be able to change its geographic range very rapidly to accommodate the abrupt and intense changes of climate. Species that were for instance too pedestrian or too rigidly attached to their latitude by photoperiodic factors, must have been eliminated at the start as large scale climatic oscillations began to play a significant role during the upper Tertiary. Those that survived the first onslaughts of the climatic changes were well fitted to withstand subsequent ones. This is reflected in the fossil record of Pleistocene insects that show little or no evidence for global extinction of species during this period in spite of the extreme hostility of the climatic changes. At first sight this may seem rather surprising but I have argued elsewhere (Coope 1978) that it was the frequent changes in geographical ranges imposed by the climatic oscillations that made evolution almost impossible by continuously breaking down the isolation between populations permitting them to merge with one another. The frequency and intensity of these climatic changes was such that adaptation by genetic adjustment to the new conditions was not a valid option and that the sole alternative to extinction was to track the acceptable climates across the continents. Thus the climatic conditions in which a species actually lived remained constant; it was the geography that changed. Species stability and the evident lack of global extinction amongst insects may therefore be a consequence of the hostility of the Pleistocene climate.

It follows from the evidence presented here that the Mediterranean region was not entirely outside the influence of the glacial/interglacial climatic fluctuations and that a component of its biota should be viewed merely as temporary residents. To know how long the various species have been present in the area, we must await the investigation of Mediterranean subfossil assemblages.

References

Atkinson, T.C., Briffa, K.R. & Coope, G.R. 1987. Seasonal temperatures in Britain during the past 22,000 years, reconstructed using beetle remains. *Nature* 325: 587–592.

Coope, G.R. 1973. Tibetan Species of Dung Beetle from Late Pleistocene Deposits in England. *Nature* (London) 245: 335–335.

Coope, G.R. 1978. Constancy of insect species versus inconstancy of Quaternary environments. In L.A. Mound & N. Waloff (eds.), *Directory of Insect Faunas*, Blackwell, Oxford, pp. 176–187.

Coope, G.R. 1979. Late Cenozoic Coleoptera: Evolution, Biogeography and Ecology. *Annual Reviews of Ecology and Systematics* 10: 247–267.

Coope, G.R. & Angus, R.B. 1975. An ecological study of a temperate interlude in the middle of the last glaciation, based on fossil Coleoptera from Isleworth, Middlesex. *Journal of Animal Ecology* 44: 365–391.

Girling, M.A. 1974. Evidence from Lincolnshire of the age and intensity of the mid-Devensian temperate episode. *Nature (London) 250: 270.*

Joachim, M.J. 1978. Late-glacial coleopteran assemblages from the west coast of the Isle of Man. Unpublished PhD Thesis, Department of Geological Sciences. University of Birmingham.

Lindroth, C.H. 1943. *Oodes gracilis* Villa. Eine thermophile Carabide Schwedens. *Notulae Entomologicae* 22: 109–157.

Pennington, W. 1986. Lags in adjustment to climate caused by the pace of soil development: Evidence from Britain. *Vegetatio* 67: 105–118.

Shotton, F.W. & Osborne, P.J. 1965. The fauna of the Hoxnian Interglacial deposits of Nechells, Birmingham. *Philosophical Transactions of the Royal Society of London.* B, 248: 353–378.

Van Geel, B., De Lang, L. & Wiegers, J. 1984. Reconstruction and interpretation of the local vegetational succession of a Late Glacial deposit form Usselo (The Netherlands), based on the analysis of micro- and macrofossils. *Acta Botanica Neerlandica* 33–34: 535–546.

13. Migratory Phenomena in European animal species

GIORGIO MARCUZZI

Abstract

Migrations in European animals are illustrated for a period going from the Miocene to the present, and their bearing upon the constitution of the present fauna is emphasized. Together with spontaneous movements of several European vertebrates (mammals and birds) recently observed, the spreading of many man-transported animals is illustrated. Some of these have become true pests.

Introduction

Migration is one of the major factors responsible for colonization of Europe since the first appearance of emergent lands. While we have detailed data about exact period and route of most recent invasions, our knowledge is more limited for more remote epochs (Pliocene, Miocene). Consideration of the land morphology of Europe and the Mediterranean Basin together with that of systematic affinities of many taxa allows us to offer some hypotheses about epoch of arrival and route followed by many terrestrial European animals. Clear examples are those of Western Mediterranean- North African elements (reptiles, beetles) and those of Western Asian elements. Well documented are the movements of Coleopterous insects during Miocene and then during Pleistocene from Balcania location to what Italy was in those epochs and vice-versa (transadriatic species). After the glacial period, a number of animals, vertebrates and invertebrates, followed the retreating ice masses both towards north and south, giving origin to the category of boreo-alpine animals. In some warm and dry epoch, after the last retreat of ices, some southern species have been able to colonize some xerothermic biotopes of southern Alps and Prealps, where they have been able to thrive up to today (xerothermic oases). Many animals have invaded Europe only in recent times, thanks to the passive transport, voluntary or not, enacted by man. With few competitors and secondary predators, some of these have reached very elevated population

F. di Castri, A. J. Hansen and M. Debussche (eds.), Biological Invasions in Europe and the Mediterranean Basin.
217–227. © 1990, *Kluwer Academic Publishers, Dordrecht.*

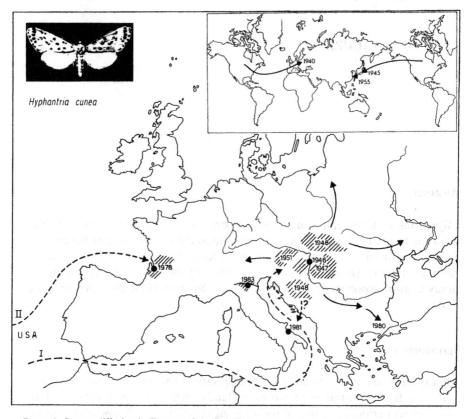

Figure 1. Recent diffusion in Europe of the Fall Webworm *Hyphantria cunea*.

densities and have become true pests (e.g., the clam *Dreissena polymorpha*, the Colorado beetle *Leptinotarsa decemlineata*, the Fall webworm *Hyphantria cunea*) (Figure 1). Finally, spontaneous occupation of new territories has also been observed in the last decades in several mammals (polecat *Putorius putorius*, northern birch mouse *Sicista betulina*, etc.) (cf. Marcuzzi, 1968a p. 282; 1979 p. 134, fig. 102–3) and birds (see Isenmann, this book). Migrations can be considered therefore as a very constant feature of animal colonization of our continent. Often it has been accompanied by evolution (speciation or sub-speciation).

Migration of European species

Ancient Western origin

A group of species migrated to Europe since Miocene times from North-West Africa. Examples include *Chalcides ocellatus,* a Scincidae lizard; *Psam-*

219

madromus algirus, a Lacertidae lizard, and the Tenebrionid beetles *Pachychila friolii, P. dejeani*. This distribution may be dated to a period going from Pontian (Late Miocene) to Pleistocene (Riss-Wurmian interglacial period), when Tunisia was united to Sicily and the present islands to Sicilian Sea. Perhaps in the same period the Scarab beetle *Onthophagus andalusicus* reached as far east as the Tremiti Islands (Adriatic Sea) and the Carabid beetle *Broscus politus* migrated from Western Mediterranean up to southern Italy.

Ancient Eastern origin

Eastern species or taxa going back to Miocene belong to the genus *Eutagenia*, which perhaps in the Miocene – when South Aegeis (southern portion of present Balcanic Peninsula) was emerged – reached Southern Puglia. There it has been found in two localities near Taranto (*Eutagenia elvirae*). The genus is perhaps original from Lesser Asia or Turkmenistan, since it shows some clear Asiatic affinities (Figure 2). *Leichenum pictum*, a psammophilous element, has Central Asiatic origin. The genus *Dendarus* has a probable eastern centre of origin, with a secondary centre, maybe more recent, in the Hiberic Peninsula*. *Opatrum obesum* reached Puglia from the east; *Ocnera philistina* has migrated as far west as Sicily. The same is true for genus *Pedinus*. The Carabid beetle *Masoreus aegyptiacus* has spread from eastern Mediterranean Basin up to Sicily, from which possibly it reached only in recent times southern Puglia, where it has been found for the first time by the present author in 1960. Among vertebrate fauna, an example of West-Asian origin is given by the horned viper, *Vipera ammodytes*, subdivided in a number of subspecies as a sign of its great antiquity (Bruno, 1968) and by the Rodent genus *Spalax* (cf. van der Brink and Barruel 1971).

Ponto-Pannonic (i.e. of the region of the Black Sea and Eastern Europe, the Roman 'Pannonia'), steppic, halophilous elements going back to Lower Pliocene in which south-eastern Europe was covered by shallow brackish water are the two hoppers *Epacromia coerulipes* and *E. pannonicus*. These are present in Europe only in brackish environments such as the so-called 'barene' of the Lagoon of Venice or the 'sansouires' of the Midi of France, bound to halophil vegetation given by *Salsola, Salicornia, Atriplex*, etc. (cf. Marcuzzi, 1979 p. 582, fig. 381). An eastern steppic element on the contrary is the Chrysomelid beetle *Tituboea macropus*, reaching in the past, xerothermic periods as far west as Northern Italy (cf. Marcuzzi, 1979 p. 328).

The retreat of several Asiatic animals which came to Western Europe when the climate was dry and cold during Pleistocene is particularly well known. *Saiga tatarica* or steppe antelope in the glacial episodes lived in Central Europe and England; in the middle of XIX century, it was present in Poland and

---

* The indication 'Tirol' in Gebien (p. 420) for *Dendarus tristis* must be attributed to Süd Tirol (or Alto Adige) since no species of *Dendarus* is indicated by Wörndle for Tirol.

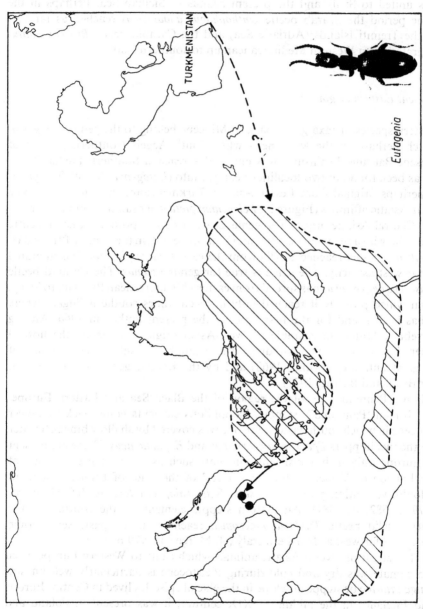

Figure 2. Distribution (schematic) of the genus *Eutagenia*, which spread as far west as the Puglie (*E. elvirae*).

the Carpathians, and in 1865, it still remained in the Kalmucks steppe between the Volga and the Don River. Some years ago, it was present in the territory extending between the Manic and the Volga, whereas it had been exterminated between the Volga and the Ural. The hamster, *Cricetus cricetus* is a rodent originating also in Asia, and reaching as far west as Belgium (Figure 3). Its spread depends on soil characteristics, being restricted to moderately solid, dry and fertile soils (loess, loam). It is present in Wurttemberg, Svevian Alb, Central Germany, Brandenburg and Rhenania. At the end of the last century, it was living also on the Vosges (a mountain chain on the left of the Rhine River, near Strasbourg); as a fossil it is known from Central France (Brehm, 1914). The Columbiform *Syrrhaptes paradoxus* is unique in its migratory behaviour, characterized by invasions from time to time to Western and Central Europe. A great eruption took place in 1888, when many birds reached the extreme west of Ireland in May and June, and some individuals bred in England. In 1863, a similar migration also occurred, though perhaps not on such a vast scale. During last great invasion in 1908, the bird made its appearance in several Italian localities such as Padova, Emilia, Lazio and Barletta (Puglia). The latter is the southernmost part of Italy (and Europe) in which the bird has ever been found. It is said to have bred in Padova in 1888 (Arrigoni degli Oddi). Sporadic invaders coming from Asia are the pin-tailed sandgrouse *Pterocles alchata* and the waxwing *Bombycilla garrulus*.

Pleistocene origin: the effect of glaciation on European fauna

The boreo-alpine migration of animals and plants, associated with the retreat of ice, is represented by two currents: one towards north (Arctic region) and one towards the Alps and other southern European mountains. Examples include ptarmigan *Lagopus mutus*, three-toed woodpecker *Picoides tridactylus*, ring ouzel *Turdus torquatus*, snow hare *Lepus timidus*, and Odonata *Somatochlora alpestris* and *S. arctica*. Many beetles show the boreo-alpine disjunction which witnessess this kind of two-directional migration (Holdhaus).

Western Asia insects which reached Puglia (and in some instances the whole of Italy) include a number of hoppers belonging to Chelifera (or Acrididae), that have a practically continuous are of distribution – evidence in favour of the recent type of diffusion partly holopalearctic, partly eurosibiric. We may mention *Calliptamus italicus, Oedipoda coerulescens, Omocestus ventralis, O. petraeus, Chorthippus mollis, C. brunneus* and *C. dorsatus* (in Puglia *d. garganicus*). The latter is one of the few examples within our fauna or a pleistocenic differentiation of a subspecies). Three species are present also in the steppic situations of Hungary (*Calliptamus italicus, Oedipoda coerulescens* and *Chorthippus brunneus*). They belong to what has been called 'Kultursteppe'. This includes the vegetation constituted by cultivated cereals whose soil and climate are somewhat analogous to those prevailing in steppes, characterized by elements following cultivation, appeared for the first time during Neolithic

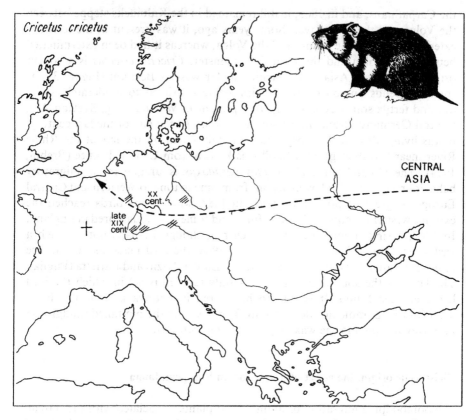

Figure 3. Present and past distribution (schematic) of the hamster, *Cricetus cricetus* (figure of animal from Van den Brink & Barruel) the cross (+) represents a fossil finding.

(La Greca).

In the Hypsothermic period some Mediterranean elements have been able to migrate up to Northern Italy, where they remained in some particularly warm xeric micro-environments, the so called 'xerothermic oases'. Examples are the beetles *Crioceris paracenthesis, Chrysomela grossa, Scybalicus oblongiusculus, Licinus silphoides, Carterus dama* and *Acinopus picipes,* as well as the Lepidopteron *Gonepteryx cleopatra* (Magistretti & Ruffo).

Recent migrations in Europe

A well-known case in which eco-climate (more precisely the climatic conditions of litoral sandy dunes) is the main factor responsible for the diffusion (and presence) concerns some Tenebrionid beetles. An analysis of northwards migration of six thermo-halopsammophilous elements (i.e. species requiring warm, salty, sandy situations) can be made on the ground of the climatic characteristics of the biotope they have been able to reach from the south (Southern Italy and/or the Western Mediterreanen Basin). We may speak here of different ecological valence as far as temperature is concerned. A parellel examination of the action of rainfall would prove that aridity is also responsible for this distribution. The species according to latitude (and average annual temperature) are from South to North as follows: 1, *Gunarus parvulus*; 2, *Pachychila frioli*; 3, *Pimelia rugulosa*; 4, *Pseudoseriscius helvolus adriaticus*; 5, *Erodius siculus*, as far north as the Lagoon of Comacchio near Ravenna; 6, *Xanthomus pallidus*, as far north as the Lagoon of Venice.

In recent times, several steppic animals came from Eastern Europe to occupy the newly expanded cultivated areas in Central Europe. There they found optimal conditions, and generally, empty ecological niches. Examples are the rodents *Cricetus cricetus* or hamster and *Citellus citellus*, or European souslik, and the two birds *Otis tarda* and *Perdix perdix*. With an exception for the latter, all of them retreated when intensive agriculture began to develop in Central Europe (Tischler, 1979 p. 248). In more ancient times felling of trees and drainage of swampy areas to permit the diffusion of agriculture in Central Europe brought about the settlement in the new open spaces of two vertebrates which proved to be harmful to crops, namely, the rook (*Corvus frugilegus*) and the field mouse (*Microtus arvalis*) (Tischler, 1979 p. 249). This can be considered as a case of invasive immigration from another biotope, due exclusively to the presence of man.

A quite peculiar example of short-range, periodic, migration is known in lemming (*Lemmus lemmus*), living in the arctic part of Eurasia. The migration, due to an increase of population density, sometimes results in the death of animals which move into the sea, as a consequence of their migratory drive. The phenomenon is cyclic (4 years-cycles) (Marcuzzi, 1979 p. 261). We should not disregard in this short discussion of migrations in European fauna the two species of rats reached Europe in various, but not well known epochs. Both had enormous negative effects on man and his stored products. *Rattus norvegicus* comes from east or north, while one subspecies of *Rattus rattus (r. rattus)* comes from the East only during the Middle Ages. The other subspecies (*r. alexandrinus*) came to Europe in ancient times and is 'indigenous' in southern countries (Brohmer 1957). The epidemics of plague which struck Athens in 429 B.C. (causing the death of Pericles) were probably due to this subspecies (cf. Marcuzzi, 1979 p. 705-6).

Migrations in aquatic fauna

Very characteristic are the migrations of several euryhalinous (tolerant of a great variation in salt content of water) aquatic species, belonging to fishes, cyclostomes and mammals. Letting apart the well-known amphibiotic fishes, we limit ourselves to mention the cyclostomes *Petromyzon marinus* and *Lampetra fluviatilis* (Zanandrea), and, among mammals, the dolphin known

224

in Russian as 'beluga', *Delphinapterus leuca* (Marcuzzi & Pilleri). The migration of *Phoca* (*Pusa*) from Caspian Sea to Arctic Sea must be dated to an ancient, immediately post-glacial time: the presence of *Phoca hispida* in the Baltic Sea, very akin to *Phoca caspica*, is explained by means of a migration in a time when the Caspian Sea was communicating with the Arctic Sea and when the Baltic Sea was freely communicating with the Atlantic Ocean, namely in the cold 'Litorina-Zeit', nearly 4000 B.C. (cf. Marcuzzi, 1979 p. 560, fig. 369).

The role of passive transport

Finally, migratory movements can be easily observed and studied in a number of vertebrates and invertebrates. Transported voluntarily or not by man, these animals rapidly spread in Europe where they found an empty ecological niche and no competitors. The brachyuran Decapod *Eriocheir sinensis*, originally from China was imported with water tanks by ship and appeared first in the Weser River in 1912. It then diffused to Elba, Oder and Vistula Rivers, and in a second time in several parts of North-Central Europe (cf. Marcuzzi, 1968 p. 279-280, fig. 170a, p. 282, Table XLI). The clam *Dreissena polymorpha*, originally from the Caspian Sea and the Black Sea, invaded most European rivers and in some places (e.g. Balcania) multiplied with an exceptionally high rate*. Among the fresh water fishes, *Eupomotis sol* (= *gibbosus*) the sun fish, came to Italy in 1900 coming from USA, diffused widely in Europe. *Ameiurus nebulosus* (– *Ictalurus* sp.) the cat-fish, imported from USA in 1906 (acc. to Gridelli, 1936, in 1898), is now widespread, often having replaced the indigenous species. More recently the black bass *Micropterus salmoides* has been imported from the USA and now is kept in several water-bodies for sport-fishing, though adapted only in a few European lakes. Finally, we wish to mention the guppy, *Gambusia affinis hollbrooki*, imported from USA in 1922 to fight malaria (it eats the larvae of Culicidae). This species has also eliminated some indigenous elements, as *Triturus vulgaris meridionalis* (Marcuzzi, 1968 p. 355). Also, a marine litoral scallop (Crustacea, Cirripedia), *Elminius modestus*, reached from Australia the southern shores of England in 1944, spread to all Britain and Ireland and then to the European coasts (cf. Marcuzzi, 1968 p. 279).

All these species have been able to spread with an unparalleled speed that is rarely observed in nature. These examples, therefore, cannot serve to understand how the species *normally* spread in their environment. But there are at our disposal some examples of European animals which moved from their home very recently with a well known speed. We may mention *Vanellus vanellus*, the lapwing, in the period 1899-1954, the polecat *Putorius putorius*,

---

* This is a very harmful animal to all the structures used for aquatic supply of any kind (drink water, irrigation water, hydroelectrical uses of water, etc.) (Grossu).

in the period 1879-1939. Also the passeriform *Serinus serinus* (= *S. canaria serinus* s. Zimmer & Rensch), the serin has spread northwards since the beginning of this century (cf. Marcuzzi, 1979 p. 134, fig. 102). Finally the small steppic rodent *Sicista betulina*, the northern birch mouse, in recent years has appeared in some parts of Austria for the first time.

We taken from Margalef (1984) a table in which the average advance per year for some particularly invasive animals is represented:

Species	Group	Geographical area	Aver. approximate advance per year (km)
Elminius modestus	Cirripedia	Australasia-Europe	19–52
Eriocheir sinensis	Dec. Brachyra	East. Asia-Europe	30
Leptinotarsa 10-lineata	Coleopt.	USA-Europe	70
Serinus serinus	Passerine birds	Europe-NE	10–23
Emberiza aureola	Passerine birds	Russia-West	23
Sturnus vulgaris	Passerine birds	Europe-USA	43
Fiber zibethicus	Rodents	(N. America)-Czechoslovakia	25

Conclusion

From what we have seen, we may conclude that migrations have always been present in most animal taxa. In some instances we may observe also today the speed of migration in indigenous, European species and in elements recently introduced by man, both terrestrial and aquatic. For the spreading in a given biotope of a newly arrived species the lack of competitors and secondarily of predators is of paramount importance. This is particularly valid in the study of pests. We cannot extend our experience on living species to old European elements (Miocene or even older periods): the fossil evidence existing only for a very small number of species (mammals, molluscs, etc.) is too scarce to permit us any generalization. Anyway, we may affirm that migration is a sure and constant feature in the diffusion of many animals, as it has been shown since long by many authors (for general aspects of ecology and physiology of migration, see for instance Elton 1958 for animals in general, or Johnson 1969 for flying insects). In some instances climatic changes have been sufficient to bring about a migration of some animals which have been able to stay in their new environment also after the climate had returned to its original conditions, favoured by some micro-climatic conditions. Recent movements northwards as in the case of the beetle *Masoreus aegyptiacus* can be perhaps attributed to recent changes (increase) in temperature. Migrations have been sometimes accompanied by evolution, namely speciation or sub-speciation. Examples of subspeciation in old European species, belonging to both vertebrates and invertebrates are present in *Psammodromus algirus, Onthophagus andalusicus, Pseudoseriscius helvolus*. Of course, the 'tempo' of

226

evolution was various in the different taxa and perhaps in the different epochs, according to changes in the extension of emerged lands.

References

Arrigoni degli Oddi, E. 1929. *Ornitologia italiana*, Hoepli, Milano, 1046 pp.

Atlante Fisico Economico d'Italia. 1940. Consoc. Turist. Ital., Milano.

Brehm, A. 1914. *Die Säugetiere*, Bibliogr. Institut, Leipzig & Wien, II. 654 pp.

Brehm, A. 1920. *Die Säugetiere*, Bibliogr. Institut, Leipzig & Wien, IV. 714 pp.

Brink, F.H. van den & Barruel, P. 1971. *Guide des mammifères sauvages de l'Europe Occidentale*, Delachaux & Niestlé, Paris & Bruxelles, 263 pp.

Brohmer P. 1957. Säugetiere, In: Tierwelt Mitteleuropas, VII.

Bruno, S. 1968. Sulla Vipera ammodytes (Linnaeus 1758), *Mem. Mus. St. Nat. Verona*, 15: 289.

D'Ancona, U. 1962. *Biologia generale*, Cedam, Padova.

Elton, C.S. 1958. The ecology of invasions by animals and plants, Methuen, London, 195 pp.

Gebien, H. 1938-1942. Katalog der Tenebrioniden (Col. Heteromera), II, *Mitt. Münchn. Entom. Gesellsch.* 28-32: 49-345.

Gridelli, E. 1936. *I pesci d'acqua dolce della Venezia Giulda*, Udine. 133 pp.

Gridelli, E. 1950. Il problema delle specie a diffusione transadriatica, *Mem. Biogeogr. Adriat.* 1: 7-299.

Grossu, A.V. 1962. Bivalvia, *Fauna Republ. Popul. Romine*, Moll. III, Fasc. 3; Ed. Acad. Rep. Pop. Rom.

Holdhaus, K. 1954. Die Spuren der Eiszeit in der Tierwelt Europas, *Abh. Zool. Bot. Ges. Wien* 18: 1-493.

Johnson, C.G. 1969. *Migration and dispersal of Insects by flight*, Methuen, London.

Koch, C. 1948. Die Tenebrioniden Kretas, *Mitt. Münchn. Entom. Gesells.* 34: 255-386.

Ladiges, W. & Vogt, D. 1965. *Die Süsswasserfische Europas*. Verl. Paul Parey, Hamburg & Stuttgart.

La Greca, M. 1962. L'ortotterofauna pugliese ed il suo significato biogeografico. *Mem. Biogeogr. Adriat.* 4: 33-170.

Lydekker, R. 1913-1916. *Catalogue of the ungulate mammals.* 5 vols., London.

Magistretti, M. & Ruffo, S. 1959. Primo contributo alla conoscenza delle oasi xerotermiche prealpine. *Mem. Mus. St. Nat. Verona* 7: 99.

Magistretti, M. & Ruffo, S. 1960. Secondo contributo alla conoscenza della fauna delle oasi xerotermiche prealpine. *Mem. Mus. St. Nat. Verona* 8: 223-240.

Marcuzzi, G. 1968a. *Ecologia animale.* Fetrinelli, Milano, 832 pp.

Marcuzzi, G. 1968b. Osservazioni ecologiche qualitative sull'erpetofauna della laguna veneta. *Mem. Accad. Patav. S.L.A.* 80: 333-372.

Marcuzzi, G. 1969. Contributo alla zoogeografia dei Tenebrionidi della Sicilia. *Mem. Soc. Ent. Ital.* 48: 499-518.

Marcuzzi, G. 1976. Osservazioni biogeografiche sulla tenebrionidofauna del Mediterraneo. *Quad. Ecol. Anim., Padova* 6: 3-48 pp.

Marcuzzi, G. 1979. European ecosystems, Junk ed., Dordrecht, Boston & London, 779 pp.

Marcuzzi, G. & Pilleri, G. 1971. On the zoogeography of Cetacea. *Investig. Cetacea* 3: 101-170.

Marcuzzi, G. & Turchetto Lafisca, M. 1977. Ricerche sui Coleotteri della Puglia raccolti da G. Marcuzzi (1960-63). *Quad. Ecol. Anim. Padova* 9: 1-186 pp.

Marcuzzi, G. & Turchetto Lafisca, M. 1981. Nuovi dati faunistici sulla Coleotterofauna pugliese. *Atti Ist. Ven. S.L.A.* 139: 59-79.

Marcuzzi, G. & Turchetto Lafisca, M. 1981. Eutagenia elvirae. Nuova specie di Coleoptero Tenebrionide dell'Italia meridionale (Heteromera), *Mem. Soc. Ent. Ital., Genova* 60: 235-238.

Margalef, R. 1984. *Ecologia.* Ed. Omega, Barcelona. 950 pp.

Niethammer, J. & Krapp, F. 1978. *Handbuch der Säugetiere Europas, I.* Akadem. Verlags.,

227

Wiesbaden. 476 pp.

Scharff, R.F. 1899. *The history of the European fauna*. Walter Scott, London. 364 pp.

Schweiger, H. 1960. *Das Tierleben in der Grosstadt*, Wien. 22 pp.

Tischler, W. 1979. Einführung in die Okologie, G. Fischer V., Stuttgart & New York, 306 pp.

Wagler, E. 1937. Crustacea. In: *Tierwelt Mitteleuropas, II*, 2.

Wörndle, A. 1950. *Die Käfer von Nord-Tirol*, Innsbruck, 388 pp.

Zanandrea, G. 1947. Notizie ed appunti sui Petromizonti delle Tre Venezie, *Boll. Pesca Piscic. Idrobiol.* 23: 5–8.

Zanandrea, G. 1957. Esame critico e comparativo delle Lamprede catturate in **Italia**, *Arch. Zool. Ital.* 42: 249–307.

Zanandrea, G. 1959. Le lamprede del Trentino-Alto Adige, *Natura Alpina*, 81–85.

Zanandrea, G, 1959. Lampetra fluviatilis catturata in mare nel golfo di Gaeta, *Publ. Staz. Zool. Napoli* 31: 265–307.

Zanandrea, G. 1962. Rapporti tra l'alto e il medio versante adriatico d'Italia nella biogeografia delle Lamprede, *Boll. Zool.* 29: 727–734.

Zangheri, S. 1986. Note sulla distribuzione geografica, la biologia, l'etologia dell'H. cunea Drury (Lepidoptera, Arctiidae), l'ifantria americana (Hyphantria cunea, Drury) nella realtà padana, *Atti*, Reggio Emilia, 1–8.

14. The bean beetle (*Acanthoscelides obtectus*) and its host, the French bean (*Phaseolus vulgaris*): a two-way colonization story

VINCENT LABEYRIE

Abstract

Phaseolus vulgaris and *Acanthoscelides obtectus* originated from Central America. But, outside equatorial America, where mature pods (of wild or cultivated beans) are present all the year in field, bean pods occur only during restricted periods of the year. In such regions, the bean bruchid exists as a domestic, man-dependent insect. Its ability to reproduce in stored seeds allows it to expand its range in anthropogenic habitats by developing alternatively in the field during the ripening time of pods, and during the rest of the year in stored seeds. Reproduction in the stores is allowed by the ability to oviposit in spite of suppression of the first sequences of the original egg-laying behaviour, *i.e.* flight to mature pods, prospection of pods, holing of pods by mandibles. In stores, it glues eggs directly to the seeds tegument. Growing in stores, it can live temporarily each year in fields, during the ripening of pods, where climatic conditions in the remainder of the year are inconsistent with its biological constraints.

Introduction

Since *The genetics of colonizing species* (Baker & Stebbins 1965), relatively few studies have examined the genetic and demographic characteristics of insect pests undergoing a range extension. Emigration to new areas is often considered a response to new climatic conditions, but, modification of the niche may also result in insects colonizing new habitats. Hirsch (1967) underlined the danger of typologism in behavioural studies. It was because I observed a tremendous polymorphism of the host acceptance during the egg-laying behaviour of an ichneumonid wasp that I hypothesized (Labeyrie 1964a) that females with the less rigid and stereotyped oviposition behaviour may colonize new kinds of hosts. The well-known cosmopolitan pest, the bean beetle (*Acanthoscelides obtectus*) with its recent appearance in many countries and its unique use of two habitats – beans in fields and in stores

F. di Castri, A. J. Hansen and M. Debussche (eds.), Biological Invasions in Europe and the Mediterranean Basin.
229–243. © 1990, Kluwer Academic Publishers, Dordrecht.

230

– is a very exciting model to study colonization by insects.

Phaseolus vulgaris results from selection by pre-Columbian farmers of a *Phaseolus* species complex originating in the Andean highlands. Expansion of cultivars to the north and south occurred before the Spanish invasion. During the 17th century, *P. vulgaris* was introduced also into Europe. Since then, with extension of the European colonial empires, it has been planted in Asia and in Africa (Duke 1981).

Wild *Phaseolus* is a vine and the traditional agricultural practice was to mix corn and bean crops: bean vines climbing on corn stems. Today, in Europe, intercropping of corn and beans is less and less frequent. Contemporary plant breeding is designed to obtain determinated growth and short cycles that facilitate mechanized harvest of pods eaten as vegetables; either green or when almost ripe (Pickersgill 1977). These selection programs are carried out in monocultures. With modern selection, breeders have obtained an annual legume, cultivated alone, with determinate habit and simultaneous blooming (Evans 1974).

Acanthoscelides obtectus is a holometabolous insect parasitoid (parasitically-living larva and free-living adult); its larva is a gregarious seminivore specialized on mature seeds of *Phaseolus ssp*. The number of generations produced each year is dependent on temperature. It is calssified by Loukianovitch & TerMinassian (1957) in the second group of economically important bruchids because it is able to reproduce in the field and also on stored seeds.

Like the *P. vulgaris* complex, *A. obtectus* originated in northern South America, probably at middle altitude between 1,000 and 2,000 m elevation, near the equator. It appeared in Europe two centuries later than the bean itself; the first good evidence of its presence was made by Fabre in the seventies of the last century (Fabre 1879). With an expansion so recent, it is possible to follow its anthropogenic spread throughout the world.

According to economic considerations, *A. obtectus* is abusively distinguished as a stored product insect (Singh & Van Emden 1979). Consequently, as if it were spontaneously generated in stores, its biology in the field has generally been neglected. This strange failure results from the U.S. Department of Agriculture's operational dichotomy, made in 1917, between pre-harvesting pests and post-harvesting pests (Simmonds 1964). This artificial break still persists nowadays. Therefore, its actual geographical distribution is more restricted than is described in the applied entomology books: in stores it is a very cosmopolitan insect, its limits are only the historical opportunities, given by human activities, to invade new countries. On the other hand, in fields its presence is defined by local ecological constraints (Labeyrie & Maison 1954).

In this paper, I explore the ecological and anthropogenic factors that have allowed *A. obtectus* to succesfully invade new regions and the demographic patterns associated with this range extension. I shall argue that it is very unlikely in this species that colonization corresponds to an enlargement of its niche. In fields, besides some occasional ovipositions within pods of *Vigna*

231

unguiculata (Jarry & Bonet 1982), with further larval development in seeds, *A. obtectus* lays its eggs only in *Phaseolus* pods. Rather, it appears that the spread of this species is related to its ability to thrive and multiply inside stored beans and to its ability to colonize new fields of beans.

The rigidity of the niche of *Acanthoscelides obtectus*

Rigidity of niche could be a general characteristic of phytophagous insects specialized in feeding on legume seeds. Ordinarily, this specialization is explained by its linkage with the occurrence of specific allelochemicals in the seeds of legumes (Applebaum 1964); these substances represent barriers that only some bruchids have been able to overcome. To digest these plants, not only the basic enzyme complex common to all bruchid species is needed, but also specific enzymes able to attack specific lectins, enzyme inhibitors, saponins, etc. of mature *Phaseolus* seeds. Thus, few bruchid species attack the same species of legume, and conversely, the number of legume species used by each species of bruchid also is very limited (Johnson 1981b).

It is generally not thought that this situation is the result of coevolution between a particular legume and a specialist beetle (Varaigne-Labeyrie & Labeyrie 1981), although legume allelochemicals have been used as examples to offer evidence of coevolution (Janzen 1977). The 'raison d'être' of secondary substances, according to Fraenkel (1959), can be the result of selective pressure by consumers, when the loss of the structure jeopardizes the life of the plant, or at least lowers its reproductive value.

In annual plants, it is clear that the destruction of fruit exerts a strong selective pressure; but allelochemical production is not the only response. Several other responses allow survival of the plant: redundancy of surnumary reproductive organs or induction of new inflorescences appear as other answers to the destruction of fruits by consumers (Harris 1973). I consider that coevolution between plant and insects consuming plant reproductive organs may be limited to annual plants (Labeyrie 1977b, 1978, 1987).

Circumstances are different with perennial plants, among which population survival is assured with a limited number of offspring (Hossaert & Valero 1985). The perennial way of life and vegetative multiplication can be considered as means to avoid or to limit hazards resulting from the disappearance of reproductive organs.

It must be remembered that wild *P. vulgaris* is a perennial plant; the selective work of breeders modified it into an annual plant (Evans 1974). Therefore, because seeds were not essential for perennial primitive beans, today the chemical defenses of seeds of this neo-annual plant are insufficient against bruchid larvae, even if they prevent attacks of some generalist consumers. Even if the toxic effects of chemical barriers are insufficient to explain the specialization of *A. obtectus*, they reveal the aptitude of the larvae to use some enzymes to neutralize toxic effects of allelochemicals.

232

An alternative explanation for host specificity involves behaviour. Jermy (1984) emphasized that 'Host plant selection is mainly a behavioural process,' at the very least in perennial plants. For many species the number of potential hosts suitable for larvae is greater than the number of hosts actually used (Wiklund 1975). The first factor limiting host range is the behaviour of ovipositing females. Several sequences of complex behaviour define subsequent opportunities for further oviposition (Labeyrie 1964a). This reproductive behaviour implies many signals and several kinds of sensory receptors (Pouzat 1977).

For example, *A. obtectus* adults are able to discover isolated developed bean plants among several hectares of maize (Jarry 1981); generally this first step of reproductive behaviour (long range attraction) is neglected in laboratory bioassays. Attraction by the plant favours sex-meeting, and the oviposition site is also the site of the sexual rendez-vous (Labeyrie 1971). Unfortunately, the influence of trophic relations upon sexual behaviour is generally neglected (Labeyrie 1977a), although, 'it serves as an effective isolation mechanism' (Zwölfer & Harris 1971).

Following discovery of mature pods – especially yellowish ones – the preoviposition behaviour of the female is very elaborated, involving sensory receptors of antennae, mandibular and labial palps, and both olfaction and gustation to prospect the surface of the pod (Pouzat 1981). After such inspection the female bores a hole with her mandibles through the ventral suture of the pod; even after the completion of this hole, characteristics of the carpellar cavity are analysed by sensillae of the ovipositor before laying eggs (Pouzat 1975). If desiccation of carpellar cavity is only partial, the female does not lay eggs; dehydration is a prerequisite to oviposition. Up to thirty eggs are glued side by side on the interior wall of pods. Larvae, which are chryso-meliform, wander through the carpellar cavity and perforate the teguments of mature seeds, where they will grow and metamorphose. Up to thirty adults will escape from the same mature seed (Labeyrie 1962).

The usual list of *A. obtectus* hosts (Johnson 1981a) are inaccurate. They correspond only to the list of seeds in which larvae are able to penetrate, to develop and to give emerging adults, and not to the plants in which females lay eggs and from which adults emerge.

To test the relative importance of different sequences on the rate of insect attack on seeds, we compared, in 1985, attacks on french bean (*P. vulgaris*), chickpea (*Cicer arietinum*), garden pea (*Pisum sativum*), and *Lathyrus sylvestris* in the field, with bioassays of mature pods in the laboratory.

In September 1985, the field study showed no attacks on the pods of any species except for *P. vulgaris* in which 20% and 40% of the pods contained one or more attacked seed. This is a normal value in this part of France. In the laboratory, females confined with mature pods of bean oviposited with a very high frequency in them, and the ratio of adults matured to eggs laid was also very high. The situation was totally different with pods of other species of legumes (Table 1). The ratio of adults matured to eggs laid was

Table 1. Contamination of pods and bean bruchid development in three legumes (females confined with mature pods in Petri dishes).

Plant species	Treatment[a]					
	n	A	B	C	D	E
Pisum	24	13	3	2	2	4
Cicer	25	23	0	0	1	1
Lathyrus	25	17	0	0	8	0

[a] A: eggs, larvae, adults absent; B: only larvae present; C: larvae and adults present; D: eggs and adults present; E: eggs, larvae, and adults present.
n: number of pods.

0.75 (54/72) with *P. sativum*. In *C. arietinum* the ratio was not actually quantified but was probably very near 0.75. In *L. sylvestris* no adults were matured, though a large number of eggs was laid (0/103). Thus, only *P. vulgaris* attracted *A. obtectus* adults in the field; but when females were confined with pods, results were more complex: some females bored holes in pods of *Pisum* and *Lathyrus* (25% and 32%, respectively), but only 8% did so in *Cicer*. Some larvae arising from eggs laid in boxes had perforated the pods of *Pisum*, contaminated seeds, and about 20% of the pods produced adults (5/24). The proportion of larvae surviving to adults was normal in *Pisum* and *Cicer*, but none did so in *Lathyrus* whose seeds contain cspecific non-protein amino acids (Bell 1981). The toxicity of seeds to *A. obtectus* larvae was determinant in *Lathyrus*. In all the other cases, host choice was made earlier.

Therefore, it is an oversimplification to ascribe niche specialization solely to toxicity of seeds. We can even consider that the more efficient is the adaptation of the specialized insect, the earlier are the effects of significant cues orientating its host choice.

Any enlargement of the niche implies not only allelochemical similarity between contents of earlier and further hosts, but also similarity of external traits serving as cues in reproductive behaviour. I think this is the reason why specialist insects of legume seeds do not attack cereals, however poor they may be in toxic allelochemicals. The fact that first instar larvae penetrate into pods of peas is interesting, it corroborates that physical characteristics of mature seeds do not seem to originate a barrier to larvae of this bruchid. Indeed, larvae drill also into polystyrene tubes and other materials. Absence of phagostimulants does not appear to be an obstacle to attack by larvae of bean bruchids. This being so, it is difficult to understand why corn grains, mixed with contaminated beans seeds in storage of small farms in Mexico and Colombia, were not attacked by bruchid larvae.

None the less, modification of the niche of the very specialist insect, *A. obtectus*, would imply changes in its olfactory, gustatory, mechanical, and other sensory reactions, and also in its enzymatic equipment. Being able to feed is not sufficient. It is first necessary to be attracted to the potential food

234

(i.e. to oviposit where the larvae will be able to eat). Probably, the very great range of alterations required is sufficient to explain why adaptation to new niches is very exceptional, whereas adaptation to new habitats in this species is very usual.

The rigidity of the niche of *A. obtectus* is all the more remarkable, since, as I shall show, the reproductive behaviour on stored beans is strongly truncated.

The double enlargement of the habitat of *Acanthoscelides obtectus*

The bean beetle has several behavioural traits that make the problem of colonization of new habitats (stores and new geographical areas) by this insect especially interesting. *A. obtectus* multiplies in stores with several generations a year; it contaminates the pods of cultivated bean in fields of many countries in the world, and it excapes from stores to colonize fields.

Multiplication in stored beans

Having emerged in stores from contaminated seeds, after several days (number varying with temperature), adults lay and glue their eggs at points of contact between contiguous seeds. The young larva bores into the integument at a contact point; all the growth takes place in cotyledons; the adults of the new generation emerge in their turn, and after several days they oviposit, beginning the cycle anew (Labeyrie 1962). In stores, all the first sequences of activity leading to oviposition have disappeared. Feeding is no more a prerequisite, as it seems to be in fields (Jarry 1984). Females seem very opportunistic and able to leave out all the now useless components of their preovipositional behaviour; hence, their behaviour is not stereotyped (i.e. the last sequences are not necessarily induced by the previous ones (Labeyrie 1964b).

Different observations demonstrate that oviposition in the field takes place chiefly in sunshine and during the warmer hours of the day (Labeyrie & Maison 1954); nevertheless in stored seeds, with constant darkness and without apparent thermoperiod, females lay their eggs without decreasing fecundity and without particular delay. In the same way, adults develop a strong positive phototactism; in the laboratory they are obviously attracted to windows, and it is impossible to obtain their oviposition on caged bean plants in the field because they concentrate on the cage wall directed to the sun. However, Perttunen & Häyrinen (1970) have observed 'in both males and females the individual variation in the take-off activity is considerable.' Similarly, Jarry observed that on stocks a great proportion of young adults are not attracted to light, and stay and reproduce again on the heap of seeds (1984), but in successive generations the frequency of adults migrating into the field increases, and during summer they oviposit in mature pods.

A. obtectus is clearly an example of an insect with two behaviours, the first one as a field insect, the second one as a domestic insect subjected to man's activities. Twenty years ago, I considered that these two behaviour- were the expression of a kind of behavioural polymorphism of this species. In connection with this hypothesis, and in order to know if only some females present the shortened behaviour in stores, I examined offspring of field females to discover if some of them are able to lay their eggs in empty Petri dishes. I succeeded, and after four to five generations, I obtained pure lines of 'astime females' (Labeyrie 1961). Every examined European field population generates about 15% of offspring able to oviposit without any sensory stimulation. I thought that this constant frequency of astime females resulted from the heterozygote condition of females presenting a double reproductive behaviour. Secondly, all the females of the different populations examined are able to lay their eggs directly on seeds; when they fail, they usually present some malformations. Thus, all the females reproduce on seeds inside dark boxes. Conversely, all the females bred on seeds or in empty Petri dishes are able to bore holes and to select dehydrated pods.

Since the adults of different generations, after a more or less long time of multiplication on stored seeds, can recover the complex field reproductive behaviour, it is difficult to imagine a system of alternating field and store generation, or an expression of phase phenomena without the external dimorphism observed in other bruchids, (e.g. *Callosobruchus maculatus*). In the same way, it is clear that the stores are not colonized by marginal adults with less strict cues or requirements, as can be suspected when we observe the colonization of a new niche (Labeyrie 1964a).

It is clear that *A. obtectus* exploited seeds of *Phaseolus* long before the discovery of fire and its use to cook legumes prerequisite to their eating by man, which results in detoxification of mature seeds by destruction of the thermolabile lectins (Labeyrie, 1981). Storage of mature bean seeds by man is consecutive to the enlargement of his trophic basis enabled by cooking of this legume. This would lead us to suppose a relatively recent appearance of the bean bruchid behaviour of ovipositing directly on seeds linked to the appearance of bean storage by man! This is not my opinion. I think that storage has only promoted a previous behaviour.

Hypothesis on the origin of Acanthoscelides obtectus reproduction in stores

Reproduction in stores might have been introduced through two different ways, before any human intervention:
– In Colombia, very vigorous vines of some *Phaseolus ssp.* climb on the stems of isolated trees up to five or six meters. Ripe pods, strongly lignified, are indehiscent. Therefore, it is impossible for the adult emerged from the seed with its mandibles to bore a hole through the carpellar wall. The emergence

236

hole of the young adults is prepared in the seed coat by the mature larva before pupation; then, when pods are indehiscent, new adults are kept and confined in them. Many of them reproduce *in situ*, ovipositing within the pod in which they had developed.

Must we then consider that the original hosts of *A. obtectus* had presented indehiscent pods? Maybe, but a new problem arises; indeed, if the dry pericarp of mature pods is too hard to be bored by adult, such an indehiscent host is a lethal trap for the bruchid and evolution from this kind of pod is impossible without arising a way to improve the boring of the carpellar wall.

Some other observations on indehiscent pod in bean crops show that the hypothesis of multivoltinism as an adaptation to indehiscent pods must not be necessarily rejected. In these circumstances, some adults can bore into the pod wall while cutting the hole prepared by the larva through the seed coat. To succeed in this work, the hole prepared by the larva must be just against the carpellar wall.

– An alternative hypothesis involves the storage of food by small mammals. About ten years ago, just before dusk in Mexico, in a pre-columbian agricultural site, near the pyramid of Tepoztlan (Morelos), altitude 1,900m, under a pergola of vines of *Phaseolus* intermixed with branches of trees, covered with mature and dehiscent pods, I observed a heap of seeds upon the ground. But the following morning, all the seeds had disappeared. During the night, animals had collected them. In France, the field mouse *Apodemus sylvaticus* collects, stores and eats seeds of Fabaceae. They do not refuse seeds containing larvae or pupae of bruchids; moreover, it seems that they eat the bruchid larvae before eating the seed's cotyledons (P. Haffner, personal communication). Some meso-american rodents probably present a similar behaviour. If such is the case, we consider that some species of rodents store seeds of *Phaseolus*, including seeds containing bruchid larvae. Temperature inside burrows is probably relatively constant and higher than 17°C in the Tepoztlan area; it is sufficient to allow reproductive behaviour.

So, it is not out of the question that some bruchid adults multiply directly on seeds, inside rodent burrows. With the exhaustion of the rodent store, after a limited number of generations, adults would escape and would contaminate new wild pods. According to this hypothesis, stores simulate rodent burrows, and the store-field system simulates the burrow-wild bean one. Other stored product insects might have evolved from rodent burrow insects. The antiquity of this biocenotic relation would explain why, today, all the adults present this behaviour.

In countries with a dry season – with no production of pods during this time – it is possible that the adults maintained in indehiscent pods or in rodent burrows helps the survival of *A. obtectus* populations. But, some contaminated areas in central America do not present actual dry seasons! However that may be, the colonization of stored beans illustrates previous remarks on the rigidity of niche. The critical trait seems to be the chemical stimulants included in the pods as in the seeds.

237

Range expansion by the bean beetle

It is out of the question to include in the bruchid's area of distribution the countries where this pest is observed in stores only. Exclusively the countries where pods are contaminated in the field should be included in the geographic area of this insect. Occurrence of pre-harvest attacks does not imply that we can observe the same ecological relations everywhere. Indeed, we have two kinds of situations.

In the first one, the attack of pods in the field is contingent upon the adults' escape from stores. In this category of countries, *A. obtectus* is unable to maintain populations year-round in fields; stores constitute a reservoir for field invaders. The ecological conditions for obtaining this first situation are less restrictive than in the next.

In the second situation, as in the bruchid's area of origin, the whole annual cycle takes place in the field; in this case, multiplication in stored seeds is only of subordinate importance in the annual cycle.

Oscillating infestation between fields and stores

A. obtectus lays its eggs only into ripe pods of *Phaseolus* and exceptionally into pods of *Vigna* (Jarry & Bonnet 1982). Oviposition obviously requires crops in which pods are left in the field until maturity. In contrast, areas in which *Phaseolus* are planted only as a green vegetable crop, with pods harvested before maturity, are out of danger.

When there are no more ripe pods in the fields, adults could survive inside indehiscent pods or inside rodent burrows, as previously explained. Of course, these circumstances could assure survival. This situation arises where climatic conditions in the country of cultivation prevent planting several bean crops per year, and where winters are too cold to allow survival of adults outdoors. In the laboratory, it is possible to maintain adults alive for up to 300 days at a constant temperature of 4°C, but they are subsequently unable to reproduce (Leroi 1981). Similarly, they are unable to survive in seeds fallen on the ground as some arid-zone *Bruchidius* species do (Southgate 1981). After the first rain, imbibition of the seed kills the insect (Larson 1924). Thus, survival outdoors in countries with cold and rainy seasons seems out of the question for this multivoltine species of bruchid.

However, presence of stores as a winter reservoir is not by itself sufficient to maintain an infestation hazard in field. Until a few years ago, I considered that the reproductive activity requires temperatures higher that 18–19°C. Therefore, it is possible to define countries at risk to field infestation by bruchids. Thirty years ago, I noted that the boundary of field-infested areas in Europe corresponded to the 18°C isotherm in July (Labeyrie 1957). It is not sure that today the determinism of the boundaries would always be the same. In Gascogne, France, the last several years, with late summers and

238

the first weeks of autumn warmer that in previous years, we observe later infestations of pods in September, and even sometimes in October. It is not possible to exclude, after at least a century of oscillating activities between stores and fields, that a lowering of the tempertature threshold of reproductive activity is appearing.

With the spread of extensive and mechanized agriculture, a new category of countries appeared where climatic conditions are propitious but pre-harvest attacks are not possible. For example, in the central part of Cuba, farming of dry beans in large fields followed after the clearing of bushlands. The average field spreads over several hundred hectares; there are no small fields in the surrounding area. The harvest is totally mechanized, and the yield immediately exported to silos, several kilometres away, where seeds are submitted to chemical fumigation. Farm workers use exclusively dry beans bought in shops and issued from silos. Therefore, there are no small personal stocks of untreated beans inside farms, and infestation of crops by bruchids escaping from stores is virtually impossible. I presume that the most important and the last potential sources of contaminating adults, in this part of Cuba or in similar places, would be adults issued from rodent burrows and long-distance flying insects coming from stores of small farmers living in distant villages.

I discuss this new situation because I think that the enlargement of field size has contradictory ecological effects (Labeyrie 1977a), and not only negative ones, on pest populations. In some cases, like that of *A. obtectus*, we can cut the sequences that assure the annual cycle of the insect. A very significant example of this kind of situation, is that of *Bruchus pisorum*. Larvae of this monovoltine bruchid feed on growing seeds of *Pisum sativum*, but adults emerge from mature seeds. By separation of areas of two kinds of pea crops: vegetable crops (harvested as green pods) on the one hand, and production of seeds for planting (pods harvested when mature) on the other hand, there are no more attacks in areas where it is grown as a vegetable. Indeed, this bruchid disappears from fields where pods are harvested before maturing and, therefore, before growth of the larvae is completed, since the cycle of the host-plant is disrupted. Of course, seeds are imported from other areas, where *B. pisorum* can always infest local crops of peas.

Permanent infestation in the field

In Colombia, where it is native, populations of *A. obtectus* are maintained in nature during all its annual cycle. Several conditions allow this multivoltinism in the field, but there is always a climatic prerequisite: each month must include several hours equal to or higher than the temperature threshold for reproduction (ca. 18–19°C). When climatic conditions are favorable, occurrence of attacks depends on agricultural practices and on plant community composition, and also on the biological traits of the bruchid population.

In Colombia, pods of *P. vulgaris* are contaminated only between 1,000 and

239

2,400 m altitudes. This equatorial insect can breed throughout the year because of the absence of a distinct dry season. Therefore, this is not a tropical* outdoor pest because in the Tropics, during the dry season its populations are maintained, as in Mediterranean countries, by its multiplication in stored beans. For example, in the central altiplano of Mexico, 19°N, wild *Phaseolus* are attacked, not by *Acanthoscelides obtectus*, but by a related species *Acanthoscelides obvelatus*. This North American species attacks pods of wild *Phaseolus coccineus*, the seeds of which present dormancy, and *A. obvelatus* presents diapause. In this part of Mexico, the two species of *Acanbthoscelides* live in sympatry: *A. obtectus* in pods of cultivated beans, and *A. obvelatus* in wild bean pods.

The study of geographical distribution in equatorial South America is very stimulating. In the lowlands, the climate is too hot for *A. obtectus*, but another genus attacks cultivated or wild beans in pods, *Zabrotes subfasciatus*, which glues its eggs upon seeds in partly or totally dehiscent pods of wild *Phaseolus lunatus* or on cultivated beans (Pimbert 1985). Similarly, in Punjab (India), *P. vulgaris* is attacked by *Z. subfasciatus* and not by *A. obtectus*.

In potential habitats of Equatorial America – sides of cordilleras and altiplanos – conditions of pre-harvest infestation vary with agricultural practices, but different wild plants of the *Phaseolus* complex grow between rocks, along roads, climbing on stems of trees in arborescent savannas. These wild *Phaseolus* can support, at different times of the year, some bruchid eggs and larvae in their pods. Therefore, they help maintain *A. obtectus* populations that attack cultivated beans. The annual cycle of *A. obtectus* is passed integrally in the field, and passing through stored beans is ecologically subsidiary, even if it is quantitatively important.

When the climate is suitable, two kinds of potential areas can be defined, those where wild *Phaseolus* exist and those where they are absent.

Phaseolus is an American genus of Fabaceae (Maréchal 1969); other species from Asia and Africa, very similar morphologically, belong to the related genus *Vigna*. For this reason, outside America, all the countries where climatic conditions would permit continuous multiplication, are placed in the latter category. This is the case of Rwanda and Burundi, in central Africa, where crops of *P. vulgaris* are very important and *A. obtectus* a serious pest. These equatorial countries have the same periodicities as Colombia in South America, they also have the same altitude as the infested part of Colombia, but in Equatorial Africa there are no wild *Phaseolus*, but only wild *Vigna*. Since, in some particular situations (Mexico) it is possible to observe contaminated pods of *Vigna unguiculata*, it is necessary to examine if in Rwanda or Burundi this bruchid is able to colonize wild *Vigna*; I don't think that this occurs

---

* The terms 'tropics' and 'tropical' are not used in their conventional geographical sense, in which they include all areas between 23°27'N and 23°27'S, but to refer to the part of this area that is concerned by seasonal changes in primary periodicity (Monchadsky 1958), i.e. in solar energy inputs. For this ecological reason, 'tropical' and 'equatorial' areas are distinguished.

240

today.

Succession of many crops of beans over the year, with overlapping periods of pods' maturation, or at least a great enlargement of time of this maturation, appear as the prerequisites for *A. obtectus* to maintain its population in the field during the entire year, in the absence of wild *Phaseolus*. Hence, persistence of *A. obtectus* in the fields depends either on the cultural practices (number of sowings around the year, times of sowing), or on the kind of variety: vine vs. dwarf, extended vs. synchronous maturation.

In Colombia, near the upper altitudinal boundaries of its area, we observed another situation. Many females present delayed ovarian maturation (Huignard & Biémont 1981); thus females can pass through several weeks without ripe pods in which to oviposit. However that may be, in all populations examined we observe that vitellogenesis is stimulated by mature pods or seeds of *Phaseolus* (Labeyrie 1960). Without stimulation, the reproductive power is delayed during several weeks, and spontaneous vitellogenesis is very low (Huignard 1970).

Conclusion

Today, *Acanthoscelides obtectus* is still invading new countries. Its colonization shows different degrees of integration: on the one hand an oscillating activity, with combination of one generation in the field and several generations in stores, and on the other hand passage of the entire annual cycle in the field. But, this colonization is always anthropogenic. Furthermore, we know neither examples of colonization of wild plants, nor examples of attacks of introduced *A. obtectus* by local entomophagous insects.

In all the countries it has colonized, this bruchid remains man-dependent: it is not integrated into the local biocenosis; when cultivation of beans stops, this insect must disappear. But if its presence is man-dependent, many different factors favour its colonization, some factors are linked to the characteristics of the plant: climbing, dwarfishness, determined habits, etc. Others are linked to agricultural practices, and still others to the reproductive habits of the female: direct or partially delayed vitellogenesis. So, a very great variety of variables interact to determine whether *A. obtectus* can adapt to a new country, or only to a particular field. The example of this insect reveals that it is dangerous to conclude after the discovery of one determinism; it is not necessarily the only one, but generally, one among others.

Since the migration from America of this bruchid is the result of traffic through the Atlantic since the beginning of the 17th century, it is possible to know with accuracy the paths of its expansion. For example, it is interesting to understand why two centuries separate the introduction of *Phaseolus vulgaris* in Europe from the official observation of the first attacks of *A. obtectus* in fields. It is important to study historical data to know the complex path followed by *P. vulgaris* and *A. obtectus* in getting from America to Central

Africa. I suppose that this expansion was through steps in Europe, and that colonial countries (Belgium, France, Germany, Portugal) introduced beans into their colonies in the highlands of central Africa, because dry beans were already an element of their diet. If this hypothesis is corroborated, it means that in order to reach a habitat with very similar climatic characteristics to its habitat of origin, *A. obtectus* had to transit by European countries, where ecological conditions strongly differ from conditions of both original and final countries.

For all these reasons, *A. obtectus* is a very interesting example for studying effects of propagation on the characteristics of populations, to develop genetic ecology and population genetics with a model supported by precise historical data. It is possible to know the direction of trade, quantities of beans introduced with time, their origin, and even in some cases their sanitary condition.

Karyological studies reveal chromosomal modifications in some populations of *A. obtectus* (Garaud 1984), but distributions of these alterations appear not to be correlated with historical events of colonization or with ecological conditions. I suspect that a comparative study of some polymorphic aspects of *A. obtectus* behaviour may be more enlightening than karyological studies, and will allow us to discover the selective effects of different ecological pressures.

But the most important aspect of the biology of *A. obtectus* is to offer a very interesting model for conducting experiments about the evolutionary relations between wild and domestic species, because the behaviour of this insect always includes components of both kinds of species.

Curiously enough, this very exciting biological material was not used by experimental evolutionists and ecologists; the reason may be that it is too frequently considered only as an agricultural pest. This is one more example of the noxiousness of the gap between general biology and applied entomology.

Acknowledgements

This research was supported for the last fifteen years by the National Center for Scientific Research (CNRS). G. Fabres, D. McKey and M. Thoraval gave helpful comments on the manuscript.

References

Applebaum, S.W. 1964. Physiological aspects of host specificity on the Bruchidae. *Journal of Insect Physiology* 10: 783–788.

Baker, H.G. & Stebbins, G.L. 1965. *The genetics of colonizing species.* Academic Press, New York, London.

Bell, E.A. 1981. Non-protein amino acids in the Leguminosae. In: R.M. Polhill & P.H. Raven (eds.), *Advances in Legume Systematics,* Royal Botanic Gardens, Kew, England, pp. 489–499.

Duke, J.A. 1981. *Handbook of Legumes of world economic importance.* Plenum Press, New York.

242

Evans, A.M.A. 1974. Research on the evolution and genetic improvement of grain legumes. *Journal of Applied Biology* 46: 5–11.

Fabre, H. 1879. *Souvenirs entomologiques* 8: 26–27, 47–48, du Layet, Paris.

Fraenkel, G.S. 1959. The raison d'être of secundary plant substances. *Science* 129: 1466–1470.

Garaud, P. 1984. Mise en évidence d'un polyporphisme chromosomique de translocation dans une population naturelle d'*Acanthoscelides obtectus* (Coléoptère, Bruchidae) du Burundi. *Genetica* 63: 65–71.

Harris, P. 1973. Insects in the population dynamics of plants. *Proceedings of the 6th Symposium of the Royal Entomological Society* 6: 201–209.

Hirsch, J. 1967. Behavior genetics, or 'experimental' analysis; the challenge of science versus the Cure of technology, *The Psychologist* 22(2): 118–130.

Hossaert, M. & Valero, M. 1985. Differences in population biology within the *Lathyrus sylvestris* group (Leguminosae; Papilionaceae). In: J. Haeck & J.W. Woldenburg (eds.), *Structure and functioning of plant populations. Phenotypic and genotypic variation in plant populations 2*, North Holland Publ., pp. 65–76.

Huignard, J. 1970. Influences comparées de la plante-hôte et de la copulation sur l'ovogenèse puis sur la ponte chez *Acanthoscelides obtectus. Comptes Rendus de l'Académie des Sciences* 271: 2171–2174.

Huignard, J. & Biémont, J.C. 1981. Reproductive polymorphism of populations of *Acanthoscelides obtectus* from different colombian ecosystems. In: V. Labeyrie (ed.), *The ecology of bruchids attacking legumes (pulses)*, Junk Publishers, Dordrecht, pp. 149–164.

Janzen, D.H. 1977. The interaction of seed predators and seed chemistry. In: *Comportement des insectes et milieu trophique*, C.N.R.S. Publ., pp. 415–428.·

Jarry, M. 1981. Evolution of spatial pattern of attacks by *Acanthoscelides obtectus* Say (Coleoptera, Bruchidae) of *Phaseolus vulgaris* L. pods in South West France. In: V. Labeyrie (ed.), *The ecology of bruchids attacking legumes (pulses)*, Junk Publishers, Dordrecht, pp. 131–141.

Jarry, M. & Bonnet, A. 1982. La bruche du haricot, *Acanthoscelides obtectus* Say (Coleoptera, Bruchidae), est-elle un danger pour le cowpea, *Vigna unguiculata* (L.)? *Agronomie* 2(10): 963–968.

Jermy, T. 1984. Evolution of insect-host plant relationships. *The american naturalist* 124(5): 609–630.

Johnson, C.D. 1981a. Seed bettle host specificity and the systematics of the leguminosae. In: R.M. Polhill & P.H. Raven (eds.), *Advances in Legume Systematics*, Royal Botanic Gardens, Kew, England, pp. 995–1027.

Johnson, C.D. 1981b. Relations of *Acanthoscelides* with their plant hosts. In: V. Labeyrie (ed.), *The ecology of bruchids attacking legumes (pulses)*, Junk Publishers, Dordrecht, pp. 73–81.

Labeyrie, V. & Maison P. 1954. De l'influence du microclimat sur la ponte de la Bruche du haricot (*Acanthoscelides obtectus* Say) dans la nature. *Comptes Rendus de l'Académie d'Agriculture* 40: 733–736.

Labeyrie, V. 1957. Sur les conditions de pullulation en France de la Bruche du haricot (*Acanthoscelides obtectus* Say) en culture. *Comptes Rendus de l'Académie d'Agriculture* 43: 590–593.

Labeyrie, V. 1960. Influence de l'hôte sur la fécondité d'*Acanthoscelides obtectus* Say. *Comptes Rendus de l'Académie des Sciences* 250: 615–617.

Labeyrie, V. 1962. Les *Acanthoscelides*. In: Balachowsky ed. *Entomologie appliquée à l'agriculture*, Vol. 1(1), Masson Publ., pp. 469–484.

Labeyrie, V. 1964a. Importance de l'intégration des signaux fournis par l'hôte lors de la ponte des insectes. *Année psychologique* 66(1): 1–14.

Labeyrie, V. 1964b. Action sélective de la fréquence de l'hôte utilisable (*Acrolepia assectella* Zell) sur *Diadromus pulchellus* WSM: la variabilité de la fécondité de fonction de l'intensité de la stimulation. *Comptes Rendus de l'Académie des Sciences* 259: 3644–3647.

Labeyrie, V. 1971. Trophic relations and sex meetings in insects. *Acta Phytopathologica Academiae Scientiarum Hungaricae* 6(1–4): 229–231.

Labeyrie, V. 1977a. Influence de la variabilité du comportement dans la dynamique des populations d'insectes. In: J. Medioni & E. Boesiger (eds.), *Mécanismes éthologiques de l'évolution*, Masson Publ., pp. 21–33.

Labeyrie, V. 1977b. Environnement sensoriel et coévolution des insectes. In: *Colloques Internationaux du C.N.R.S.: Comportement des Insectes et Milieu Trophique*, Tours 13–17/09/1976, C.N.R.S. Publ., pp. 15–35.

Labeyrie, V. 1978. Reproduction of insects and coevolution of insects and plants. *Entomologia Experimentalis & Applicata* 24: 496–504.

Labeyrie, V. 1981. Vaincre la carence protéique par le développement des légumineuses alimentaires et la protection de leurs récoltes contre les bruches. *Food and Nutrition Bulletin* 3(1): 24–38.

Labeyrie, V. 1987. Towards a synthetic approach to insect-plant relationships. In: V. Labeyrie, G. Farbes, D. Lachaise (eds.), Plant 86, Junk Publ., Dordrecht, pp. 3–5.

Larson, A.D. 1924. The effect of weevily seed beans upon the bean crop and upon the dissemination of weevils *Bruchus obtectus* Say and *B. quadrimaculatus* Feb. *Journal of Economic entomology* 17: 538–548.

Leroi, B. 1981. Feeding, longevity and reproduction of adults of *Acanthoscelides obtectus* Say in laboratory conditions. In: V. Labeyrie ed. *The ecology of bruchids attacking legumes (pulses)*, Junk Publ., Dordrecht, pp. 101–111.

Loukianovitch, F.K. & Ter Minassian, M.E., 1957. Jouki Zernoyki (Bruchidae) *Faune de l'U.R.S.S.* 24(1) N.S. 67: 1–210.

Maréchal, 1969. Données cytologiques sur les espèces de la sous-tribu des Papilionaceae, Phaseloae-Phaseolenae *Bulletin du Jardin Botanique National Belge* 39(2): 125–165.

Monchadsky, A.S. 1958. A propos de la classification des facteurs du milieu. *Zool. Zh.* 37: 680–692.

Perttunen, V. & Häyrinen, T. 1970. Individual variation in the take-off activity of *Acanthoscelides obtectus* Say (Col. Bruchidae) *Ann.Ent.Fenn.* 36(2): 107–110.

Pickersgill, B. 1977. Taxonomy and the origin and evolution of cultivated plant in the new world. *Nature* 268: 591–595.

Pimbert, M. 1985. A model of host plant change of *Zabrotes subfasciatus* Boh. (Coleoptera Bruchidae) in a traditional bean cropping system in Costa Rica. *Biological Agriculture and Horticulture* 3: 39–54.

Pouzat, J. 1975. Analyse expérimentale du rôle de l'ovitube dans le comportement de ponte de la bruche du haricot, *Acanthoscelides obtectus* Say. *Behaviour* 54: 258–277.

Pouzat, J. 1977. Effet des stimulations provenant de la plante-hôte, le haricot *Phaseolus vulgaris* L., sur le comportement de la ponte de la Bruche du haricot *Acanthoscelides obtectus* Say. In: *Colloques Internationaux du C.N.R.S.* 265: *Comportement des insectes et milieu trophique*. pp. 115–131.

Pouzat, J. 1981. The role of sense organs in the relations between bruchids and their host plants. In: V. Labeyrie (ed.), *The ecology of bruchids attacking legumes (pulses)*, Junk Publ., Dordrecht, pp. 61–72.

Simmonds, P. 1964. An outline of recent progress in stored-product entomology. *Journal of Economical Entomology* 57(1): 29–31.

Singh, S.R. & Van Emden, H.F. 1979. Insect pests of grain legumes. *Annual Review of Entomology* 24: 255–278.

Southgate, B.J. 1981. Univoltine and multivoltine cycles. Their significance. In: V. Labeyrie (ed.), *The ecology of bruchids attacking legumes (pulses)*, Junk Publ., Dordrecht, pp. 17–22.

Varaigne-Labeyrie, C. & Labeyrie, V. 1981. First data on bruchidae which attack the pods of legumes in Upper Volta of which eight species are man consumed. In: V. Labeyrie (ed.), *The ecology of bruchids attacking legumes (pulses)*, Junk Publ., Dordrecht, pp. 83–96.

Wiklund, C. 1975. The evolutionary relationship between adult oviposition preferences and larval host plant range in *Papilio machaon. Oecologia* 18: 185–197.

Zwölfer, H. & Harris, P. 1971. Host specificity determination of insects for biological control of weeds. *Annual Review of Entomology* 16: 159–178.

15. Some recent bird invasions in Europe and the Mediterranean Basin

PAUL ISENMANN

Abstract

In Europe and the Mediterranean Basin, where the natural environment has been dramatically altered by man, many species have vanished or became endangered whereas others have been able to expand their breeding range. This chapter deals only with a sample of some expanding species, most of them concern westward oriented expansions but also some northward expansions by Mediterranean and desertic species. Each of these range expansions reflects more a combination of several factors than the influence of a single factor. Species able to live in mostly open and eutrophic habitats shaped by man and/or in close contact with man seem to be favoured. Thus, the ever-growing impact of human activities now plays a significant role in recent bird invasions.

Introduction

Europe and the Mediterranean Basin have undergone climatic and land use changes during this century. Forest fragmentation, mechanizaton of agriculture, heavy use of pesticides, urbanization, industrial pollution, increasing wastes, recreation and hunting activities constitute the pool of major factors that have reshaped the environment. These changing patterns in food and habitats have strongly influenced the abundance and distribution of birds.

A bird's breeding range is rarely static, rather it changes under the combined pressures of environment, dispersal processes and population dynamics. In Europe and the Mediterranean Basin, where the natural environment has been dramatically altered by man, many species have vanished or became endangered whereas others have been able to expand their breeding range. This chapter deals only with the expanding species (for papers on this topic, see Kalela 1940, 1949, 1952, Niethammer 1951, Otterlind 1954, Frey 1970, Yeatman 1971, Nowak 1975a & b, Järvinen & Ulfstrand 1980, Hildén & Sharrock 1985,

F. di Castri, A. J. Hansen and M. Debussche (eds.), Biological Invasions in Europe and the Mediterranean Basin.
245–261. © 1990, *Kluwer Academic Publishers, Dordrecht.*

246

Mauersberger 1985, Hengeveld 1988). First, I present a typology of the different invasion types (invasion is understood as an expansion of a species' range into a new area), followed by a sample of some spectacular cases that have occurred during this century in the Western Palearctic. Then, I briefly discuss their most plausible causes.

Invasion types

It is easy to follow the different steps of most of the recent range expansions in Europe, thanks to an enthusiastic army of ornithologists which promptly detects any changes in species distributions. It remains much more difficult to explain most of their range shifts. A bird species may expend the limits of its breeding range due to climatic and environmental changes, changes in the bird's tolerance to climate or habitats, and changes in dispersal or migration patterns. If reproduction is successful in the new habitats, then a permanent range expansion becomes likely. Obviously, all of these changes are interrelated. Thus, each particular range expansion reflects more a combination of several factors than the influence of a single factor. The exact weight of each factor, however, often remains impossible to evaluate. Nevertheless, I will try to distinguish the following invasion types according to the main reasons I assume to be involved in their dynamics:
(1) creation of new habitats or/and new natural feeding resources: Pochard (*Aythya ferina*), Woodpigeon (*Columba palumbus*), Fieldfare (*Turdus pilaris*) and Common Rosefinch (*Carpodacus erythrinus*);
(2) increasing artificial foor resources (rubbish tips, etc.): Black-headed Gull (*Larus ridibundus*) and Herring Gull (*Larus argentatus*):
(3) response to climatic fluctuations: Fan-tailed Warbler (*Cisticola juncidis*) and Cetti's Warbler (*Cettia cetti*);
(4) introductions: Pheasant (*Phasianus colchicus*), Ring-necked Parakeet (*Psittacula krameri*) and Red Avadavat (*Amandava amandava*);
(5) unknown causes: Cattle Egret (*Bulbucus ibis*), Collared Dove (*Streptopelia decaocto*), Melodious Warbler (*Hippolais polyglotta*) and Penduline Tit (*Remiz pendulinus*).

Case studies

Pochard (Aythya ferina)

During the last 50 years, this duck colonized southern Sweden and western Europe (Netherlands, Belgium, south-western Germany and France); it increased in Great Britain (200–400 pairs in the mid-1970s, Sharrock 1976) and reached also Ireland and Iceland. In France, it reached the westernmost parts between 1960 and 1965. It has been argued that this westward expansion

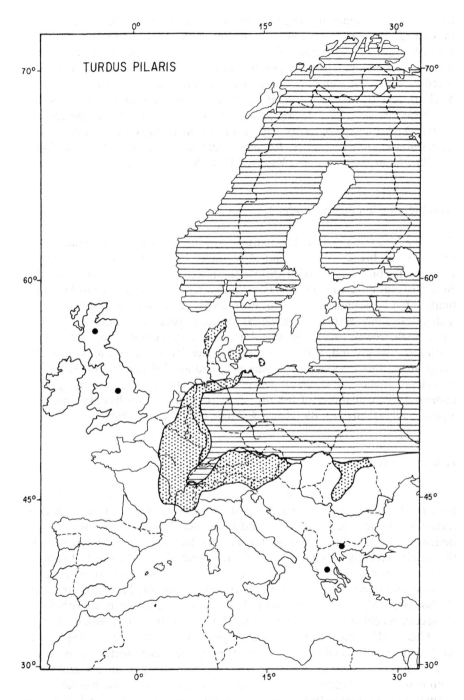

Figure 1. Breeding range of the Fieldfare (*Turdus pilaris*) in 1960 (horizontal lines) and range expansion until 1987 (dotted area and full circles) (after Glutz von Blotzheim & Bauer 1988).

had its origin in some drying processes occurring in the species natural range in Central Asia (Kalela 1949). It seems more likely, however, that the multiplication of water reservoirs and barrages, as well as a marked eutrophication of many European ponds, allowed this species to winter and to breed in increasing numbers in sites not existing before (Bezzel 1969, Bauer & Glutz von Blotzheim 1969). It is obvious that the spatial expansion has also been favoured by a demographical boom. A closely related species, the Tufted Duck (*Aythya fuligula*), has shown a similar range expansion into the westernmost parts of Europe for nearly the same reasons (Bauer & Glutz von Blotzheim 1969, Cruon & Vielliard 1975, Nowak 1975a & b).

Woodpigeon (Columba palumbus)

Since the second half of the last century, this pigeon has shown an important expansion in breeding range and population. Originally native to Central Europe, the bird has invaded the northern and south-western parts of the British Isles and Ireland (also the Faeroes and Iceland) and pushed its breeding limits in Fenno-Scandia towards the north by about 250 km. In Central Europe, it drastically increased its numbers by invading parks and gardens in cities. Even small treeless islands in the North Sea were colonized by individuals nesting on the ground, a rather unfamiliar site for a tree-nesting species. In many areas, this pigeon has become a farmland bird and the development of crops such as maize, clover, rape seed, cabbage, turnip, peas, etc. has greatly improved the availability of feeding resources. The species has even become a pest locally (Glutz von Blotzheim & Bauer 1980, Cramp 1985).

Fieldfare (Turdus pilaris)

This species' westward expansion began during the previous century and has been particularly active during the past 50 years (Figure 1) (for a general review, see Lübcke & Furrer 1985 and Glutz von Blotzheim & Bauer 1988). Besides some breeding points as far west as Iceland and even Greenland, the species recently reached Denmark, the Netherlands, Belgium (Leprince 1985) and sporadically the British Isles (Frost & Shooter 1983). In France, the western boundary reached Lille, Paris, Burgundy, the Massif Central and the Southern Alps (Isenmann 1986). A southward expansion is also obvious, the species reaching Italy (Brichetti 1982), Yugoslavia (Geister 1980), Romania, and Greece (Hölzinger 1986). It was suggested that the species' expansion has been favoured by climatic changes, its poor site attachment, high breeding success owing to colonial nesting and increasing feeding resources, and wide habitat tolerance within a range of more or less open moist habitats and cultivated land, ranging from poplar plantations, parks, gardens and orchards to light forests.

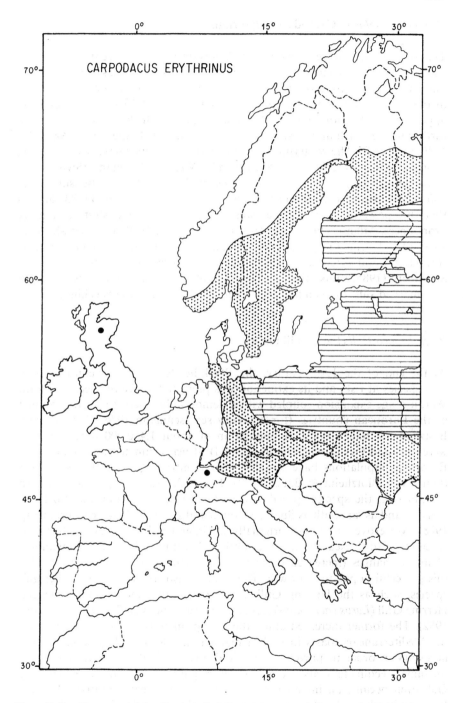

Figure 2. Breeding range of the Common Rosefinch (*Carpodacus erythrinus*) in 1960 (horizontal lines) and range expansion until 1986 (dotted area and full circles).

250

Common Rosefinch (Carpodacus erythrinus)

This species first showed a westward expansion in the 19th century, reaching southern Norway, Denmark, Eastern Germany, Czechoslovakia and Austria, but this was soon followed by a recession. A second expansion slowly began in the 1930's, and, more significantly, after 1960 and 1970. This expansion appears to be the more important because it pushed the breeding limits much further to the north and west (Czikeli 1976, Jung 1983, Stjernberg 1985, Hill 1986) (Figure 2). The westernmost points hitherto attained have been Norway, Scotland (Mullins 1984), the Netherlands, Western Germany, Switzerland (Mosimann 1988) and Yugoslavia (Geister 1980). The Asiatic subspecies *kubanensis* also spread during the same time in Turkey. Jung (1983) argued that this powerful and rapid expansion coincides with a shift to a more pronounced Atlantic climate in Western Europe after 1930 and especially after 1960–1970. During this period, more and more observations were made during springtime, and often were soon followed by breeding records. However, Stjernberg (1985) thinks that, at least in Finland, it has actually been caused by a marked shift to open habitats, where the breeding success is higher.

Black-headed Gull (Larus ridibundus)

At the end of the 19th century, the Black-headed Gull was considered to be a species at risk, and its extinction was predicted. Since the Second World War, a range expansion coupled with a demographic increase started in most of the European countries (for a review see Isenmann 1976–77, Glutz von Blotzheim & Bauer 1982). A population of about 1,400,000 breeding pairs were censused ther (Soviet Union exluded). It was estimated that the Central European populations have doubled their numbers between 1960 and 1975 (Glutz von Blotzheim & Bauer 1982). From Central Europe and southern Scandinavia, the species spread also as far north as Iceland and Northern Norway and as far south as Spain, Sardinia and since 1987 Greece (Jerrentrup 1988). Morerover, although originally an inland gull, this species has also spread toward the seacoasts. It is doubtless that it must have taken advantage of many changes (increased protection and artificial food items) to exhibit this powerful population growth. The same is also true for several other gull species, such as the Herring Gull (*Larus argentatus*) and the Yellow-legged Herring Gull (*Larus cachinnans*) (for reviews, see Glutz von Blotzheim & Bauer 1982). The former increased along the European coasts and the latter along the Mediterranean coasts in such a manner that local culling soon became necessary in order to prevent a drastic decrease of other species that share the same breeding habitats as these predatory gulls. The Yellow-legged Herring Gull even became an inland breeder in a few European countries (France, Italy, Switzerland, West-Germany). Among the seabirds, the spread of the Fulmar (*Fulmaris glacialis*) is a further example in which a combined action of protection and increasing food resources (first offal from whalers and then from fish trawlers) allowed a large spread in the eastern North Atlantic area for a species whose breeding range was formerly restricted to a single Scottish island (St. Kilda) (Cramp *et al.* 1974, Cramp & Simmons 1977).

Fan-tailed Warbler (Cisticola juncidis)

During this century, this species showed no less than three waves of expansion outside its Mediterranean breeding range (Géroudet & Lévêque 1976). The last one, which begun at the end of the sixties and reached a maximum in 1976, pushed the breeding limits northward to the Netherlands, Switzerland, Northern Italy and Northern Yugoslavia. After 1978 and especially since the harsh winter 1982–83, the Fan-tailed Warbler populations crashed and disappeared, even from most of its northern Mediterranean breeding points and, evidently, north of them. Despite a high fecundity (up to two generations per year) and efficient dispersal that takes individuals far beyond their traditional breeding area limits, the success of the Fan-tailed Warbler's spread has been limited by local winter conditions (long frost or snow coverage periods are lethal for this small Warbler). A similar story is that of the Cetti's Warbler (*Cettia cetti*) which also showed a series of expansion waves with recent breeding records as far north as England (Bonham & Robertson 1975), the Netherlands (Teixeira 1979), Western Germany (Becker 1975), Switzerland (Schifferli *et al.* 1980) and Bulgaria (Robel & Königstedt 1980). Other climatically-induced shifts of breeding ranges have been demonstrated by the Desert Wheatear (*Oenanthe deserti*) and the desertic White-crowned Black Wheatear (*Oenanthe leucopyga*), whose ranges were extended northward in northwestern Africa after several recent drought years (François 1975, Ledant *et al.* 1981). A similar expansion is also exhibited by the House Bunting (*Emberiza striolata*) that invaded northward a great part of Morocco (it reached Rabat, Fès and Oujda during the 1960–1987 period) (Courteille & Thévenot 1988).

Pheasant (Phasianus colchicus)

This is the earliest known bird species successfully introduced to Europe. In Germany, the earliest evidence comes from the 6th and the 9th centuries. The records increased in many other countries between the 14th and 17th centuries. But it is only since the end of the 19th and the beginning of the 20th century that this bird has bred and thrived in the temperate parts of Europe, despite heavy hunting pressures and winter adversities. The present European populations are a mixture of birds of various Asiatic origins. Nevertheless, the natural populations are permanently enriched by individuals born in aviaries. Another introduced pheasant species has been less successful: the Reeves Pheasant (*Syrmaticus reevesii*) has only a sporadic and patchy

252

distribution (especially in France, Yeatman 1976) and seems not to be able to persist without the help of regular introductions (Glutz von Blotzheim, Bauer & Bezzel 1973). The same is also true for another Phasianidae, the Bobwhite (*Colinus virginianus*), of North American origin. Hunters regularly introduced this species with the hope that it would become established. Most, if not all, of these introductions failed despite local breeding in France (Yeatman 1976, Cruon & Nicolau-Guillaumet 1985) and Northern Italy (Fasolo & Gariboldi 1987).

Ring-necked Parakeet (Psittacula krameri)

This introduced or escaped Afrotropical parakeet has developed, mostly since 1970, with a few persistant populations in suburban parks and gardens in Great Britain (Greater Manchester and London), Belgium (Brussels and Antwerp), Netherlands (Amsterdam, Rotterdam, The Hague), Western Germany (Lower Rhine valley), France (near Lyon in 1984), Spain (Barcelona since 1982) and Italy (since 1980, 4 breeding places in the northern part and probably also in Sicily) (Cramp 1985, Spano & Truffi 1986). Some of them constitute self-supporting populations that curiously survive in temperate climates. For example, the breeding population near the Zoological Garden of Köln, Western Germany includes about 50 pairs. Another introduced tropical species (originating from southwest Asia), a small-sized passerine (9g), the Red Avadavad (*Amandava amandava*), has spread since 1973–74 and mainly after 1978 between Badajoz and Madrid in Central Spain, where it now abundantly breeds (Lope *et al.* 1984 and 1985). In northern Italy (Treviso province), about 80–90 breeding pairs were also recorded in 1983 and in 1984 (Mezzavilla & Battistella 1987). This species has also bred in Great Britain, Germany and France (Yeatman 1976, Mezzavilla & Battistella 1987) but, owing to harsh winter conditions, it rapidly became extinct in these countries.

Cattle Egret (Bulbucus ibis)

In 1950, this egret successfully colonized parts of North and South America after a spectacular transatlantic crossing from Africa (for reviews see Handtke & Mauersberger 1977, Franchimont 1986a and Arendt 1988). Stragglers coming from Southeast Asia successfully invaded Australia in 1954 and many thousands pairs now breed there (McKilligan 1985). In the Western Paleartic, its breeding range long remained restricted to northwest Africa, where it is known to have been breeding since the 19th century, and to southern Spain. The species showed a recent strong expansion there, especially in Morocco (Franchimont 1986a) and southern Spain. After several attempts after 1957, it has bred since 1968 in the Camargue (southern France) (Hafner 1970) and at least since 1972 in the Ebro Delta in north-eastern Spain where it reached 1136 pairs in 1978 (Muntaner *et al.* 1984). In the Camargue, its population amounted 468 pairs in 1982, but dropped to 74 pairs in 1985 after some severe winters (Wallace *et al.* 1987). In 1985, one breeding pair was found in Sardinia (Grussu & Secci 1986); this seems to announce a colonization of an area east of southern France. The bird has also been found breeding in very low numbers in Western France: 1981 in the Loire Atlantique (Marion & Marion 1982), in 1984 in the Charente Maritime (Bredin 1985) and in the Pyrénées Atlantiques (Carlon 1985). The species has bred since 1973 in Alsace (North-Eastern France) and also in Baden (West Germany) where birds from Morocco have been introduced (Dronneau & Wassmer 1985). The Cattle Egret's spread has been favoured by an extended breeding season and a high fecundity (it can breed twice a year) and a highly opportunistic breeding and feeding behaviour (it frequently uses anthropogenic habitats and the vicinity of cattle herds) (Franchimont 1986b, Arendt 1988). The species also shows an impressive dispersal ability demonstrated by transcontinental jumps and spectacular recoveries of ringed birds (i.e., ringed as a nestling 17 May 1986 Cadiz/ South Spain – recovered 19 July 1987 Stockholm/ Sweden, Stolt 1988). However, being chiefly sedentary, hard winters considerably slow its expansions outside its tropical and southern mediterranean breeding range.

Collared Dove (Streptopelia decaocto)

The most spectacular recent invasion is that of the Collared Dove (*Streptopelia decaocto*). This dove literally has flooded nearly the whole of temperate Europe and north-western Africa in 30–50 years (Figure 3). Between 1933 and 1972, its breeding range increased by 2,500,000 km^2 (Glutz von Blotzheim & Bauer 1980). After 1972, the speed of expansion considerably decreased, the species having colonized by that time nearly all areas favourable to it in Europe. The remaining parts of Northern Europe are too sparsely inhabited by man or climatically too harsh to allow a further rapid expansion. In the Mediterranean area, other constraints must have hindered the spatial expansion. It is only recently that new areas have been colonized: it has bred for the first time in 1979 in Sardinia (Brichetti *et al.* 1986). Although first breeding has been recorded in 1974 in northern Spain and in northern Portugal, the Iberian Mediterranean zone has only been colonized since 1980 (Bernis *et al.* 1985, Barcena & Dominguez 1986) and Morocco in 1986 (Franchimont 1987). It must be remembered that for at least 200 years the species showed no tendency to spread. When the Turks, who protected this dove, had been evicted from a large part of southeastern Europe, the species became less numerous or even extinct due to severe persecution by local populations wishing to destroy all the symbols of the Ottoman occupation period. But, unexpectedly, at the end of the twenties, a massive spread began, the reasons for which are not well known. It is suspected that climatic changes which permitted a long breeding period associated with a high fecundity (3 to 5 broods per

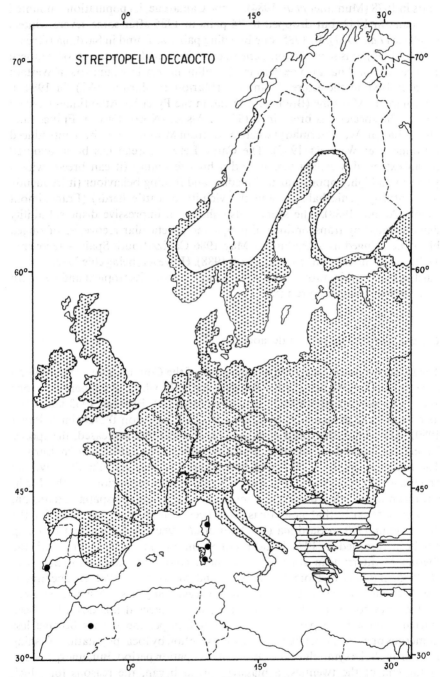

Figure 3. Breeding range of the Collared Dove (*Streptopelia decaocto*) in 1932 (horizontal lines) and range expansion until 1986 (dotted area and full circles) (after Glutz von Blotzheim & Bauer 1980 modified).

year). Moreover, the possibility of reaching maturity when 3 to 4 months old allowed this species an avalanche-like spread throughout most of Europe (Hengeveld 1988). Finally, it preferably lives in man-made habitats in and around towns (Glutz von Blotzheim & Bauer 1980). It is not by chance that its spread in the Mediterranean area was less active than elsewhere or was delayed, as the species found much less favourable conditions here (less urbanization?, lack of protection?).

Melodious Warbler (Hippolais polyglotta)

The breeding range of the Melodious Warbler long remained restricted to southwestern Europe and northwestern Africa, with a narrow hybrid zone and some mixed pairs with the more northern and eastern sister-species, the Icterine Warbler (*Hippolais icterina*). A first small range expansion began westward after 1960 in Brittany (Guermeur & Monnat 1980). A much larger expansion began in 1970–75 when the species invaded the Icterine Warbler's range (reviewed by Tombal 1980, Landenbergue & Turrian 1982, François 1983, Jacob et al. 1983, Fernex 1985, Preiswerk 1985, Jacob 1986, Hayo & Weyers 1986) (Figure 4). In Switzerland, where breeding was first found 1960

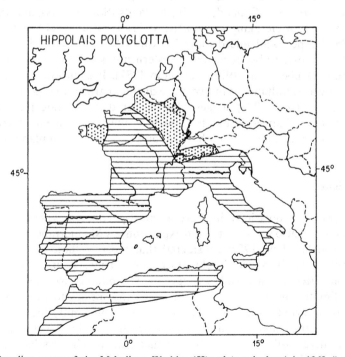

Figure 4. Breeding range of the Melodious Warbler (*Hippolais polyglotta*) in 1969 (horizontal lines) and range expansion until 1987 (dotted area).

256

in Tessin, other breeding records were made in 1974 in Wallis and 1977 near Geneva. This warbler reached Belgium in 1981 (in 1985 the breeding populations amounted to about 100 pairs), Luxemburg in 1986 (Melchior 1988), northeastern France in 1981–1984 and Western Germany in 1983–84. Some local population increases were impressive: thus, in Belgium (see above) and in Saarland (Western Germany) the species was first recorded in 1983, found breeding in 1984, and by 1985 60 singing males were censused (Hayo & Zannini 1986). The reasons for the sudden northward spread of this Mediterranean warbler are unexplained: the Icterine Warbler, despite some local declines, is still present, the habitat requirements of both species are different (the Icterine W. is rather arboreal and the Melodious W. lives in shrubby areas), and the habitats of the colonized areas are not new. Contrary to the Fan-tailed Warbler and the Cetti's Warbler, whose expansion waves are limited by severe winter conditions, the Melodious Warbler is a summer resident present only from May to August.

Penduline Tit (Remiz pendulinus)

Since 1930 and especially after 1965, the Penduline Tit has spread to the west. As shown by Flade *et al.* (1986), the species now reaches southeastern Sweden and southern Finland, the Netherlands (Sovon 1988), and the Rhine Valley in Western Germany and northeast France (Lutz 1988). Between 1965 and 1985, the species advanced about 250 km to the west (Figure 5). A westward expansion occurred also at its southwestern limits in Mediterranean France and Spain (Delibes *et al.* 1978, Isenmann 1987). It is curious that the range expansion was simultaneous in two quite distinct parts of the species' breeding range. Flade *et al.* (1986) suggest that higher dispersal rate plays a major role, but some changes in habitats, such as the development of gravel-pits with poplars and willows, could have favoured many local implantations.

Conclusions

The recent bird invasions in Europe mostly concern westward oriented expansions, but also some northward expansions by Mediterranean and desertic species. Kalela (1949, 1952) considered that most of these expansions were directly or indirectly induced by climatic changes. These changes could influence migration patterns, which lead these species to encounter new potential breeding areas. Climatic changes not only allowed some species to breed abroad; if they also occurred within the traditional breeding limits, they could push birds outside their breeding limits, forcing them to find new suitable breeding sites elsewhere. Although it is obvious that climatic changes have markedly influenced birds' breeding range in the past as well as in present times, the evergrowing impact of human activities now plays a significant role (Järvinen

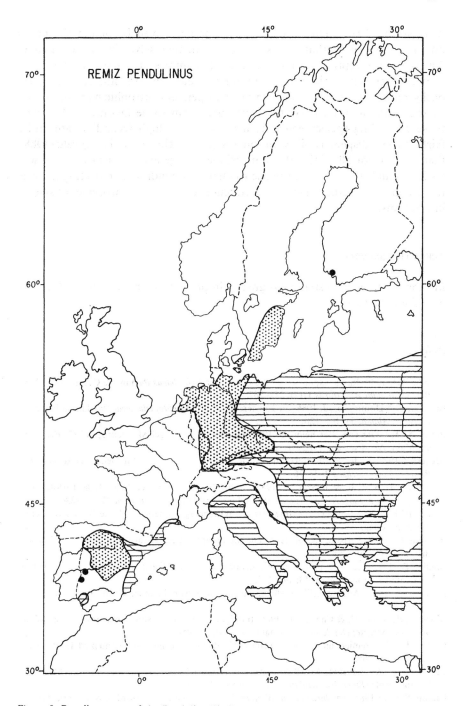

Figure 5. Breeding range of the Penduline Tit (*Remiz pendulinus*) in 1965 (horizontal lines) and range expansion until 1985 (dotted area and full circles).

258

& Ulfstrand 1980). This latter factor is perhaps even becoming the major one, if we recognize that man's activities can themselves influence climates. Albeit a few exceptions in the above cited examples, it is striking that man's activities through various kinds of intermediary factors still appear in the background. Species able to live in mostly open and eutrophic habitats shaped by man and/or in close contact with man seem to be favoured. Moreover, those expanding species generally also possess a high fecundity rate and a fairly efficient dispersal ability. In conclusion, we believe with Hengeveld (1988), that most if not all of the above mentioned range expansions can be viewed 'as individualistics species response to changing conditions, reflecting internal range dynamics' rather than as consequences of changes in processes of species interactions.

Acknowledgements

My thanks to J. Bronstein who greatly improved this paper and A. Carrière who drew the figures.

References

Arendt, W.J. 1988. Range expansion of the Cattle Egret (*Bulbucus ibis*) in the Greater Caribbean Basin. *Colonial Waterbirds* 11(2): 252–262.
Barcena, F. & Dominguez, J. 1986. La Tourterelle turque (*Streptopelia decaocto*) dans la Péninsule Ibérique. *Alauda* 54: 107–120.
Bauer, K. & Glutz von Blotzheim, U. 1969. Handbuch der Vögel Mitteleuropas 3/2. Frankfurt am Main, Akademische Verlagsgesellschaft.
Becker, P. 1975. Erster Brutnachweis des Seidensängers (*Cettia cetti*) für Deutschland. *Vogelkdl. Berichte Niedersachsen* 7: 73–76.
Bernis, F., Asensio, B. & Benzal, J. (1985). Sobre la expansion y ecologia de la Tortola turca (*Streptopelia decaocto*) con nuevos datos del interior de España. *Ardeola* 32: 279–294.
Bezzel, E. 1969. Die Tafelente (*Aythya ferina*). Wittenberg-Lutherstadt, Neue Brehm Bücherei/Ziemsen Verlag.
Bonham, P.F. & Robertson, J.C.M. 1975. The spread of the Cetti's Warbler in north-west Europe. Brit. Birds 68: 393–408.
Bredin, D. 1985 Première preuve de nidification du Héron Garde-Boeufs (*Bulbucus ibis*) en Charente-Maritime. *Alauda* 53: 144–145.
Brichetti, P. 1982. Atlante delli uccelli nidificanti sulle Alpi Italiane, I. *Riv. ital. Orn.* 52: 3–50.
Brichetti, P., Saino, N. & Canova, L. 1986. Immigrazione ed espansione della Tortora dal collare orientale (*Streptopelia decaocto*) in Italia. *Avocetta* 10: 45–49.
Carlon, J. 1985. Première nidification du Héron Garde-boeufs (*Bulbucus ibis*) au pied des Pyrénées. *Alauda* 53: 64–65.
Courteille, C. & Thévenot, M. 1988. Notes sur la répartition et la reproduction au Maroc du Bruant striolé (*Emberiza striolata*). *L'Oiseau* 58: 320–349.
Cramp, S. (ed.) 1985. *The birds of the Western Palearctic*, Vol. IV. Oxford & New York, Oxford University Press.
Cramp, S., Bourne, W.R.P. & Saunders, D. 1974. *The seabirds of Britain and Ireland.* London, Collins.

259

Cramp, S. & Simmons, K.E.L. (eds.) 1977. *The birds of the Western Palearctic*, Vol. I. Oxford & New York, Oxford Univ. Press.

Cramp, S. (ed.) 1985. *The birds of the Western Palearctic*, Vol. IV. Oxford & New York, Oxford Univ. Press.

Cruon, R. & Nicolau-Guillaumet, P. 1985. Notes d'ornithologie française XII. *Aluada* 53: 34–64.

Cruon, R. & Veilliard, J. 1975. Notes d'ornithologie française XI. *Alauda* 43: 1–21.

Czikeli, H. 1976. Die Ausbreitung des Karmingimpels (*Carpodacus erythrinus*) in Österreich und seinen Nachbarländern. *Egretta* 19: 1–10.

Delibes, M., Costa, L., Gisbert, J., Llamas, O. & Tirados, I. 1978. Sobre la expansion reciente del Pajaro moscon (*Remiz pendulinus*) en la Peninsula Iberica. *Ardeola* 25: 193–206.

Dronneau, C. & Wassmer, B. 1985. L'introduction du Héron gardeboeuf (*Bulbucus ibis*) en Alsace: historique, conditions de maintien de la population et déplacements. *Ciconia* 9: 123–146.

Fasola, M. & Gariboldi, A. 1987. Il Colino della Virginia (*Colinus virginianus*) in Italia. *Riv. ital. Orn.* 57: 3–13.

Fernex, M. 1985. Nidification de l'Hypolaïs polyglotte (*Hippolais polyglotta*) en Alsace. *Nos Oiseaux* 38: 25–36.

Flade, M., Franz, D. & Helbig, A. 1986. Die Ausbreitung der Beutelmeise (*Remiz pendulinus*) an ihrer nordwestlichen Verbreitungsgrenze bis 1985. *J. Orn.* 127: 261–287.

Franchimont, J. 1986. a. Aperçu de la situation du Héron gardeboeuf (*Bulbucus ibis*) en Afrique du Nord dans le contexte de l'expansion actuelle de l'espèce. *Aves* 23: 121–134.

Franchimont, J. 1986. b. Les causes de l'expansion géographique du Héron garde-boeuf (*Bulbucus ibis*). *Cahiers d'Ethologie appliquée* 6: 373–388.

Franchimont, J. 1987. A propos de l'installation de la Tourterelle turque (*Streptopelia decaocto*) au Maghreb. *Aves* 24: 150–151.

François, J. 1975. Contribution à la connaissance de l'avifaune d'Afrique du Nord. *Alauda* 43: 279–293.

François, J. 1983. L'Hypolaïs polyglotte (*Hippolais polyglotta*) dans le nord-est de la France. Analyse de sa répartition. *Ciconia* 7: 151–162.

Frey, H. 1970. Tiergeographische Untersuchungen über säkuläre quantitative und qualitative Veränderungen im Brutvogelbestand der Oberrheinische Tiefebene und der Wetterau. *Decheniana-Beihefte Nr.* 16: 1–177.

Frost, R.A. & Shooter, P. 1983. Fieldfares breeding in the Peak District. *Brit. Birds* 76: 62–65.

Geister, I. 1980. Über Ausbreitung von Wacholderdrossel (*Turdus pilaris*) und Karmingimpel (*Carpodacus erythrinus*) in Jugoslawien. Proc. IV Meeting European Orn. Atlas Committee. Göttingen 1979.

Géroudet, P. & Lévêque, R. 1976. Une vague expansive de la Cisticole jusqu'en Europe centrale. *Nos Oiseaux* 33: 241–256.

Glutz von Blotzheim, U. & Bauer, K. 1980. *Handbuch der Vögel Mitteleuropas 9*. Wiesbaden, Akademische Verlagsgesellschaft.

Glutz von Blotzheim, U. & Bauer, K. 1982. *Handbuch der Vögel Mitteleuropas 8*. Wiesbaden, Akademische Verlagsgesellschaft.

Glutz von Blotzheim, U. & Bauer, K. 1988. *Handbuch der Vögel Mitteleuropas 11*. Wiesbaden, Aula Verlag.

Glutz von Blotzheim, U., Bauer, K. & Bezzel, E. 1973. *Handbuch der Vögel Mitteleuropas 5*. Frankfurt am Main, Akademische Verlagsgesellschaft.

Grussu, M. & Secci, A. 1986. Prima nidificazione in Italia dell'Airone guardabuoi (*Bulbucus ibis*). *Avocetta* 10: 131–136.

Guermeur, Y. & Monnat, J.Y. 1980. *Histoire et géographie des oiseaux nicheurs de Bretagne*. Centrale Ornithologique de Bretagne, Ar Vran.

Handtke, K. & Mauersberger, G. 1977. Die Ausbreitung des Kuhreihers (*Bulbucus ibis*). *Mitt. Zool. Mus. Berlin* 53, Suppl. Ann. Orn. 1: 3–78.

Hafner, H. 1970. A propos d'une population de Hérons garde-boeufs (*Ardeola ibis*) en Camargue.

260

Alauda 38: 249–254.

Hayo, L. & Weyers, H. 1986. Zum Brutvorkommen des Orpheusspötter (*Hippolais polyglotta*) im Saarland. *Lanius* 24: 15–43.

Hayo, L. & Zannini, G. 1986. Orpheusspötter (*Hippolais polyglotta*) im Saarland. *J. Orn.* 127: 244.

Hildén, O. & Sharrock, J.T.R. 1985. *A summary of recent avian range changes in Europe*. Moscow, Acta XVIII Congr. Intern. Ornithologici, pp. 716–736.

Hengeveld, R. 1988. Mechanisms of biological invasions. *J. Biogeogr.* 15: 819–828.

Hill, A. 1986. Einwanderung des Karmingimpels (*Carpodacus erythrinus*) in die Bundesrepublik Deutschland. *Orn. Mitteilungen* 38: 72–84.

Hölzinger, J. 1986. Die Wacholderdrossel (*Turdus pilaris*) neuer Brutvogel für Griechenland. *Ökol. Vögel* 8: 113–115.

Isenmann, P. 1976–77. L'essor démographique et spatial de la Mouette rieuse (*Larus ridibundus*) en Europe. *L'Oiseau et R.F.O.* 46: 337–366 and 47: 25–40.

Isenmann, P. 1986. Le point sur la nidification de la Grive litorne (*Turdus pilaris*) en 1984–85 en France. *Alaude* 54: 100–106.

Isenmann, P. 1987. Zur Ausbreitung der Beutelmeise (*Remiz pendulinus*) in Westeuropa: Die Lage an der südwestlichen Verbreitungsgrenze. *J. Orn.* 128: 110–111.

Jacob, J.P., Van Der Elst, D., Schmitz, J.P., Parquay, M. & Maréchal, F. 1983. Progression de l'Hypolaïs polyglotte (*Hippolais polyglotta*) en Belgique et au Grand-Duché de Luxembourg. *Aves* 20: 92–102.

Jacob, J.P. 1986. 1984, 1985: Poursuite de la progression de l'Hypolaïs polyglotte (*Hippolais polyglotta*) en Wallonie. *Aves* 23: 1–11.

Järvinen, O. & Ulfstrand, S. 1980. Species turnover of a continental bird fauna: Northern Europe, 1850–1970. *Oecologia* 46: 186–195.

Jerrentrup, H. 1988. Die Lachmöwe (*Larus ridibundus*) neuer Brutvogel für Griechenland. *Kartierung mediterr. Brutvögel* 1: 5–6.

Jung, N. 1983. Struktur und Faktoren der Expansion des Karmingimpels (*Carpodacus erythrinus*) in Europa und Kleinasien. *Beitr. Vogelkunde* 29: 249–273.

Kalela, O. 1940. Zur Frage der neuzeitlichen Anreicherung der Brutvogelfauna in Fennoskandien mit besonderer Berücksichtigung der Austrocknung in den früheren Wohngebieten der Arten. *Ornis fennica* 17: 41–59.

Kalela, O. 1949. Changes in geographic ranges in the avifauna of Northern and Central Europe in relation to recent changes in climate. *Bird-Banding* 20: 77–103.

Kalela, O. 1952. Changes in the geographic distribution of finnish birds and mammals in relation to recent changes in climate. *Fennia* 75: 38–51.

Landenbergue, D. & Turrian, F. 1982. La progression de l'Hypolaïs polyglotte dans le Pays de Genève. II. *Nos Oiseaux* 36: 306–324.

Ledant, J.P., Jacob, J.P., Jacobs, P., Malher, F., Ochando, B. & Roché, J. 1981. Mise à jour de l'avifaune algérienne. *Gerfaut* 71: 295–398.

Leprince, P. 1985. La Grive litorne (*Turdus pilaris*) en Wallonie. Progrès récents et choix des milieux de reproduction. *Aves* 22: 153–168.

Lope, F. de, Guerrero, J., De la Cruz, C. 1984. Une nouvelle espèce à classer parmi les Oiseaux de la Péninsule Ibérique: *Estrilda* (*Amandava*) *amandava* (*Ploceidae*, Passeriformes). *Alauda* 52: 312.

Lope, F. de, Guerrero, J. and De la Cruz, C. and Da Silva, E. 1985. Quelques aspects de la biologie du Bengali rouge (*Amandava amandava*) dans le Bassin du Guadiana (Extrémadoure, Espagne). *Alauda* 53: 167–180.

Lübcke, W. & Furrer, R. 1985. *Die Wacholderdrossel* (*Turdus pilaris*). Wittenberg Lutherstadt, Neue Brehm Bücherei 569/ Ziemsen Verlag.

Lutz, A. 1988. Nouvelle nidification de la Mésange rémiz (*Remiz pendulinus*) en Alsace. *Ciconia* 12: 175–176.

McKilligan, N.G. 1985. The breeding success of the Indian Cattle Egret (*Ardeola ibis*) in eastern Australia. *Ibis* 127: 530–536.

Marion, L. & Marion, P. 1982. Le Héron garde-boeuf (*Bulbucus ibis*) niche dans l'ouest de la France. Statut de l'espèce en France. *Alauda* 50: 161–175.

Mauersberger, G. 1985. *Analysis of different factors causing dynamics of birds ranges.* Moscow, Acta XVIII Congr. Intern. Ornithologici, pp. 757–762.

Melchior, E. 1988. Erster Brutnachweis des Orpheusspötter (*Hippolais polyglotta*) in Luxemburg. *Regulus, Beilage* 9: 53–58.

Mezzavilla, F. and Battistella, U. 1987. Nuove ricerche sulla presenza del Bengalino comune (*Amandava amandava*) in Provincia di Treviso. *Riv. ital. Orn.* 57: 33–40.

Mosimann, P. 1988. Die bisherigen Beobachtungen des Karmingimpels (*Carpodacus erythrinus*) in der Schweiz. *Orn. Beobachter* 85: 179–181.

Mullins, J.R. 1984. Scarlet Rosefinch breeding in Scotland. *Brit. Birds* 77: 133–135.

Muntaner, J., Ferrer, X. & Martinez-Vilalta, A. 1984. *Atlas dels ocells nidificants de Catalunya i Andorra.* Barcelona, Ketres Editora.

Niethammer, G. 1951. Arealveränderungen und Bestandsschwankungen mitteleuropäischer Vögel. *Bonner zool. Beiträge* 2: 17–54.

Nowak, E. 1975a.*The range expansion of animals and its causes.* Publ. Smithsonian Institute, Washington D.C.

Nowak, E. 1975b. *Ausbreitung der Tiere.* Wittenberg Lutherstadt, Neue Brehm Bücherei 480/ Ziemsen Verlag.

Preiswerk, G. 1985. Zum Brutvorkommen des Orpheusspötters (*Hippolais polyglotta*) in der Badischen Rheinebene bei Basel. *Orn. Beobachter* 82: 124–125.

Otterlind, G. 1954. Migration and distribution. A study of the recent immigration and dispersal of the Scandinavian avifauna. *Vaar Faagelvärld* 13: 1–99.

Robel, D. & Königstedt, D. 1980. Über die Ausbreitung des Seidensängers (*Cettia cetti*) in Bulgarien. *Larus* 31–32: 371–375.

Schifferli, A., Géroudet, P. & Winkler, R. 1980. *Atlas des oiseaux nicheurs de Suisse.* Sempach, Station ornithologique suisse.

Sharrock, J.T.R. 1976. *The atlas of breeding birds in Britain and Ireland.* Tring, Brit. Trust Ornithology.

SOVON, 1988. Penduline Tit (*Remiz pendulinus*) colonization in the Netherlands up to 1987. *Limosa* 61: 145–149.

Spano, S. & Truffi, G. 1986. Il Parrocchetto dal collare (*Psittacula krameri*) all stato libero in Europa, con particolare referimento alle presenze in Italia. *Riv. ital. Orn.* 56: 231–239.

Stjernberg, T. 1985. *Recent expansion of the Scarlet Rosefinch (Carpodacus erythrinus) in Europe.* Moscow, Acta XVIII Congr. Intern. Ornithologici, pp. 743–753.

Stolt, B.O. 1988. Cattle Egret (*Bulbucus ibis*) ringed in Spain recovered in Sweden. *Vaar Faagelvärld* 47: 374–377.

Teixeira, R.M. 1979. *Atlas van de Nederlandse broedvogels.* 's-Graveland, Ver. Behoud Natuurmonumenten Nederland.

Tombal, J.C. 1980. L'Hypolaïs ictérine (*Hippolais icterina*) et l'Hypolaïs polyglotte (*H. polyglotta*) dans le nord de la France: le point de la situation en 1980. *Le Héron* n°4: 50–54.

Wallace, J., Hafner, H. & Dugan, P. 1987. Les hérons arboricoles de Camargue. *L'Oiseau et R.F.O.* 57: 39–43.

Yeatman, L. 1971. *Histoire des oiseaux d'Europe.* Paris-Montréal, Bordas Découverte.

Yeatman, L. 1976. *Atlas des oiseaux nicheurs de France.* Société Ornithologique de France, Paris.

16. Of mice and men

JACQUES MICHAUX, GILLES CHEYLAN and HENRI CROSET

Abstract

Natural immigrations, invasions linked to the influence of man and some consequences of these range shifts are examined within rodents of the family Muridae. The Neogene and Pleistocene history of faunas from the periphery of the Mediterranean Sea as shown by murids makes clear the role of paleogeography and climate in determining immigrations. A few species of rats and mice show spectacular invasions under the influence of man. Various aspects of invasions (morphological change, genetic variability, ecological amplitude) are discussed. A short review of the plague shows how the commensal rat is involved in the human cycle of the disease.

Introduction

Murids offer several examples of natural immigrations into Europe and the Mediterranean Basin (Table 1). The most recent immigrations involve some species of mice (*Mus*) and rats (*Rattus*). These species are commensal with man and man played a prominent role in their rapid expansions. In this case, these expansions can be called invasions. Immigrations occurred previously which are, on the contrary, true natural phenomena. The first part of the chapter deals with natural immigrations of the murids. The geographical frame will be Europe and the Mediterranean Basin which yield a large number of fossiliferous localities for the last 15 million years. The documentation on small mammals is relatively accurate and immigration events can be recognized. The second part of the chapter deals with invasions linked to the influence of man; two genera are involved: *Mus* and *Rattus*. Special attention is paid to *Rattus* and to the consequences of rapid evolution of the members of this genus. The third part is a discussion about the contact between man and the invaders, the negative aspect of the epizootia.

F. di Castri, A. J. Hansen and M. Debussche (eds.), Biological Invasions in Europe and the Mediterranean Basin.
263–284. © 1990, *Kluwer Academic Publishers, Dordrecht.*

264

Table 1. Genera of murids immigrating into western and southern Europe, North Africa.

Western and southern Europe	North Africa
Progonomys	*Progonomys*
Paraethomys	*Paraethomys*
Castillomys	*Pelomys*
Apodemus (small size species)	*Castillomys*
Micromys	*Apodemus* (large and small size species)
Mus	*Mus*[a]
Rattus	*Arvicanthis*
	? *Praomys*
	Mastomys
	Lemniscomys
	Acomys

[a] Several immigrations brought small murids of the genus *Mus* to North Africa (Jaeger, 1975).

Natural immigrations

Natural immigrations involve the successful settlement of a species in a region usually occupied by other mammals: the colonizing species has to settle and to establish itself. Rodents called murids belong to the family *Muridae* Gray, 1821 which is a highly diversified taxon. Many murids are fundamentally adapted to a tropical environment of high grass and mixed forest-grass habitats (Misonne 1969). Zoologists believe the center of evolution of the murids is in tropical southeast Asia (including or not – see below – the Indian subcontinent) (Misonne, ibid). Murids spread over the Old World out of southeast Asia at the beginning of the Upper Miocene. The radiation of the group dates back at least to the middle Miocene (Jacobs 1978, Jaeger *et al.* 1986a). In western and southern Europe, North Africa and the Middle East (Turkey) they are traced back to the beginning of the Upper Miocene. The immigration of the murid *Progonomys* (Hartenberger *et al.* 1967) and the tridactyl horse *Hipparion* are the two most important events in mammalian faunal history which occurred at this moment in the Old World.

The first invasion: Progonomys, beginning of the Upper Miocene (about – 11,5 Ma)

The Neogene biogeographical pattern of the Old World faunas has been progressively worked out by paleomammalogists who recognize several provinces. The time span in relation to the first expansion of the murids in Europe and the Mediterranean Basin covers the Middle and Upper Miocene, –15 to –5 Ma. Following Bernor (1983) the area under the scope of this study includes three provinces: the western and southern European province, the North African province and a part of the Sub-Paratethyan province. The

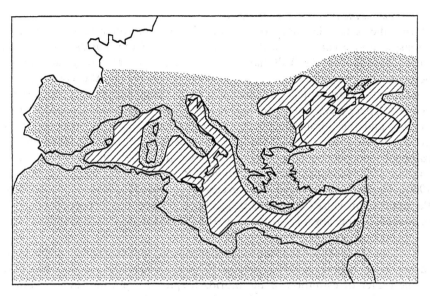

Figure 1. Maximum expansion of Bernor's 'evergreen woodland biome' in Upper Miocene (redrawn from Bernor 1983). Paleogeography (the restricted Tethys area – hatched – during Uppermost Miocene) and widespread similar climatic conditions (– dotted – as inferred from biogeography) explain why some species had a very large distribution for at least a short time. Tectonics in the Gibraltar zone may have also facilitated migrations before the salinity crisis (see Biju-Duval et al. 1977, and Biju-Duval & Montadert 1977, for further paleogeographic information).

composition of the faunas from these areas shows a definite modification around –12 Ma, particularly in the western and southern European provinces.

Environmental context during Middle and Upper Miocene
The mammalian faunas of the Astaracian stage (–15 to –12 Ma) in the latter three provinces indicate a closed and forested environment. The North African and the Sub-Paratethyan provinces, in contrast, supported more open-country forms. With the next stage, the Vallesian (–12 to –9,5 Ma), an enrichment of the fauna occurred in Europe with the addition of new species indicative of a shift towards more open woodlands. This trend continued in the succeeding stage named the Turolian (–9,5 to –5 Ma) and a maximum was rapidly reached during the second half of the stage (–8 to –5 Ma) (Figure 1). The faunas of the North African and Sub-Paratethyan provinces in the Vallesian showed a similar trend. The two faunas became still more alike due to their more open-country habitat which was even more open than the habitats of the contemporaneous faunas of the western and southern European Provinces. During the Turolian stage the faunas of these three areas were still similar in their ecological adaptations. This is correlated to what Bernor called the spread of the Old World evergreen woodland biome (Bernor 1983).

266

Progonomys, the first immigrant murid

During the Astaracian none of the rodents from the three faunal provinces discussed above can be regarded as an ancestor of the murids, so they are clearly immigrants. The Vallesian shows a decrease in the ecological contrast between the previously cited mammalian provinces but the contrast was certainly still sharp with south-eastern Asia where the murids have supposedly evolved. There were only a limited number of species which migrated, one or two, and the immigrants were able to spread rapidly all over the three provinces. One species is known in the North African province (*Progonomys cathalai*) (Chabbar Ameur *et al.* 1976), two are known in the western and southern European provinces (*P. cathalai, P. woelferi*) (Bachmayer & Wilson 1970). During the Vallesian one species is known (*P. debruijni*) in the Sub-Paratethyan province (Unay & de Bruijn 1984). This species has been previously recovered in the Siwalik province in the Northern part of the Indian subcontinent and described by Jacobs (1978).

The low diversity of the murids in the Siwalik and elsewhere in the west, causes us to believe that this part of the Indian subcontinent was not part of the center of evolution of the murids despite the fact that in the Middle Miocene there were representatives of a supposed ancestor of the murids, genus *Antemus* Jacobs, 1978. Jacobs (1978) recognized that the characters of this rodent are more primitive than in *Progonomys*. But these characters can also be taken as proof for a sister-group of the murids because *Antemus* looks rather like a dendromurid. These rodents at present living in Africa south of the Sahara are known in the Lower Vallesian of North Africa (Ameur Chabbar 1979; Ameur 1984) and the group may have evolved from a common ancestor with the murids (Jaeger *et al.* 1985, 1986). The rapid expansion of *Progonomys* (geologically instantaneous) allows to infer that this murid was a pioneer and was preadapted to an environment yet characterized by seasonality.

The first murid radiations

A chronological succession of Vallesian localities in southern Europe gives indication of how the murids evolved since they immigrated. In southern France, for example, the first known species was *P. cathalai*. Soon after there was another species, *P. woelferi*. A little later, we find two other species recently described as *P. clauzoni* and *Occitanomys faillati* (Aguilar *et al.* 1986). In Spain *P. cathalai* was known first, followed by *P. hispanicus* which probably derived from *P. cathalai* (Michaux 1971). A local diversification or radiation occurred with the immigrant as ancestor. In France, drillings in Bresse yielded fossils (Farjanel et Mein 1984) which give a picture of the beginning of murid evolution similar to the one obtained in southern France. Fewer localities in the North African province and the Sub-Paratethyan give interesting information. In the North African province *P. cathalai* remained unchanged and alone until at least –9.7 +/–0.5 Ma. (Jaeger *et al.* 1973). Data from Turkey document a small radiation and a regression or stasis in the diversity of murids: during the Turolian only one genus was found (Unay & de Bruijn 1984). From these data we infer that the immigrant species encountered better conditions (climatic and biotic) in the western and southern European province than in the North African and Sub-Paratethyan provinces, which were probably drier than the former.

The later Upper Miocene invasions

Another immigration followed *Progonomys*, that of *Paraethomys*, a species more evolved than *Progonomys*. Its ancestor could be, following Thomas *et al.* (1982), *Karnimata darwini* of the Siwalik, which was found in a level correlated with the late Vallesian or early Turolian levels of Europe. The large mammal faunas from Northern African, western and southern European and Sub-Paratethyan provinces have clear open-country characteristics at the time of the immigrations of the species of *Paraethomys*, so *Progonomys* and *Paraethomys* were adapted to different environments.

Systematics of Paraethomys and its bearing on migration events
Pareathomys was first described in North Africa and was known there from the Upper Miocene to the end of the Pleistocene. The oldest members of this genus have been referred to as the same species, *P. miocaenicus* (Jaeger *et al.* 1975) and fossils were found in contemporaneous levels in Spain and Morocco. The immigration into the Mediterranean Basin antedated the Moroccan locality of Khendek el Ouaich and indirectly correlated with a formation dated -7.4+/-1.2 Ma (Jaeger *et al.* 1973). A detailed study of the fauna discovered at Salobrena in southern Spain (Aguilar *et al.* 1984) recently discussed the occurrence of the same species of *Paraethomys* on both sides of the western Mediterranean Sea in the Upper Miocene, as admitted by Jaeger *et al.* (1975). Morphological observations on the teeth of several populations of the Spanish *Paraethomys* led Aguilar *et al.* (op. cit.) to propose the hypothesis that the immigrant in Spain was not *P. miocaenicus*, but a different species close to *P. anomalus*, a species previously described from the Lower Pliocene of Rhodes (de Bruijn *et al.* 1970). Aguilar *et al.* (1984) propose to name the Upper Miocene Spanish *Paraethomys*, earlier referred to *P. miocaenicus*, *P.* cf., as '*anomalus*.' This new interpretation, if correct, modifies the picture of the immigration toward the West of *Paraethomys* which is then slightly different from the one of *Progonomys* because different species migrated North and South of the Mediterranean Sea.

There is another interesting remark in the paper about Salobrena (Aguilar *et al.*, ibid) dealing with the pattern of murid immigrations. The authors think that in at least one locality of uppermost Miocene age in southern Spain, La Dehesa, the *Paraethomys* is different from *Paraethomys* cf. '*anomalus*' and nearly identical with *Paraethomys miocaenicus* of North Africa. This obser-

268

vation gives information about faunal exchanges between North Africa and southern Spain at the end of the Miocene. The end of the Miocene offered in the Mediterranean area a remarkable occasion for migrations when the Mediterranean Sea evaporated (Hsü 1972). This event called the 'salinity crisis' lasted about 0.5 Ma. It occurred during the Upper Messinian. It would explain the discovery of European taxa in North Africa and of African taxa in Europe. In the case of murids, species of the first group are *Apodemus* cf. *jeanteti* at Aïn guettara – Upper most Miocene, Morocco – (Brandy *et al.* 1980) and *Apodemus* cf. *jeanteti* together with *Stephanomys numidicus* and *Castillomys crusafonti* at Argoub Kemellal – Lower Pliocene, Algeria – (Coiffait *et al.* 1985). In the second group the murid is *Paraethomys* cf. '*miocaenicus*' of la Dehesa.

The review of Upper Miocene murid immigration events shows that they were more or less distributed over the whole period. Some murids came from the East, first *Progonomys*, then *Paraethomys*. Other murids may have migrated owing to the salinity crisis between Europe and North Africa. Some data indicate that migrations between southern Spain and North Africa have occurred before the salinity crisis. The fauna of Salobrena which includes a lophiomyine cricetid is considered older than the salinity crisis: respectively -7 to -6 Ma (Aguilar *et al.* ibid.) and -5.6 to -4.9 Ma (Vai 1988). Europe and Africa at the time of Salobrena were separated by straits but tectonics was still in process. This may have allowed accidental crossings.

Immigrations are the consequences of the abilities of species to move and take advantage of new ways or breakdowns of geographical boundaries. As ecological conditions were roughly the same over the area (see Figure 1 for Upper Turolian times) the immigration species have been able to spread very far. A discussion of the problems related to the mammalian settlement in Northern Africa can be found in a paper by Thomas *et al.* (1982). Several of the involved rodent species have not settled down nor stayed for a long time. So the introduction of some species into the communities of the autochtonous rodents have not been equally successful (Jaeger *et al.* 1987).

Two more immigrations of murids are recognized in western and southern Europe after the immigration of *Paraethomys*: that of *Castillomys crusafonti*, known from Caravacca, an Upper Miocene locality of southern Spain and that of *Apodemus dominans*, collected in the fauna of La Tour referred to the Upper Miocene (Aguilar *et al.*, 1982). *Castillomys* is also found in the Lower Pliocene of Turkey and Greece. Morphological data show that this *Castillomys* differs from the ones found in western Europe and northern Africa. As the *Castillomys* found in Algeria is close to the *Castillomys* from western Europe, we can infer either that this species migrated through the Betice-Rifan region, or that it came from the Middle East and had a convergent evolution with the western species. The wide distribution of this genus is probably connected with the maximum of the expansion of the evergreen woodland biome during the second half of the Upper Miocene. *Stephanomys*, already mentioned from Argoub Kemellal offers a simpler case because of its restricted distribution during the beginning of the Upper Miocene: it entered North Africa from the west.

The Plio-Pleistocene invasions

Since the great marine Pliocene transgression, faunas of western and southern Europe and North Africa evolved independently. The Plio-Pleistocene history of the murids in North Africa was still characterized by several immigrations: a *Mus* is known in the Middle Pliocene (a mouse of the *Mus musculus* type). Several African murids which evolved south of the Sahara are also found: *Arvicanthis* and *Lemniscomys* and two species of *Mus* (one belongs to the *Nannomys* subgroup) and the other is of uncertain affinity (Jaeger 1975). We have to point out that the Pleistocene *Praomys* of North Africa probably belongs to an autochtonous lineage (Jaeger 1975). *Pelomys*, which is known in the Pliocene of North Africa, is an immigrant but it may have either an Asiatic origin (the fossil *Pelomys* of Algeria stays close – Coiffait *et al.* 1981 – to *Parapelomys* Jacobs 1978 known from Asia) or an African origin (the present *Pelomys* are tropical and live south of the Sahara).

For species of tropical African origin, there were two main migration routes (Figure 2): (1) the Nile Valley and then the Mediterranean coast towards the West, and (2) the Atlantic coast of Africa. A relict population of *Mastomys* (*Mastomys erythroleucus*) still present in southwestern Morocco probably

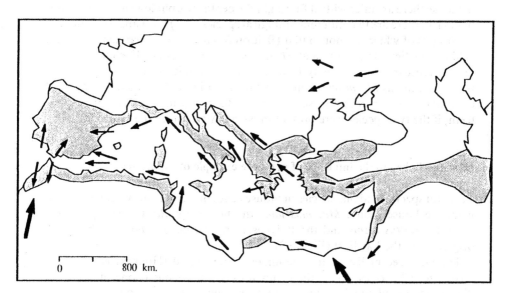

Figure 2. Migration paths in the Mediterranean region for the tropical Africa taxa during the Pleistocene (large arrows) and arrows for *Mus* (after Orsini 1982) during the last thousands years (small arrows). (Dotted area = bioclimatic Mediterranean region).

entered Africa by the latter way. The most recent immigrations of murids in North Africa are that of *Apodemus sylvaticus* and that of *Acomys cahirinus* at the end of the Pleistocene. *Acomys* is adapted to arid habitats and *Apodemus sylvaticus* is adapted to forest, so they probably entered North Africa at different times and in response to different constraints. Several immigrants of Plio-pleistocene age disappeared, with the exception of *Lemniscomys*, *Acomys*, *Mus* (of the *Mus musculus* morphology) and *Apodemus sylvaticus*.

After the Upper Miocene immigrations of *Paraethomys*, *Castillomys* and *Apodemus* (small size species) into Europe we have to notice the immigration of *Micromys* in the Lower Pliocene. Then came *Mus* for a short time in the Middle Pleistocene of Hungary (Janossy 1961), and this genus is given for the fauna of the Vallonet near Nice of Middle Pleistocene age (Chaline 1984). At last *Mus* entered Europe about 2000 B.C. (Vigne & Alcover 1985 and a recent note on this subject may be found in the thesis of Alcade-Gurt 1986).

Mus musculus is present in North Africa since the Upper Pliocene (at least -3 Ma) but it seems that its presence is not continuous. It entered Europe only very recently. This heterochrony probably has an ecological explanation. North Africa, western and southern Europe and the Middle East since the end of the Miocene were again under rather different climates. The rodents of the more forested Europe offered difficulties to the settlement of *Mus* which is more adapted to a relatively open country environment. It is only under the double action of the post-Würmian climatic change and of man with agriculture and pastoralism that the Mediterranean environment become more open and allowed the successful settlement of *Mus*. Under the common name of mouse there are mice with at first sight the dental morphology of *M. musculus*, but in fact there are two different biological species: *M. spretus* and *M. musculus*, the origins of which are not similar (Britton & Thaler 1978, Darviche & Orsini 1982). Genetical data show that *M. spretus* in southwestern Europe probably has an African origin. It may have crossed the strait of Gibraltar at nearly the same time as *M. musculus* entered Europe (Figure 2 shows different ways for *Mus*). It is impossible to say according to dental morphology of isolated teeth, if the two species were present in North Africa since the Pliocene.

Invasions linked to the influence of man: the example of rat origins

Three rat species, and one species of mouse have spread over the world following man: the house mouse *Mus musculus*, the brown rat *Rattus norvegicus*, the black rat *Rattus rattus* and the Polynesian rat *Rattus exulans*. All these rats originated in the Far East (Figure 3).

The most recent list of the mammals of the world (Honacki *et al.* 1982) recognizes 63 species within the genus *Rattus*. This genus is one of the most diverse in all mammals. This estimate is well below previous taxonomist revisions, but will probably increase again by further genetic studies. The genus *Rattus* clearly originated in southeast Asia where most of the species

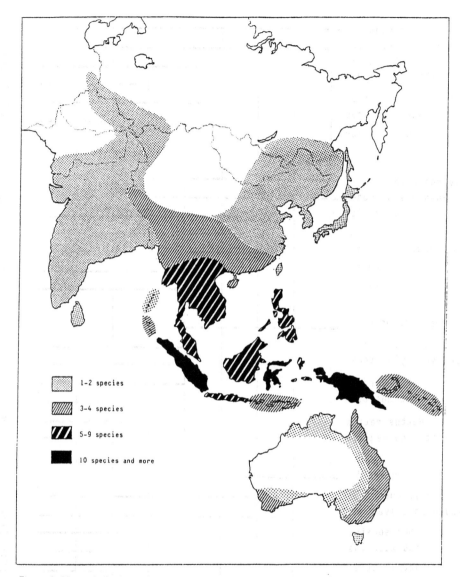

Figure 3. Natural distribution of the genus *Rattus* (from Honacki *et al.* 1982). The distribution of *R. rattus*, *R. norvegicus* and *R. exulans* outside this range is not shown.

are found. This genus displays a remarkable adaptative radiation on islands. The most diverse assemblages of species are found on three islands: Sumatra, Celebes and New Guinea, where more than ten species, most of them endemic, coexist sympatrically. Similarly, many endemic species have been discovered on small islands of the Sunda shelf.

The black rat and the Polynesian rat originated in Indochina, the Norway

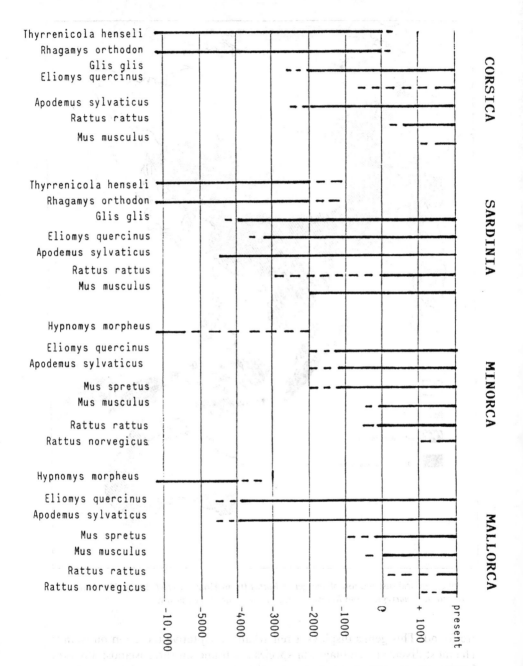

Figure 4. Rodent fauna changes on Mediterranean islands during the Holocene (from Vigne and Alcover 1985).

rat is indigenous to southern China, while the house mouse origin is from somewhere in the Middle East. The house mouse and the black rat expansions to Europe took place around the first or second millenium B.C. (Figure 4) (Vigne & Alcover 1985). On the Mediterranean islands, the black rat and two mice species appeared during the Roman empire, except in Sardinia where these species were found two or three thousand years before Christ (Sanges & Alcover 1980).

The guild of rodents found today on these islands is richer than during the Pleistocene. Before man invaded these islands, only endemic rodents were found: a Cricetid and a Murid in Corsica and Sardinia, and a Glirid in the Balearic. This depauperate fauna went extinct shortly after the colonizations of these islands by modern Murids and Glirids, but coexisted for at least two thousand years in Corsica.

Due to their huge geographical range covering all kinds of climates and biotas, mice are particularly appropriate for the study of evolution. So far, two species have been subjected to intensive genetic studies: the house mouse and the black rat; the karyological evolution of the black rat is well known and was intensively studied by Yosida and co-workers (1971, 1974, 1980) in Asia.

They found that in its primitive range in Indochina and northern India, the species had a diploid number of forty-two chromosomes, which reduced later on to a thirty-eight chromosomal complement by mean of two centric fusions. These translocations took place in India, where the modern 2N=38 karyotype has replaced the other chromosomal races (Niethammer, 1975) (Figure 5). At the same time, probably during Holocene, the new chromosomal race invaded the new habitats created by deforestation and climatic changes. We must note that this species, like most of the other rats, was originally found mainly in woodlands and tropical forests. No doubt that this habitat shift was very effective in the rapid dispersal of the species across the Middle East and the Mediterranean.

In New Guinea, where rats have undergone a marked adaptative radiation, seven endemic species are recognized. These species fall into two phylums: one of probable Australian origin, the *sordidus* group, and one independant from Australian origin, the *leucopus* group. The latter is remarkable by the presence of four or five centric fusions which reduced the diploid number from 42 chromosomes to 32 or 34 (Dennis & Menzies 1978). This high rate of translocation is absent in eleven species from southeast Asia whose primitive complement is, or was, 42 chromosomes, like New Guinea *Rattus* (Yosida 1973). Among these species, only three display centric fusions, namely *Rattus bowerssi, Rattus berdmori* and *Rattus rattus*. The karyotype of *Rattus norvegicus, Rattus exulans* and *Rattus diardii*, three widespread commensal species, is primitive, possessing 42 chromosomes. In Australia, seven endemic species are found, each having a unique karyotype with diploid numbers ranging from 32 to 50 (Baverstock *et al.* 1977).

One characteristic of populations inhabiting edges of the species' range

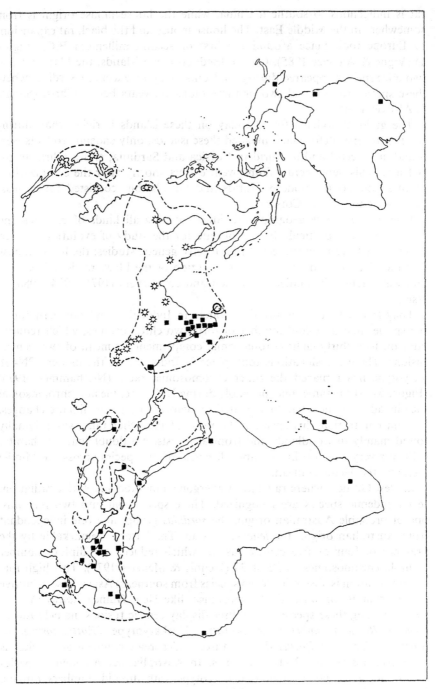

Figure 5. Distribution of free-living populations of *R. rattus* in the Old World. Stars: 2N=42 karyotype; Circle: 2N=40 karyotype (Ceylon); Squares: 2N=38 karyotype (from Yosida 1980).

deserves attention: that is, marginal populations are usually less variable than populations living in the centre of the area of distribution. This is illustrated by short-tailed mouse *Mus spretus* distribution, whose French populations are much less polymorphic than its African ones (Jacquart 1986). The same pattern is true for the black rat. From these results, one may infer that marginal populations are less fit than central ones. And, in fact, rats are strictly commensals in Northern Europe, while in the Mediterranean they are mainly found in riverine forests.

Interestingly, Spanish workers have found a chromosomal polymorphism unknown elsewhere in Europe in some populations inhabiting the southern tip of the peninsula (Pretel & Diaz 1978, Diaz *et al.* 1979, Ladron & Diaz 1981). This polymorphism involves three pericentric inversions and the presence of supernumerary 'B' chromosomes first discovered in Asia. The origin of this polymorphism is unknown, but is probably unrelated to its presence in Asiatic populations with a diploid number of 42 chromosomes.

A similar chromosomal polymorphism was not detected in several Mediterranean islands located along the coasts of Tuscany, Corsica and Sardinia (Capanna & Civitelli 1971). Nevertheless, a study of protein variation at 26 genetic loci revealed unexpected results on nine small islands of the Bonifacio straits between Corsica and Sardinia (Figure 6) (Cheylan 1986). We first note a loss of variability correlated with the size of the island. This relationship is true for morphological characters as well as genetic variability and both regressions give similar results. The loss of variability is severe on islands whose area is smaller than one hectare, where a monomorphic population was found. As these micro-populations number less than 100 individuals, we can predict a high rate of extinction and the deleterious effects of repeated bottle-necks. Similar results were obtained by several authors working on vertebrates on very isolated islands.

More surprisingly, the variability estimates for the largest islands, whose area is about a thousand hectares, show an increasing variability, when these populations are compared to mainland ones. Five alleles are unique in small islands, where they probably appeared by mutation, and two biochemical groups, reproductively isolated, have been discovered on several islands of the Lavezzi archipelago in the Bonifacio straits.

As stated above, these populations are a few thousands years old. In Corsica, the black rat appeared in a sixth century A.D. garbage pit (Vigne & Marinval-Vigne 1985), while in Sardinia, the species appeared in layers of a cave deposit around 3,500 B.C. (Sanges & Alcover 1980). The age of this population falls between those studied by Schmitt (1978) in Australia on *Rattus fuscipes* and by Patton *et al.* (1975) in the Galapagos. In southern Australia, *Rattus fuscipes* populations were isolated by sea-level transgressions 6,500 to 14,000 years ago. Most of the populations are monomorphic at 16 loci, except those living on the largest islands. On the Galapagos, the black rat was introduced by Europeans between the 18th century and 1945, depending on the island size. This result clearly shows that a long period of time is not a condition

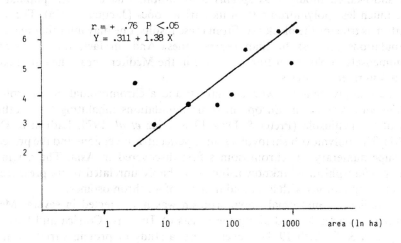
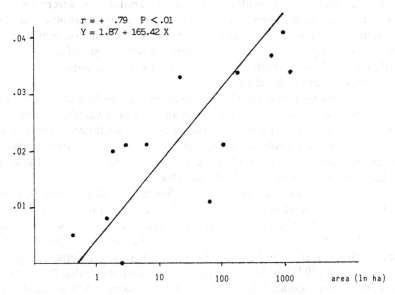

Figure 6. Variability-area relationships for *Rattus rattus* in the small Mediterranean islands (Provence, Corsica and Sardinia). Top: Head + body length variation; Bottom: genetic variability (allelic diversity at 26 genetic loci). (From Cheylan 1986).

prerequisitive to the loss of variability.

Similarly, the rodent fauna of these islands exhibit a broad range of morphological adaptations dealing mainly with size and teeth changes. A comparison between continental and Corsican populations of four species of rodents (the house mouse, the wood mouse, the black rat and the garden dormouse) was undertaken using 17 biometrical characters (Orsini & Cheylan 1983) (Figure 7). One species, the Wood mouse, is larger on the island whatever the character analyzed. The black rat is smaller on the island, except for

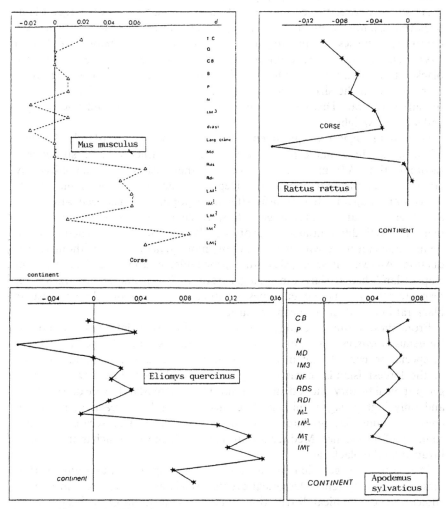

Figure 7. Biometrical differentiation between continental and insular populations of Rodents: *Apodemus sylvaticus* (wood mouse) and *Rattus rattus* (black rat) (*Muridae*); *Eliomys quercinus* (garden dormouse) (*Gliridae*). Continental populations are used as standard (vertical bar); deviation from the standard (Corsican populations) is expressed as d=Log x − Log Standard for each character used. (From Orsini & Cheylan 1983).

the tooth row whose length is the same as that of continental populations. The two last species, the house mouse and the dormouse, exhibit a longer tooth row but a similar size to that of the mainland populations. These changes are probably related to the presence or absence of competitors (Orsini 1982). In Algeria and southern France, where two mice species occur sympatrically (the house mouse and the short-tailed mouse), tooth row lengths do not overlap; in the absence of the other mouse species, we note an increase in tooth row length in the house mouse in Corsica and Sardinia, the values fall between those observed on mainland populations of *Mus spretus* and *Mus musculus*.

As stated above, the community of rodent found today on larger Mediterranean islands is much richer than during the Pleistocene, and only few continental species are missing in insular Mediterranean habitats (Cheylan 1984). This is not the case on smaller islands where only one species, the black rat, is usually found. On these islands, the size of the rat is negatively correlated with the size of the island (Cheylan 1986): the smaller the island, the larger the rat. This increase amounts to 10% for head and body length and 40% for weight.

Some data suggest a rapid phenotypic evolution on these islands. The fate of most of these micro-populations is quick extinction and subsequent recolonization. We have tested how fast these islets are recolonized by eradication of their rat population. Depending upon isolation, some islands are colonized again a few months after their population has been wiped out. So, it seems that continuous gene flow reach these islands, although they maintain well differentiated phenotypes. We do not know whether these immigrants can mate with native rats or, more generally, die without reproduction. Anyway, it seems clear that propagules establishing themselves on empty islands become larger in a few generations. Strong undirectional selection seems to act upon these phenotypes; whatever the gene pool, all micro-insular black rat in the Mediterranean are large.

From these results, we suggest that speciation by isolation is common in the genus *Rattus*, especially on the Sunda shelf islands. Nevertheless, this process of speciation cannot explain the unusual adaptive radiation found on some of the largest islands, namely the Celebes and New Guinea. Due to their geographical history, the rodent communities of these islands were depauperate and only highly vagile animals, like rats, were able to reach them. These colonists found few predators and competitors there. This recalls the environment of the smallest Mediterranean islands, which were much more recently invaded by the black rat.

The patterns of genetic diversity observed among the genus *Rattus* within these islands may reflect historical events associated with changes in habitat choice and geographical distribution. On these islands, this genus fits the niche width-variation hypothesis, which implies an adaptive relationship between ecological amplitude and genetic variability.

The role of rodents as reservoirs of diseases

The association of small mammals with certain human diseases is supported by innumerable reports which note that rodent so-and-so has been observed in nature to be infected with a pathogenic agent; a lot of them show that the species is truly involved with the transmission to man. The former are too abundant to summarize and even the latter are too numerous to quote (partial lists may be found in Arata 1975, Cox 1979, W.H.O. 1967).

The fact is that the *Muridae* family alone is a host reservoir for Plague, Tularemia, Listerosis, Erysipelas, Leptospirosis, Pseudotuberculosis, Salmonellosis, Brucellosis, Rickettsial pox, murine Typhus, Q fever, scrub Typhus and other Rickettsioses, Histoplamosis, Lassa fever, Rabies, asian Schistosomiasis, Chagas' disease, etc.

The example of the Plague is certainly the most famous and the most didactic for the understanding of the functioning of such metazoonis.

Plague is an infection of rodents caused by a bacillus, *Yersinia pestis* transmitted by fleas. According to Wo Lien-Teh *et al.* (Mollaret 1972), 40 outbreaks of Plague have decimated human populations before the time of Christ and Procope tells that the Plague of Justinien (6th century A.D.) destroyed almost all humanity (i.e. the Mediterranean peoples, the only ones known in this time).

According to Mollaret (1972), the epidemics of the Middle Ages have killed 25 million people between 1346 and 1353, that is 1/4 to 1/2 of the European population. The disease lasted for three centuries marked out by the mournfully famous episodes of Venice (1575–1577), Lyon (1628), Nijmegen (1635), London (1665), Marseille (1720).

The last large outbreak began in Yunnam and reached Hong Kong in 1894 where Yersin found the bacillus responsible for the affection on both human and rat blood. Four years later, P.L. Simond proved the transmission by the flea.

Carried by ship, the rats and their disease spread around the world: the Suez was reached in 1897, Madagascar in 1898, Japan, east Africa and Portugal in 1899, Manille, Sydney, Glasgow and San Francisco in 1900, Honolulu in 1908, Java in 1911, Ceylon in 1914, Paris in 1918 and Marseille in 1920.

The present situation is quiet, but the disease still exists in natural foci (Figure 8) where it behaves like a primary metazoonosis. For instance, Plague is endemic on the Iranian Kurdistan where two Meriones, *M. vinogradovi* and *M. persicus* are particularly involved in its sylvatic cycle (Golvan & Rioux 1961): the former is sensitive to *Yersinia pestis* and occupies the corn fields of the high plains (means altitude 2,000 m); the latter is resistant to the bacillus and lives in the rocky hills that surround these high plains; it is also a corn-eating species able to store 5 to 10 kg of grain in their burrow. *M. vinogradovi* is a segetal species favoured by human agriculture. This species is able to multiply rapidly. On such occasions, its populations meet the germ-carrying ones of *M. persicus*: the epidemic is kindled; it will kill almost every *M. vinogradovi*

280

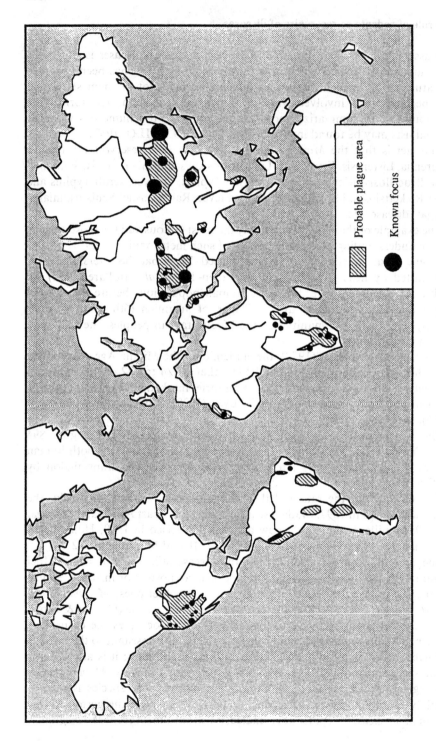

Figure 8. World map of known plague foci (from W.H.O. 1967 and 1969).

and reach the still healthy *M. persicus* individuals so that the endemic way protracts the epidemic ones until the next outbreak.

The secondary metazoonosis begins by accident: either children are contaminated by playing with dead meriones or men get inoculated by digging the burrows of *M. persicus* to get corn; indeed the outbreak of *M. vinogradovi* popultions through the corn fields leads to the loss of a large part of the crops and to famine; peasants recover the grains stolen by resistant rodents and get infected. Then, the commensal rat *Rattus rattus* and its fleas are involved in the cycle of the human bubonic form of the disease.

Before dying, infected men can transmit directly the pulmonary form: indeed, in certain circumstances (high humidity, temperature under 15 °C) the transmission is direct by sputtering. The pulmonary form is fatal after an incubation period of two or three days: the human epidemic flames out.

Conclusion

The Paleontological analysis shows clearly that Murids are able to migrate without the help of man: his recent activities have only enlarged and accelerated the inherent potentialities of this family.

Cultivation by man rejuvenates natural ecosystems: in doing so, he provides the most prolific species which are the most able to profit from such a situation; with an increase of food, their populations reach densities unknown in natural conditions. These outbreaks in number actuate the natural foci of infection such as the Plague or Cutaneous Leishmaniasis.

When he enters such natural foci, man can also initiate secondary metazoonosis in which commensal rodents are then involved.

Moreover, the adaptation to commensalism explains the worldwide distribution of *Rattus rattus*, *Rattus norvegicus* and *Mus musculus* and the local genetic evolution of their populations.

But for the moment, the reason why such species possesses these adaptations is unknown. Our studies on commensal populations of *Rattus and Mus* suggest that this capability is probably linked to their behavioural and physiological plasticity.

References

Aguilar, J.P., Dubar, M. & Michaux, J. 1982. Nouveaux gisements à Rongeurs dans la formation de Valensole: La Tour près de Brunet, d'âge Miocène supérieur (Messinien) et le Pigeonnier de l'Ange près de Villeneuve, d'âge Pliocène moyen. Implications stratigraphiques. *C.R. Acad. Sc., Paris*, 295(2): 745–750, 2 pl.

Aguilar, J.P., Brandy, L.D. & Thaler, L. 1984. Les rongeurs de Salobrena (Sud de l'Espagne) et le problème de la migration messinienne. *Paléobiologie continentale, Montpellier*, 14(2): 3–17.

Aguilar, J.P., Calvet, M. & Michaux, J. 1986. Découvertes de faunes de micromammifères aux Miocène supérieur: espèces nouvelles et réflexion sur l'étalonnage des échelles continentales

282

et marine. *C.R. Acad. Sc., Paris*, 30 (2-88): 755-760, 2 pl.

Alcade-Gurt, M. 1986. Les faunes de rongeurs du Pléistocène supérieur et de l'Holocène de Catalogne (Espagne) et leurs significations paléoécologiques et paléoclimatiques. Diplôme E.P.H.E. (3° section), Paris, Avril 1986.

Ameur Chabbar, R. 1979. Biochronologie des formations continentales du Néogène et du Quaternaire de l'oranie. Contribution des Micromammifères. Thèse Doctorat de 3° cycle, Univ. Oran, 77 p.

Ameur, R. 1984. Découverte de nouveaux rongeurs dans la formation de Bou Hanifia (Algérie occidentale). Géobis, Lyon, 17(2): 167-175, 10 fig.

Arata, A.A. 1975. The importance of small mammals in public health. In: Golley, Petrusewicz & Ryszkowski (eds.), *Small mammals: their productivity and population dynamics*, Cambridge University press, pp. 349-359.

Bachmayer , F. & Wilson, R.W. 1970. Small mammals (Insectivora, Chiroptera, Lagomorpha, Rodentia) from the Kohfidisch fissures of Burgenland, Austria. *Ann. Naturch. Mus., Vienne*, 74: 533-587, 13 pl.

Baverstock, P.R., Watts, C.H.R., Hogarth, J.T., Robinson, A.C. & Robinson, J.F. 1977. Chromosome evolution in Australian rodents. II The *Rattus* group. *Chromosoma* 61: 243-257.

Bernor, R.L. 1983. Geochronology and zoogeographic relationships of Miocene Hominoida. In: R.L. Ciochon & R.S. Corruccini (eds), *New interpretations of ape and human ancestry*, Plenum press, Advances in primatology, 21-64.

Biju-Duval, B. Dercort, J. & Le Pichon, X. 1977. From the Tethys ocean to the Mediterranean seas: a plate tectonic model of the Western Alpine system. In: Technip (ed.), *Structural History of the Mediterranean Basins*, Paris, pp. 1-12.

Biju-Duval, B. & Montadert, L. 1977. Introduction to the structural History of the Mediterranean Basins. In: Technip (ed.), *Structural History of the Mediterranean Basins*, Paris, pp. 143-164.

Brandy, L.D. & Jaeger, J.J. 1980. Les échanges de faunes terrestres entre l'Europe et l'Afrique nord-occidentale au Messinien. *C.R. Acad. Sc., Paris*, 291 (sér. D): 465-768, 1 pl.

Britton, J. & Thaler, L. 1978. Evidence for the presence of two sympartic species of mice (genus *Mus* L.) in Southern France based on the biochemical genetics. *Bioch. Genet.* 16: 213.

Bruijn, H. de, Dawson, M. & Mein, P. 1970. Upper pliocene Rodentia, Lagomorpha and Insectivora (Mammalia) from the isle of Rhodes (Greece). I, II and III. *Proc. Koninkl. Neder. Akad. van Wetensch. sér. B, ser. B*, 63(5): 535-584, 9 pl.

Capanna, E. &, Civitelli, M.V. 1971. On the chromosome polymorphism of *Rattus rattus. Experientia* 27: 583-584.

Chabbar Ameur, R., Jaeger, J.J. & Michaux, J. 1976. Radiometric age of early *Hipparion* fauna in North West Africa. *Nature* 261: 38-39.

Chaline, J. 1984. Echelles biostratigraphiques: les Micromammifères. In Synthèse géologique du Sud-Est de la France. *Mém. B.R.G.M., Fr.* 125: 524-529.

Cheylan, G. 1984. Les Mammifères des îles de Méditerranée occidentale: un exemple de peuplement insulaire non équilibré? *Revue d'Ecologie* 39: 37-54.

Cheylan, G. 1986. Facteurs historiques, écologiques et génétiques de l'évolution de populations méditerranéennes de *Rattus rattus*. Unpublished thesis U.S.T.L. Montpellier, 94 p.

Coiffait, B. & Coiffait, P.E. 1981. Découverte d'un gisement de micromammifères d'âge pliocène dans le bassin de Constantine (Algérie). Présence d'un muridé nouveau: *Paraethomys athmeniae* n. sp. Palaeovertebrata, Montpellier, v. 11(1): 1-15, 2 pl.

Coiffait, B., Coiffait, P.E. & Jaeger, J.J. 1985. Découverte en Afrique du Nord des genres *Stephanomys* et *Castillomys* (Muridae) dans un nouveau gisement de microvertébrés néogènes d'Algérie orientale: Argoub Kemellal. Proc. Koninkl. *Neder. Akad. van Wetensch. sér. B*, 88(2): 167-183, 1 pl.

Cox, F.E.G. 1979. Ecological importance of small mammals as reservoirs of disease. In: Stodaart, Chapman & Hall (eds.), *Ecology of small mammals*, London, pp. 213-238.

Darviche, D. & Orsini, P. 1982. Critères de différenciation morphologique et biométrie de deux espèces de souris sympatriques: *Mus spretus* et *Mus musculus domesticus. Mammalia* 46: 205-217.

283

Dennis, E. & Menzies, J.L. 1978. Systematics and chromosomes of New Guinea *Rattus*. *Australian Journal of Zoology* 26: 197–206.

Diaz de la Guardia, R., Ladron de Guevara, G.R.G. & Pretei Martinez, A. 1979. Chromosomal polymorphism caused by pericentric inversions in *Rattus rattus frugivorus* from the South-East of the Iberian peninsula. *Genetica* 51: 103–106.

Golvan, Y.L. & Rioux, J.A. 1961. Ecologie des Mérions du Kurdistan iranien. Relations avec l'Epidémiologie de la Peste rurale. *Ann. Parasitologie* 36(4): 449–588.

Hartenberger, J.L., Michaux, J. & Thaler, L. 1967. Remarques sur l'histoire des Rongeurs de la faune à *Hipparion* en Europe sud-occidentale. *Coll. Intern. C.N.R.S., Paris*, 163: 503–513, 1 fig., 2 tabl.

Honacki, J.H., Kinman, K.E. & Koeppl, J.W. 1982. *Mammal species of the world*. (Eds Allan Press and the Association of systematics collections. Lawrence, Kansas, U.S.A.) 694 p.

Hsü, K.J. 1972. When the Mediterranean dried up. *Scientific american*, 227, 6, 27–36.

Jacobs, L.L. 1978. Fossils rodents (Rhizomyidae et Muridae) from neogene deposits, Pakistan. *Mus. Northern Arizona Bull. (sér. 52)* 164 p.

Jacquart, T. 1986. Etude de la variabilité des gènes nucléaires et mitochondriaux chez l'espèce *Mus spretus* (Lataste, 1883) sur l'ensemble de son aire de réparition. Unpublished Thesis U.S.T.L. Montpellier. 122 p.

Jaeger, J.J. 1975. Les Muridae (Mammalia, Rodentia) du Pliocène et du Pléistocène du maghreb. Origine, évolution; données biogéographiques et paléoclimatiques. Unpublished Thesis, U.S.T.L. Montpellier (2): 148 p.

Jaeger, J.J., Coiffait, B., Tong, H. & Denys, C. 1987. *Rodent extinctions following Messinian faunal exchanges between Western Europe and Northern Africa*. Bull. Soc. géol. de France. N.S. (150), pp. 153–158.

Jaeger, J.J., Michaux, J., & David, B. 1973. Biochronologie du Miocène moyen et supérieur continental du Maghreb. *C.R. Acad. Sc., Paris*, 277 (sér. D). 2477–2480, 1 pl.

Jaeger, J.J., Michaux, J. & Thaler, L. 1975. Présence d'un rongeur muridé nouveau, *Paraethomys miocaenicus* nov. sp., dand le Turolien supérieur du Maroc et d'Espagne. Implications paléogéographiques. *C.R. Acad. Sc., Paris*, 280, (sér. D) 1673–1676, 1 pl.

Jaeger, J.J., Tong, H., Buffetaut, E. & Ingavat, R. 1985. The first rodents from the Miocene of Northern Thailand and their bearing on the problem of the origin of the Muridae. *Revue de Paléobiologie, Genève* 4(1): 1–7.

Jaeger, J.J., Tong, H. & Denys, C. 1986. Age de la divergence *Mus – Rattus*; comparaison des données paléontologiques et moléculaires. *C.R. Acad. Sc, Paris*, 917–922.

Janossy, D. 1961. Die Entwicklung der Kleinsäugerfauna Europas im Pleistozän (Insectivora, Rodentia, Lagomorpha). *Zeitschrift für Säugertierkunde* 26: 40–50.

Ladron de Guevara, R.G. & Diaz de la Guardia, R. 1981. Frequency of chromosome polymorphism for pericentric inversions and B-chromosomes in Spanish populations of *Rattus rattus frugivorus*. *Genetica* 57: 99–103.

Michaux, J. 1971. Muridae (Rodentia) néogènes d'Europe sud-occidentale. Evolution et rapports avec les formes actuelles. *Paléobiologie continentale, Montpellier* 2(1): 67 p., 12 pl.

Misonne, X. 1969. African and indo-australian Muridae. *Ann. Mus. roy. de l'Afrique centrale, Tervuren, Sc. zool.* 172: 219 p., 27 pl.

Mollaret, H. 1972. La Peste; In Encyclopaedia universalis, 12: 846–848.

Montenat, C. 1976. Chronologie et principaux évènements de l'histoire paléogéographiques du Néogène récent. *Bull. Soc. géol. de France* (7), 19 (3): 577–583.

Niethammer, J. 1975. Zur Taxonomien und Ausbreitungsgeschichte der Hausratte *Rattus rattus*. *Zoologische Anzeiger* 194: 405–415.

Orsini, P. 1982. Facteurs régissat la répartition des souris en Europe. Unpublished Thesis U.S.T.L. Montpellier, 134 p.

Orsini, P. & Cheylan, G. 1983. Les rongeurs de Corse: modifications de taille en rapport avec l'insularité. Bulletin d'Ecologie.

Patton, J.L., Yang, S.Y. & Myers, P. 1975. Genetic and morphologic divergence among introduced rat populations *Rattus rattus* of the Galapagos archipelago, Ecuador. *Systematic Zoology*

24: 296–310.

Pretel, M.A. & Diaz de la Guardia, G.R. 1975. Chromosomal polymorphism caused by supernumerary chromosomes in *Rattus Rattus frugivorus* (Rafinesque, 1814) (Rodentia, Muridae). *Experientia* 34: 325–328.

Sanges, M. & Alcover, J.A. 1980. Noticia sobre la microfauna holocenica de la grotta Su Guano o Gonagosula (Oliena, Sardenya). *Endins* 7. 57–63.

Schmitt, L.H. 1978. Genetic variation in isolated populations of the Australian Bush-Rat *Rattus fuscipes*. *Evolution* 32: 1–14.

Thomas, H., Bernor, R., & Jaeger, J.J. 1982. Origines du peuplement mammalien en Afrique du nord durant le miocène terminal. *Géobios* 15(3): 283–297.

Unay, E. & Bruijn, H. de 1984. On some Neogene rodent assemblages from both sides of the Dardanelles, Turkey. *Newsl. Stratigr., Berlin*, 13(3): 119–132, 3 fig.

Vai, G.B. 1988. A field trip guide to the Romagna Apennine geology. The Lamone valley. In Fossil vertebrates in the Lamone valley Romagna Apennines. Field trip guidebook. Continental faunas at the Miocene/Pliocene boundary. International Workshop, Faenza, Italy, March 28–31, pp. 7–37.

Vigne, J.D. & Alcover, J.A. 1985. Incidence des relations historiques entre l'homme et l'animal dans la composition actuelle du peuplement amphibien, reptilien et mammalien des îles de Méditerranée occidentale. 110° congrès national Soc. savantes Montpellier 2: 79–91.

Vigne, J.D. & Marvinval, M.C. 1985. Le Rat en Corse du VI° siècle après J.C.? *Mammalia* 49: 138–139.

Weerd van De, A. 1976. Rodents faunas of the Mio-pliocene continental sediments of the Teruel – Alfambra region, Spain. Micropal. S.P., Utrecht, 2. 185 p.

World Health Organization. 1967. Joint FAO/WHO Export Committee on Zoonoses. Third Report. World Health Organization Technical Report Series N° 378.

World Health Organization. 1973. Expert Committee on Rabies. Sixth Report. World Health Organization Technical Report Series N° 553.

Yosida, H.S. 1980. Cytogenetics of the Black Rat. Univ. Park Press Baltimore, U.S.A. ed., 256 p.

Yosida, H.S., Tsuchiya, K. & Moriwaki, K. 1971. Katyotipic differences of Black Rats collected in various localities of East and South-East Asia and Oceania. *Chromosoma* 33: 252–267.

Yosida, T.H. 1973. Evolution of karyotypes and differentiation in 13 *Rattus* species. *Chromosoma* 40: 285–297.

Yosida, T.H., Tsuchiyak, K., Sagal, T. & Moriwaki, K. 1974. Cytogenetical survey of Black Rat *Rattus rattus* in South-West and Central Asia with special regard to the evolutional relationship between three geographical types. *Chromosomes* 45: 999–109.